EARLY EDUCATION AND CARE, AND RECONCEPTUALIZING PLAY

ADVANCES IN EARLY EDUCATION AND DAY CARE

Series Editor: Stuart Reifel

Recent volumes:

Volume 8:	Theory and Practice in Early Childhood Teaching – edited by Judith A. Chafel and Stuart Reifel
Volume 9:	Family Policy and Practice in Early Child Care – edited by C. Dunst and M. Wolery
Volume 10:	Foundations, Adult Dynamics, Teacher Education and Play – edited by Stuart Reifel

ADVANCES IN EARLY EDUCATION AND DAY CARE
VOLUME 11

EARLY EDUCATION AND CARE, AND RECONCEPTUALIZING PLAY

EDITED BY

STUART REIFEL
University of Texas at Austin, USA

MAC H. BROWN
University of South Carolina, Columbia, USA

2001

JAI
An Imprint of Elsevier Science

Amsterdam – London – New York – Oxford – Paris – Shannon – Tokyo

ELSEVIER SCIENCE Ltd
The Boulevard, Langford Lane
Kidlington, Oxford OX5 1GB, UK

© 2001 Elsevier Science Ltd. All rights reserved.

This work is protected under copyright by Elsevier Science, and the following terms and conditions apply to its use:

Photocopying
Single photocopies of single chapters may be made for personal use as allowed by national copyright laws. Permission of the Publisher and payment of a fee is required for all other photocopying, including multiple or systematic copying, copying for advertising or promotional purposes, resale, and all forms of document delivery. Special rates are available for educational institutions that wish to make photocopies for non-profit educational classroom use.

Permissions may be sought directly from Elsevier Science Global Rights Department, PO Box 800, Oxford OX5 1DX, UK; phone: (+44) 1865 843830, fax: (+44) 1865 853333, e-mail: permissions@elsevier.co.uk. You may also contact Global Rights directly through Elsevier's home page (http://www.elsevier.nl), by selecting 'Obtaining Permissions'.

In the USA, users may clear permissions and make payments through the Copyright Clearance Center, Inc., 222 Rosewood Drive, Danvers, MA 01923, USA; phone: (+1) (978) 7508400, fax: (+1) (978) 7504744, and in the UK through the Copyright Licensing Agency Rapid Clearance Service (CLARCS), 90 Tottenham Court Road, London W1P 0LP, UK; phone: (+44) 207 631 5555; fax: (+44) 207 631 5500. Other countries may have a local reprographic rights agency for payments.

Derivative Works
Tables of contents may be reproduced for internal circulation, but permission of Elsevier Science is required for external resale or distribution of such material.
Permission of the Publisher is required for all other derivative works, including compilations and translations.

Electronic Storage or Usage
Permission of the Publisher is required to store or use electronically any material contained in this work, including any chapter or part of a chapter.

Except as outlined above, no part of this work may be reproduced, stored in a retrieval system or transmitted in any form or by any means, electronic, mechanical, photocopying, recording or otherwise, without prior written permission of the Publisher.
Address permissions requests to: Elsevier Science Global Rights Department, at the mail, fax and e-mail addresses noted above.

Notice
No responsibility is assumed by the Publisher for any injury and/or damage to persons or property as a matter of products liability, negligence or otherwise, or from any use or operation of any methods, products, instructions or ideas contained in the material herein. Because of rapid advances in the medical sciences, in particular, independent verification of diagnoses and drug dosages should be made.

First edition 2001

Library of Congress Cataloging in Publication Data

Early education and care, and reconceptualizing play/edited by Stuart Reifel, Mac H. Brown.
    p. cm. – (Advances in early education and day care; v. 11)
  ISBN 0-7623-0810-9 (hardcover)
  1. Early childhood education. 2. Child care. 3. Play. I. Reifel, Robert Stuart. II. Brown, Mac H. III. Series.

LB1139.35.P55 E37 2001
372.21–dc21   2001040674

British Library Cataloguing in Publication Data
A catalogue record from the British Library has been applied for.

ISBN: 0-7623-0810-9
ISSN: 0270-4021 (Series)

∞ The paper used in this publication meets the requirements of ANSI/NISO Z39.48-1992 (Permanence of Paper).
Printed in The Netherlands.

CONTENTS

LIST OF CONTRIBUTORS   vii

INTRODUCTION
Stuart Reifel   ix

PART I: EARLY CHILDHOOD EDUCATION AND CARE

GLOBALIZATION AND ITS DISCONTENTS: EARLY
CHILDHOOD EDUCATION IN A NEW WORLD ORDER
    Sally Lubeck, Patricia A. Jessup and Abigail M. Jewkes   3

CHILD CARE QUALITY: A MODEL FOR EXAMINING
RELEVANT VARIABLES
    Eva L. Essa and Melissa M. Burnham   59

PROFESSIONAL DEVELOPMENT AND THE QUALITY OF
CHILD CARE: AN ASSESSMENT OF PENNSYLVANIA'S
CHILD CARE TRAINING SYSTEM
    Joyce Iutcovich, Richard Fiene, James Johnson,   115
    Ross Koppel and Frances Langan

PROFESSIONAL CARING AS MOTHERING
    Noelene McBride and Susan Grieshaber   169

THE THEMATIC UNIT: OLD HAT OR NEW SHOES?
    C. Stephen White, Greta G. Fein, Brenda H. Manning and   203
    Anne Daniel

"AIR IS A KIND OF WIND": ARGUMENTATION AND THE
CONSTRUCTION OF KNOWLEDGE
    Sue Dockett and Bob Perry   227

PART II: RECONCEPTUALIZING PLAY

"WE DON'T PLAY THAT WAY AT PRESCHOOL": THE MORAL AND ETHICAL DIMENSIONS OF CONTROLLING CHILDREN'S PLAY
    Mac H. Brown and Nancy K. Freeman   *259*

THE DANGEROUSLY RADICAL CONCEPT OF FREE PLAY
    David Kuschner   *275*

PLAY AND DIVERSE CULTURES: IMPLICATIONS FOR EARLY CHILDHOOD EDUCATION
    Jaipaul L. Roopnarine and James E. Johnson   *295*

UNDER THE LENS: THE PLAY-LITERACY RELATIONSHIP IN THEORY AND PRACTICE
    Kathleen A. Roskos and James F. Christie   *321*

THE PLAY FRAME AND THE "FICTIONAL DREAM": THE BIDIRECTIONAL RELATIONSHIP BETWEEN METAPLAY AND STORY WRITING
    Jeffrey Trawick-Smith   *339*

AUTHOR INDEX   *357*

SUBJECT INDEX   *367*

LIST OF CONTRIBUTORS

Mac H. Brown	College of Education, University of South Carolina, USA
Melissa M. Burnham	Department of Human Development and Family Studies, University of Nevada, Reno, USA
James F. Christie	Department of Curriculum & Instruction, Arizona State University, USA
Anne Daniel	Department of Human Development, University of Maryland, College Park, USA
Sue Dockett	School of Education and Early Childhood Studies, University of Western Sydney, Australia
Eva L. Essa	Department of Human Development and Family Studies, University of Nevada, Reno, USA
Greta G. Fein	Department of Human Development, University of Maryland, College Park, USA
Richard Fiene	Department of Curriculum & Instruction, The Pennsylvania State University, USA
Nancy K. Freeman	College of Education, University of South Carolina, USA
Susan Grieshaber	School of Early Childhood, Queensland University of Technology, Brisbane, Australia
Joyce Iutcovich	Keystone University Research Corporation, Erie, PA, USA
Pat Jessup	School of Education, University of Michigan, USA

Abby Jewkes	School of Education, University of Michigan, USA
James E. Johnson	Department of Curriculum & Instruction, The Pennsylvania State University, USA
Ross Koppel	University of Pennsylvania and Social Research Corporation, USA
David Kuschner	College of Education, University of Cincinnati, USA
Frances Langan	Department of Education, Keystone College, PA, USA
Sally Lubeck	School of Education, University of Michigan, USA
Brenda H. Manning	College of Education, University of Georgia, USA
Noelene McBride	School of Early Childhood, Queensland University of Technology, Brisbane, Australia
Bob Perry	School of Education and Early Childhood Studies, University of Western Sydney, Australia
Jaipaul L. Roopnarine	Brooklyn College and the Graduate Centre, The City University of New York and Syracuse University, USA
Kathleen A. Roskos	Department of Education, John Caroll University, USA
Jeffrey Trawick-Smith	Department of Education, Eastern Connecticut State University, USA
C. Stephen White	Graduate School of Education, George Mason University, USA

INTRODUCTION

Volume 11 of the *Advances in Early Education and Day Care* series continues the tradition of providing a forum for current thought about the field of early education and care. Multi-disciplinary and inter-disciplinary work are at the heart of the field, reflecting scholarly contributions from established disciplines as well as emerging views. New forms of traditional scholarship meet non-traditional approaches to inquiry in this volume, as they did in earlier volumes of the series (e.g. Reifel, 1999). Irrespective of disciplinary home or scholarly orientation, all of this work provides rigorous perspective on early childhood practice and the ideas associated with it.

Part I of this volume includes chapters on a number of issues related to Early Childhood Education and Care. Sally Lubeck, Pat Jessup and Abby Jewkes begin this section with an international perspective on early education. Their chapter on "Globalization and Its Discontents: Early Childhood Education in a New World Order" reveals the range of ways that programs are viewed in different countries. They begin with a review of efforts around the world to educate young children, showing how policies, standards, and conceptions of early education vary. The profound differences in conceptions of early education from country to country are illustrated in Lubeck, Jessup, and Jewkes' case studies of the United Kingdom, the Czech Republic, and Sweden. The details of different policies, administrative approaches, regulations, curricula, parent involvement patterns, and research have profound implications for how early education and care can be considered. Global perspectives could alter research and practice in the field.

Much research has been conducted, particularly in the United States, on issues related to the quality of child care programs. Two chapters address this research, beginning with "Child Care Quality: A Model for Examining Relevant Variables" by Eva Essa and Melissa Burnham. Essa and Burnham create an ecologically based model for organizing empirical findings on the outcomes of child care. Proximal variables, including center characteristics, family characteristics, and child characteristics, combine with distal variables, such as community and social characteristics, to explain demonstrated outcomes of child care experience. An analysis of the layers of influences on children's development adds to our understanding of policy and preparation for programs. A second chapter on child care presents an evaluation of one state's programs, "Professional Development and the Quality of Child Care: An Assessment of Pennsylvania's Child Care Training System" by Joyce Iutcovich, Richard Fiene, James Johnson, Ross Koppel and Frances Langan. The details of their intensive study of child care quality, in centers and homes, analyzes the state's effort to improve quality. These authors report major findings on caregivers' interest and appreciation of training, needs for training, and use of environmental evaluation as a basis for determining needs. This comprehensive study provides a complementary picture to the review of child care quality provided by Essa and Burnham.

Evaluations of child care quality provide one form of evidence about early childhood care. How one becomes a caregiver, and what that means in a society, is presented by Noelene McBride and Susan Grieshaber. "Professional Caring as Mothering" presents a cultural feminist perspective on caregiving. These scholars review how life experience(s) relates to professional action, in particular how women are socialized through their relationships with their mothers to become caring adults. The many philosophical, psychological, sociological, and cultural roots of caring are presented, then used as a framework for analyzing child care practices. A case study provides a complex picture of how personal experience both empowers and limits relationships in child care. Implications for training and evaluation are vast.

Two chapters in Part I address topics in curriculum. Each presents a framework for understanding teaching practice. "The Thematic Unit: Old Hat or New Shoes?" by Stephen White, Greta Fein, Brenda Manning and Anne Daniel links constructivist theories with project approaches to teaching young children. Integrating children's experiences with teacher's purposes is described from an historical perspective, as well as in terms of other influences on curricular thought. Also from a social constructivist perspective, Sue Dockett and Bob Perry look to the social transactions in the classroom to see when and how children use arguments as a source of knowledge acquisition. "'Air Is a Kind of Wind': Argumentation and the Construction of Knowledge" provides evidence that demonstrates models of interaction that teachers can observe and build on. Argumentation operates in some settings to promote expression and thinking.

Part II of this volume addresses Reconceptualizing Play. Mac Brown has served as co-editor of this set of papers, a project he initiated with an idea for a symposium at the 2000 meeting of the American Educational Research Association. That interactive symposium was titled "Reconceptualizing Play in Early Childhood Eduction," and it attracted a large crown of early childhood educators with diverse perspectives on the role of play in early childhood practice. The purpose of the session was to deconstruct play by examining research and theory from a variety of perspectives. Participants in the session presented a truly diverse set of ideas. Some ideas were traditional, developmentally based views of play; others reflected disciplines that do not often appear in the early childhood play literature. It was apparent that the field of early education and care, like all other fields, was undergoing a paradigm shift from modernist, positivist views of the world to post-modern, post-positivist views. Although this shift was not universally welcomed and was, in fact, resisted in many quarters, the symposium was intended to contribute to dialog among these different views.

Exactly what the shift means to the field is yet to be determined. Many modernist views which have dominated the literature on children's play have been challenged and seen as problematic. One supposedly self-evident truth that early education has held as almost sacred deals with the centrality of play in children's development and learning; seminal authorities such as Piaget, Vygotsky, Bruner, Bronfenbrenner, Freud and Erikson placed play in a central role for young children's development. But these modern-era scholars and their work are now characterized as problematic by post-modernists (Cannella, 1997; Cannella & Bailey, 1999; Cannella & Grieshaber, 2001). This challenges the field's very definition of play as central to best practices, which has been codified in the statement on Developmentally Appropriate Practices (DAP; Bredekamp & Copple, 1997). Major questions facing the field revolve around this challenge to DAP (Jipson & Johnson, 2000). Does DAP represent best practice, and for whom? Are DAP practices really appropriate only for Euro-centric, middle class children? What comprises best practice for children of color, children of non-middle class families, or children from a variety of cultures, and who makes these decisions? Are behaviors like play, which may be inappropriate for children of one culture appropriate for children from another? Questions about gender styles in play and the freedom of "free play" may be constrained by teacher beliefs and classroom materials (Cannella, 1997).

The chapters that appear in Part II are a beginning sample of a larger set of papers that attempt to reconsider classroom play. Other papers will appear in a later volume in this series. The current chapters raise questions about the social context wherein we consider play as part of the early childhood curriculum. Mac Brown and Nancy Freeman explore ethical issues related to play in "'We Don't Play That Way at Preschool': The Moral and Ethical Dimensions of Controlling Children's Play". Theory and practice are revisited in light of modern and postmodern views of play, and in terms of ideologies that shape professional thought. Play confronts us with moral and ethical considerations, when we see it as an activity where teachers make decisions about freedom and control; decisions about toys, play themes, privacy, and autonomy complicate the way we observe and plan for play. Questions about play become problematic when they are filtered through our standards of professional ethics (Feeney & Freeman, 1999).

The question of freedom and its relationship to development is explored by David Kuschner in his historical look at "The Dangerously Radical Concept of Free Play". Many of the ideas we may take for granted, like the idea of play, have had controversial roots. As kindergartens spread across the United States and "scientific" ideas emerged from the child study movement, it seems that a clash between practices, ideology, and theory was inevitable (Frost, Wortham & Reifel, 2000). Kuschner shows how one study, published in the early decades of the 20th century, caused consternation in the early childhood community. Frederic Burk's study of kindergarten practices raised questions about Froebel's notion of classroom play as a means of understanding children's needs and interests. Burk analyzed kindergarten activities such as circle time and pretend in a manner that highlighted the deeper meanings of the play curriculum, thereby challenging extant dogma on the topic. Kuschner shows how historical analysis can be relevant to contemporary discussions about play and practice.

Just as ethics and history can provide new ways of considering play, so can cultural comparisons move us in new directions. "Play and Diverse Cultures: Implications for Early Childhood Education", by Jaipaul Roopnarine and James Johnson, reviews empirical and theoretical research in light of philosophies, models of development, parent-child play, teacher-child relationships, and their implications for practice and policy. Their review includes a view of play from diverse cultures, where a variety of ideas may differ from predominant views in Western societies. Roopnarine and Johnson look at play as it relates to children's competence, early childhood practices, child training, parental beliefs, contexts, and social position. They provide a data-rich analysis of the diversity of play around the world, an analysis that informs practice in our diverse settings. Notions about play that appear in many standard texts are challenged.

Ideas about play and context are delineated by Kathleen Roskos and James Christie in "Under the Lens: The Play-Literacy Relationship in Theory and Practice". From their perspectives as established scholars of the literacy-play connection, these authors provide a nuanced analysis of the contributions of existing and emerging theories to our understanding of classroom play. Piaget and Vygotsky have become standard theorists for play and literacy researchers,

Introduction xiii

and their theories are inspected for their contributions. The work of Bronfenbrenner is added to this discussion, to provide a broader picture for contrasting ideas and evidence. Roskos and Christie revisit these theorists to provide a new understanding for teachers, as they take roles in the "dynamic activity system" of literacy play. Jeffrey Trawick-Smith expands our debate on literacy play with a look at "The Play Frame and the 'Fictional Dream': The Bidirectional Relationship Between Metaplay and Story Writing". The developmental context for literacy takes on a new meaning as he describes play stories in context in terms of "story, play, and metaplay." To understand writing, print, and play stories, Trawick-Smith provides non-developmental play theories. He analyzes evidence to show play as a kind of composition that can be understood with Bateson's idea of "frames." He argues that teachers can make literacy and metaplay explicit to children, building on their facilities to create stories to become literate.

Early childhood education is adjusting to the cultural paradigm shift which is challenging established notions of developmentally appropriate practices and child development. The reconceptualization of play that we provide in Part II demonstrates how diverse disciplines, such as ethics, history, culture studies, and others, can interface with developmental views of play to create new understandings. This effort builds on the work of Sutton-Smith (1997, 1999) and others who show us multiple ways, or rhetorics with which we can understand children's play. Volume 10 of *Advances in Early Education and Day Care* (Reifel, 1999) contributed to this effort. We anticipate further explorations in a forthcoming *Advances* volume. Thanks to Mac Brown for his work on this volume. We hope that its deconstruction of play will be of assistance by providing some of the issues involved, as well as examples of how thoughtful people struggle with the complexities of this discourse.

Both parts of this volume reflect the increasing diversity of thought that is contributing to early education and care. Traditional disciplinary inquiry is moving in new directions and contributing to new depth of understanding, while postmodern research is opening new vistas. The complexity of the field is being described in ways that will nourish research, speak to practice, and inform teacher education. It is my hope that this volume, like earlier volumes in the *Advances in Early Education and Day Care* series, will serve as a resource for research, practice, and professional preparation.

Many have collaborated with me to prepare and produce this book. Our new production staff at Elsevier Science/JAI has been most helpful as we make the transition to our new presentation for the series. I want to thank Gerhard Boomgaarden for his continuing support for this series, and for his encouragement with this particular volume. At The University of Texas, Laura

Havlick has been a constant source of assistance with document preparation and with managing the flow of manuscripts. Helpful blind reviews were provided by numerous colleagues who deserve thanks, including the following: Sandra Briley, The University of Texas at Austin; Mac Brown, University of South Carolina; Christine Chaille, Portland State University; Celia Genishi, Teachers College, Columbia University; Lisa Goldstein, The University of Texas at Austin; Priscilla Hoke, Southwest Texas State University; June Yeatman, Austin Community College.

Stuart Reifel, Series Editor

REFERENCES

Bredekamp, S., & Copple, C. (1997). *Developmentally appropriate practice early childhood programs serving children from birth through age 8*. Washington, D.C.: National Association for the Education of Young Children.

Cannella, G. S. (1997). *Deconstructing early childhood education: Social justice and revolution*. New York: Peter Lang.

Cannella, G. S., & Bailey, C. D. (1999). Postmodern research in early childhood education. In: S. Reifel (Ed.), *Advances in Early Education and Day Care* (Vol. 10, pp. 3–39). Greenwich, CT: JAI Press.

Feeney, S., & Freeman, N. K. (1999). *Code of professional ethics and statement of commitment*. Washington, D.C.: National Association for the Education of Young Children.

Frost, J. L., Wortham, S., & Reifel, S. (2001). *Play and child development*. Upper Saddle River, NJ: Merrill/Prentice Hall.

Grieshaber, S., & Cannella, G. S. (2001). *Embracing identities in early childhood education: Diversity and possibilities*. New York: Teachers College Press.

Jipson, J., & Johnson, R. (2000). *Resistance and representation: Rethinking childhood*. New York: Peter Lang.

Reifel, S. (1999). *Advances in early education and day care* (Vol. 10). Greenwich, CT: JAI Press.

Sutton-Smith, B. (1997). *The ambiguity of play*. Cambridge, MA: Harvard University Press.

Sutton-Smith, B. (1999). The rhetorics of adult and child play theories. In: S. Reifel (Ed.), *Advances in Early Education and Day Care* (Vol. 10, pp. 149–161). Greenwich, CT: JAI Press.

PART I:

EARLY CHILDHOOD EDUCATION AND CARE

GLOBALIZATION AND ITS DISCONTENTS: EARLY CHILDHOOD EDUCATION IN A NEW WORLD ORDER

Sally Lubeck, Patricia A. Jessup and Abigail M. Jewkes

ABSTRACT

This chapter discusses features of globalization and reviews international case and comparative studies of early childhood education and care (ECEC) policy. The chapter has four purposes: (1) to provide an international context for discussing ECEC policy reforms related to globalization, including international efforts to forge a shared vision of children's rights; (2) to review cross-national studies of ECEC policy; (3) to use case study examples as a way to highlight how historical precedent and contextual factors influence responses to globalization; and (4) to suggest a values-based, contextual framework for international ECEC policy research. Recent ECEC policy concerns and initiatives are evident in the areas of governance and regulation, funding, access, curriculum, staff recruitment and retention, and parent involvement.

INTRODUCTION

Children are the lifeblood of every society, their value expressed in biological, social, cultural, and economic terms. Children ensure the perpetuation of the species and of society; children become the bearers of tradition, and children ensure that the work force will be regenerated. For most of recorded history, children have been reared in families and communities. Between the 16th and 19th centuries, formalized group care in the form of child-care centers, preschools, nursery schools, crèches, kindergartens, and primary classes developed in Europe and spread throughout the world under the impetus of the Industrial Revolution and colonial expansion (Woodill, 1992).

Today early childhood education and care (ECEC) is an international phenomenon, but it is by no means a universal one. Despite the ubiquity of factors that have increased the need for extra-familial provision (increased rates of female labor force participation, increased numbers of single-parent families, etc.), early care and education services, in general, are most prevalent in high income nations and least prevalent in nations that struggle to meet more fundamental needs for food, shelter, and health care (UNICEF, 2000). Even primary schooling is beyond the reach of more than 100 million of the world's children (UNESCO, 1990). Thorvald Moe (2001), Deputy Secretary General of the Organization for Economic Co-operation and Development (OECD), recently suggested that there are three primary reasons for the increasing international interest in early childhood education at the turn into the new millennium. First, it has become a means to strengthen short- and long-term educational, emotional, and social outcomes. Secondly, it can be used to foster equity and achieve social objectives. And, finally, it is a response to increasing rates of female labor force participation.

This chapter describes recent developments in early childhood education and care (ECEC) policy at a time when the sheer rapidity of local and global change is unprecedented. Recent efforts to establish common "indicators" of child well being have been constrained by economic disparities, national interests, and by what Adamson (1996) refers to as a "minefield of cross-cultural value judgments" (p. 2). This becomes apparent when nations are to be compared on specific criteria.

> If the indicator used is the rate of child survival, or adequate nutrition, or access to primary health care, then the implied value judgments, although certainly present, pose few problems because they enjoy a high level of acceptance in all societies. In other words, there is broad consensus on the meaning of progress and therefore on what it is that ought to be measured. But beyond such basics it is possible, indeed, desirable, to imagine a much greater degree of diversity in concept and definition of progress, in underlying values, and

therefore in choice of what it is that should be measured and by what means (Adamson, 1996, p. 1).

At the same time, Western nations, especially the United States, have exerted strong influence on early education and care internationally (e.g. Lamb, Sternberg, Hwang & Broberg, 1992; Cochran, 1993). In her description of early childhood education in Korea, Lee (1997) specifically mentions the major influence of the American individualized approach to education. The Project for Early Childhood Education in Jamaica was heavily influenced by the U.S. Head Start model (Morrison & Milner, 1997). Trinidad and Tobago have used *The Guidelines for Developmentally Appropriate Practice* (DAP) (Bredekamp, 1987), developed by the U.S.-based National Association for the Education of Young Children (Logie, 1997), and, in Hong Kong also, DAP has been influential (Mellor, 2000). Starnes (2000) describes how an early childhood learning center was established in Azerbaijan in 1997 based on U.S. early childhood education programs, and, in Zimbabwe, "Western preschool values" are reported to have been adopted by early childhood educators (Cleghorn & Prochner, 1997, p. 349). In a review of Caribbean countries, Davies (1997b) notes that Western industrialized nations have also been influential in the recent emphasis on standards.

Some Western writers argue that Minority World values have had undue influence in the Majority World (e.g. Dahlberg, Moss & Pence, 1999; Davies, 1997b; Penn, 2000; Woodhead, 1997, 2000). Penn (1999), for example, maintains that the DAP guidelines reflect American cultural norms that stress values of individualism, independence, choice, and material ownership, values that might be out of sync with lived realities in other societies. Along these same lines, Woodhead (1997) suggests that DAP be joined with an equally important principle, Contextually Appropriate Practice or CAP, in order to develop practices appropriate to the context of early development in diverse societies. Yet he cautions that such efforts are "a ripple against the tidal wave of globalization" (p. 79). LeVine et al. (1994) illustrate just how divergent contextual child rearing practices can be in their description of the Gusii "pediatric model" as compared to the American "pedagogical model" (pp. 248–256).

This chapter has four purposes: (1) to provide an international context for discussing ECEC policy reforms related to globalization, including international efforts to forge a shared vision of children's rights; (2) to review cross-national studies of ECEC policy; (3) to use case study examples as a way to highlight how historical precedent and contextual factors influence responses to globalization; and (4) to suggest a values-based, contextual framework for international ECEC policy research.

EARLY CHILDHOOD EDUCATION IN A NEW WORLD ORDER

Globalization

Globalization is often described as a relatively recent phenomenon, although its origins can be traced to the birth of capitalism in the 16th century. Carnoy (1999) distinguishes between a world economy and a global economy. Commerce and cultural sharing across national borders has existed for centuries, but what makes globalization distinctive is what Webster (1998) calls "the growing interdependence and inter-penetration of human relations alongside the increasing integration of the world's socioeconomic life" (p. 141).

Globalization is characterized by dramatic changes in the economic, political, and cultural realms. Elements of economic restructuring include: a shift from manufacturing to high-tech information and services, from Fordist mass production to post-Fordist flexible production, increasing internationalization of trade, a growing disparity between a small number of highly-skilled and well-paid workers and a large number of low-skilled and low-paid workers nationally, and an increasing economic, technological, and cultural gap between high-income and low-income nations (Burbules & Torres, 2000, pp. 5–7).

In political terms, globalization has had somewhat contradictory effects, evident both in tightening government control of education in an effort to gain competitive advantage, or of decentralizing or privatizing responsibility. As Apple (2000) explains, neo-liberal policies result in a weak nation state, economic rationality, explicit connections between education and business, and decreases in public sector spending (although public spending on education can be viewed as warranted when the goal is to increase productivity). By contrast, neo-conservative policies affirm a strong state, a national curriculum and national testing, a "return" to standards, and increased state control of teachers and teaching. While neo-liberals support policies like vouchers and tax credits in the belief that the market, left to its own devices, will solve economic and social ills, neo-conservatives believe in the efficacy of a strong state. Apple argues that efforts to increase choice, raise standards and improve quality – what he calls a re-privatizing discourse – can function to justify inequalities. For example, if people can "choose" to send their children to better schools, it is their fault if they fail to access something better. Economic rationality has come to dominate educational discourse in English-speaking countries, promoting values of efficiency and cost-benefit and re-defining children as human capital and future workers. There is, however, no single way in which institutions such as education are being affected. For example, austerity measures may lead to a reduction of spending on education in one country, while, in another, it will lead to an increase in hopes of becoming more competitive (Burbules & Torres, 2000).

Cultural changes include the globalization of media and communication, the rise of a commercial culture, an increase in movement from one locale to another, and the internationalization of religious movements, sports, and entertainment. It has been predicted that globalization will lead to increasing homogenization and standardization, or, alternatively, to the co-existence of heterogeneity and homogeneity in an uneasy tension (Burbules & Torres, 2000, p. 11). What is clear is the advent of mass culture and the commodification of culture itself through satellite, cable, and digital technologies.

Although he does not address early childhood education and care directly, Carnoy (1999) maintains that there are five major ways in which globalization is having an impact on education. First, globalization is affecting the organization of work and the type of work people do. Increasingly, workers need to have a high level of skill and the flexibility to engage in multiple tasks at once or to change jobs easily. The average level of education thus needs to be raised world wide. Governments are now exerting increased control over what and how young children learn. In England, for example, the Government has instituted universal preschooling for four year olds and expects to provide universal preschooling for threes in the next few years. In addition, Early Learning Goals (QCA, 1999a) and guidance for teachers (QCA, 2000) lay newfound stress on literacy and numeracy. A national inspectorate is now responsible for all early years settings, and teacher training schemes are being consolidated into a national "climbing frame" (e.g. QCA, 1999b). Recent reforms in the United States likewise focus on literacy (e.g. Neuman, Copple & Bredekamp, 2000), numeracy (Copley, 2000), and teacher education (NAEYC, 1996, 2000). In the U.S., however, these reforms are primarily being advanced by professional associations (NAEYC, IRA, and NCTM) and by committees of the National Research Council that have written reports commissioned by Congress (e.g. National Research Council, 1998a, b, National Research Council, 2001; see Lubeck & Jessup, in press).

Second, with the need to raise the level of education, Carnoy contends that developing countries need to increase spending on education in order to attract foreign capital through the development of a more educated work force. However, this is occurring at the same time that many countries need to curtail public sector spending, in order to repay debts owed to foreign investors and governments. As Fig. 1 illustrates, only 9% of the world's population live in

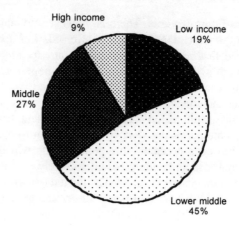

Fig. 1. Population of the World's People by Income Level of Country.
Source: World Bank, 1999.

"high income" nations and, in general, these countries have the most extensive educational – and ECEC – systems.

Developing countries are being advised to concentrate public resources on primary education (over secondary and tertiary) and to increase class size in order to accommodate more children. Carnoy maintains that governments increasingly are moving away from welfare-state policies and looking to the private sector. However, in a majority of the 12 OECD countries that participated in the recent *Thematic Review of Early Childhood Care and Education Policies,* responsibility for young children has increasingly shifted from private to public responsibility (OECD, 2001).

Third, nations are being compared on indices of child well being and educational achievement. Thus, there has developed an increasingly pervasive use of the language of standards and outcomes. The World Bank, UNICEF, and UNESCO collect data that allow for cross-national comparison, and the International Association for the Evaluation of Educational Achievement (IEA) conducts studies such as the recent Third International Math and Science Study (TIMSS) that compares educational achievement across nations. UNICEF (2000) primarily monitors for disparity reduction in terms of malnutrition, illiteracy, and disease, using explicit criteria to judge progress or regression. Katz (1989) has written that, "once basic survival risks have been addressed within a society, issues concerning children's quality of life in general, and their care and education in particular, begin to command attention" (p. 401).

Yet constant comparison also serves to underscore the fact that some nations are doing well and others not. In an increasing number of nations, international comparisons of achievement are affecting the curriculum, especially in the areas of mathematics, science, English, and communication skills. Cross-national comparison has also prompted an increased focus on literacy and numeracy in the early years.

Fourth, information technology is being exploited to increase the quantity of education (through distance learning) and the quality of education (through the Internet and computer-assisted instruction). Access to such technology remains limited, and the cost to schools to initiate and maintain technology can be burdensome. However, distance learning is increasingly being used to train early education staff in remote areas, for example, in Australia and Fiji, as well as to offer on-site training to Head Start staff in the U.S. In addition, computers have become commonplace in many preschool classrooms in the West.

Fifth and finally, globalization changes local culture and has resulted in groups feeling marginalized. Some groups, and here Carnoy mentions fundamentalist Muslims and postmodern environmentalists, are decidedly anti-market and resist the values implicit in global capitalism. But global capitalism is "resisted" in other ways. This is dramatically illustrated when looking at early education and care cross-nationally. The United States has a largely privatized, market-driven approach to child care and early education, and unpaid parental leave is granted only to those who work in companies of 50 employees or more. This provides the context for a recent television special celebrating the fact that more and more businesses are allowing mothers to bring their infants to work! Public monies, in the form of Head Start, state-funded pre-kindergarten, and Block Grants to states, are targeted at children who are vulnerable due to poverty, disability, or other risk factors, but funding limitations mean that all children who qualify are not served. By contrast, it is common in Europe for nations to ensure that mothers have an extended period of paid maternity leave, fathers paid paternity leave, and all or most children access to preschool education (see, for example, the case studies that follow).

Nonetheless, Carnoy contends that the United States, through its influence in the International Monetary Fund (IMF) and the World Bank, has had enormous influence on international educational reform efforts. The current dominant view is based on the free play of the market. "Globalization enters the education sector on an ideological horse, and its effects on education and the production of knowledge are largely a product of that financially-driven, free-market ideology, not of a clear conception for improving education" (Carnoy, 1999, no. 1354, p. 59). He argues that finance-driven reforms, which increase inequities in education, dominate reform efforts. Thus, he calls for a coherent approach to educational improvement, one that is not based simply on financial objectives but rather on "educational process and practice *within the context of globalization* rather than on globalization's financial imperatives themselves" (p. 60; emphasis in the original). As this discussion suggests, globalization is having profound, albeit varying, effects on education world wide. In the next section, we describe international efforts to address children's issues, including their care and education, in a more concerted way.

Children in a "Global Village"

The latter decades of the twentieth century have witnessed heightened attention to children's well being and, indeed, to children's rights. The 1980 International Year of the Child has been described as a "turning point" (Davies, 1997a, p. 7). From that date, the nations of the world began collecting basic statistical data on child well being, and national governments and international agencies committed to improving the conditions that militate against children's life chances. In 1989, the United Nations Convention on the Rights of the Child was held. The Convention created the possibility of becoming "the source of a new concern with the personhood of children and a world-wide revolution in children's policy, [one that offered] legislatures guideposts that could have spectacular effects on children and families" (Melton, 1993, pp. 517, 520). Children were no longer viewed as the property of their parents or as objects of charity. Instead there was an "understanding that the child is an independent human being and should therefore be treated as a *subject* with his or her own inalienable rights" (Meljeteig, 1994, no. 1449, p. 1; emphasis in the original). Nonetheless, there is a risk that children's needs and rights will be overlooked in a time of global restructuring (Melton, 1993).

Four articles of the Convention are of particular relevance with respect to early childhood education and care (ECEC). The first is Article 18 which addresses the responsibility of parents to bring up their children and of the state to assist parents through the development of services for the care of children. Article 18 also specifically addresses child care stating that "State Parties shall take all appropriate measures to ensure that children of working parents have the right to benefit from child-care services and facilities for which they are eligible" (UNICEF, 1989). Articles 28 and 29 address children's right to education, and Article 27 the rights of children to an adequate standard of living. States are directed to assist parents in the implementation of this right. Since all but two countries (the U.S. and Somalia) have ratified the Convention, it can be assumed that these emphases on education, child care, an adequate standard of living, and the role of the State in ensuring these rights for children have influenced developments in ECEC over the last decade.

In 1990, the World Conference on Education for All (EFA) also drew attention to the education of the world's children. The Conference, jointly sponsored by the United Nations Development Programme (UNDP), the United Nations Educational, Scientific, and Cultural Organization (UNESCO), the United Nations Population Fund (UNFPA), the United Nations Children's Fund (UNICEF), and the World Bank, recognized "the need to prepare young people to lead productive lives in rapidly changing societies" (World Bank, 1999, p. 36). It resulted in the "World Declaration on Education for All: Meeting Basic Learning Needs," which included six target areas. The first pertains to early childhood, focusing on "expansion of early childhood care and developmental activities, including family and community interventions, especially for poor [and] disadvantaged [children] and children with disabilities" (Johnson, 2000, p. 1). The Convention also initiated the EFA 2000 Assessment, producing detailed reports on the status of education in over 180 countries (see the UNESCO web site: unesco.org).

Cross-National Studies of ECEC Policy

Cross-national comparisons of child well being and delivery of children's services are conducted for a number of reasons. One is to provide information for policy making. The World Bank, UNICEF, and other international organizations state that they use data as a basis for planning and intervention (e.g. World Bank, 1999; UNICEF, 2000). Similarly, the stated purpose of research undertaken by the International Association for the Evaluation of Educational Achievement (IEA), a non-profit consortium of research institutions in 55 countries, is to provide data to nations to enable them to base policy decisions on research findings.

A second reason for cross-national research is that it provides access to alternatives that do not exist within a particular national context (e.g. Robinson, Robinson, Darling & Holm, 1979; LeVine & White, 1984). As LeVine and White (1984) write:

> Alternative cultural inventions and differing directions of historical change constitute the great neglected resource for evaluating educational policies, because they raise fundamental questions about the status quo and provide empirical contact with possible futures (p. 20).

Cross-national research provides a way to "trouble" taken-for-granted assumptions. For example, Melhuish (1993) argues that "simple, global, generalisations about the effects of out-of-home, preschool experience on later development can not be sensibly justified" (p. 19) and "lead to too narrow a view of human development" (p. 29).

Still others see commonality rather than distinction. Some writers hold that the study of other countries is valuable as a way "to discover common challenges that unify early childhood educators as they strive to put the world's best educational theories and research into daily practice" (Jalongo, Hoot, Pattnaik, Cai & Park, 1997, p. 173).

A list of reasons for conducting cross-national research would not be complete without also pointing out some of the difficulties inevitably encountered. These include: (1) the absence of a universal definition of early childhood; (2) varying philosophies and pedagogies; (3) differing political, economic, social, and cultural contexts; and (4) the conflicting ideologies in which early childhood education is embedded in each country (Woodill, Bernhard & Prochner, 1992). Levin (2001) suggests that "the study of education reform is complicated even in a single setting, so trying to make comparisons across settings is fraught with additional difficulties. One inevitably risks drawing comparisons without full knowledge of local circumstances, and seeing as similar what would, with closer analysis, look quite different" (p. 16).

In the subsections that follow, we briefly describe case and comparative policy studies of early childhood education and care published before and after 1995. Case studies describe ECEC policy within particular national contexts, while comparative studies compare across nations. Due to space constraints, this review is, of necessity, selective. We do not review studies of single policy issues nor do we review international studies of pedagogical and other issues pertaining to young children. The review is also primarily limited to relatively recent studies and reports published in books and journals written in English. Because of rapid and, in many cases, dramatic social change, studies only a few years old will not be current, although they still provide useful historical information.

Case Studies Prior to 1995

The "cases" reviewed in this section focus on early childhood education and care in particular countries. Cases make it possible to understand how values regarding children and child care are rooted in particular ways of life and how context makes certain forms and approaches possible. Sheila Kamerman and her colleague, Alfred Kahn, have done the most in-depth international case and comparative studies of family and child-care policies over time (see, for example, Kamerman & Kahn, 1981, 1995; Kamerman, 1991).

One of the major efforts to study early childhood policy issues in a number of countries was proposed by IEA in the late 1970s. The High/Scope Educational Research Foundation became the international co-ordinating center for the Pre-Primary Project, with each participating nation setting up its own research center responsible for securing funds and carrying out research within the country.

The Project's first publication, an edited volume, was entitled *How Nations Serve Young Children* (Olmstead & Weikart, 1989). The authors use information from public records to draw profiles of early care and education in 14 nations. The original 14 nations participating included: Belgium, the Federal Democratic Republic (before unification of East and West Germany), Finland, Hong Kong (before it became a part of China), Hungary, Italy, Kenya, Nigeria, the People's Republic of China, the Philippines, Portugal, Spain, Thailand, and the United States. Each chapter describes ECEC services in a particular nation. In her review of this work, Katz (1989) describes three themes that weave through the chapters: the lack of availability of pre-primary services, the lack of co-ordination and cohesion in service delivery, and the lack of a skilled workforce, due to the low salaries and low status attributed to work with young children.

Another edited volume, *Day Care for Young Children: International Perspectives* (Melhuish & Moss, 1991) provides two chapters on each of five countries: France, Sweden, the (former) German Democratic Republic, the United Kingdom, and the United States. Leave policies, forms of extra-familial provision for children under three, factors that influence use, and developments over time are described in the first chapter on each nation; the second reports on research on day care.

In 1992, Woodhill, Bernhard, and Prochner published the *International Handbook of Early Childhood Education* (1992) and Feeney published a source book entitled *Early Childhood Education in Asia an the Pacific* (1992). The former is an overview of early care and education in 45 countries on five continents. Each of the country chapters has information on the history of early childhood education and current provision. Some are relatively brief, while others compile comprehensive information on legislation, curriculum, special education provision, teacher training, and parent support and involvement. The latter provides histories of early childhood education and descriptions of policies and provision in countries in Asia and the Pacific.

In addition to books that include case studies of a number of countries, there are also case studies that focus in depth on only one country. For example, Piscitelli, McLean and Halliwell (1992) and Clyde et al. (1994) provide broad overviews of child care policy, provision, and research in Australia, and Brennan (1994) traces the transition of child care from the periphery to the mainstream of the Australian political agenda, focusing on the role women in this transition.

Comparative Studies Prior to 1995

Some writers have sought to develop typologies, dimensions, or frameworks by which to compare across nations. In a relatively early effort, Robinson, Robinson, Darling and Holm (1979) joined with other professionals from 12 countries to form the International Study Group for Early Child Care. The authors were concerned with the lack of coherent policy for ECEC in the United States and looked to systems in other industrialized nations for alternatives. Robinson and her colleagues clustered these countries into four models of care based on issues such as the rationale, funding, oversight, and implementation of ECEC programs. The four models were: (1) Latin-European, (2) Scandinavian, (3) Socialist, and (4) Anglo-Saxon. Since this framework was developed, ECEC in many of the countries studied has changed somewhat or been quite radically transformed.

At the time, however, the authors argued that the Latin-European approach characterized ECEC in France, Belgium, French-speaking Switzerland, and Italy. They saw a clear distinction between the crèche system for those under three, was overseen by the Ministry of Health and Welfare, and the educationally focused preschools for those ages 3 to school age administered by the Ministry of Education. The state set policy, paid teachers' salaries, and participated in teacher training, but programs were locally administered. There was no standard national curriculum. The child-care system functioned within the context of family policies in these countries – policies that included such benefits as maternity leave, children's allowances, and housing allowances.

The Scandinavian model was based on "the individual's right to assistance from the community when in need" (Robinson et al., 1979, p. 98) and the idea that child rearing should be shared by parents and society. Social supports were viewed as socially and economically beneficial to the community as well as the individual. There was also an emphasis on sexual equality as more women entered the labor force. Thus, there were extensive family support policies such as parental leave, child allowances, and publicly provided child care, that, in concept, provided "comprehensive, integrated programs of developmentally oriented care for children from birth to school age" (p. 99). Program guidelines from the central government were flexibly implemented at the local level and were concerned more with social and emotional development than academics.

Family support policies also underpinned child-care policies in the Socialist model. The age-separated model of crèches and kindergartens, similar to the Latin-European model, was seen as moving toward an integrated care model with attention to cognitive and social development as well as physical care. Services were highly centralized but also broadly available. A highly developed curriculum emphasized structured work, spontaneous play, and the development of an awareness of the group as opposed to the individual. School readiness and creativity were emphasized. Although the USSR and former Communist Bloc no longer exist, residues of this approach can still be seen in former Socialist nations.

The Robinson team defined the Anglo-Saxon model as a "far less tidy affair" than the other three. Here they saw a strong emphasis on self-determination, independence, and a limited role for government. Voluntary charity was seen as the way to care for those in need rather than government intervention. Consequently, there was not one system of care but multiple systems, some public, some private; some regulated, some not. "The general picture in the Anglo-Saxon countries is one of a highly decentralized, fragmented collection of child-care arrangements, most involving little coordination with other social services affecting children or families" (Robinson et al., 1979, p. 108). There was support, however, for parental involvement and for diversity in programming.

Drawing on a cross-national survey of legislative provision conducted by the International Labor Office (ILO), Lubeck (1989) compared child-care policies in 13 industrialized societies. Family-based policies included national health insurance, job protection, parental leave, and cash benefits during the leave period; extra-familial provisions, in terms of publicly-funded preschools, were also examined.

Cochran (1993) used case studies from 29 countries for cross-national comparison. Each case study outlines: (1) extant early childhood education policies and programs; (2) the historical, cultural, political, societal, and ideological context in which the program and policies are embedded; (3) the links between early childhood and broad policy goals of the country; and (4) the strengths and challenges of the early childhood program. From these case studies, Cochran identified more than 60 themes and developed a three-part conceptual framework consisting of causal factors that stimulate particular responses, mediating influences that affect how societal needs lead to particular kinds of policies and programs, and resulting emphases that are incorporated into national policies. He maintains that similar causal conditions will not necessarily result in similar policies. "Outcomes will differ if the economic, political, cultural, and social contexts surrounding those objectives – the mediating influences – are dissimilar" (p. 8).

Lamb, Sternberg, Hwang and Broberg (1992) place the development of child care in 18 countries in its sociocultural and historical contexts. They also discuss four major ideological dimensions on which these societies vary. These ideologies concern: (1) equality between men and women, (2) child care as public or private responsibility, (3) child care as social welfare or an early educational program, and (4) basic conceptions of childhood and the developmental process (pp. 8–9). In light of these ideologies, child-care systems can be seen as "manifestations of the wider social structure" (p. 11), and the development of these systems as a " multifaceted and richly cadenced process" (p. 22) that must account for historical, sociocultural, and ideological dimensions.

Case Studies Published After 1995
In 1998, the Organization for Economic Cooperation and Development (OECD) launched a 12-country study known as the *Thematic Review of Early Childhood Education and Care Policies.* The participating countries included Australia, Belgium (Flemish and French Communities), the Czech Republic, Denmark, Finland, Italy, the Netherlands, Norway, Portugal, Sweden, the United Kingdom, and the United States. The overall project had four components: (1) Background Reports prepared by in-country researchers and guided by a common framework; (2) external review team intensive case study visits; (3) Country Notes written by members of the review team; and (4) a final Comparative Report released in Stockholm in June, 2001 (OECD, 2001). The primary areas of interest for these studies were governance, regulation, funding and financing, program content/curriculum, staff recruitment and retention, and family engagement and support. Background Reports and Country Notes can be accessed on the OECD web site (www.oecd.org).

Other recent investigations have also focused on specific countries. These vary in coverage and detail, with the majority reporting on ECEC policy in Western industrialized nations. To frame our discussion of this section, we identified policy issues currently being addressed in nations throughout the world. These include: (1) governance and regulation, (2) funding and financing, (3) access, (4) curriculum, (5) staff recruitment and credentialing, and (6) parent involvement. We conclude with challenges to the development of ECEC revealed in these studies.

Governance and Regulation. Governments have played varying roles in developing and supporting ECEC programs. In general, developing countries can be characterized by minimal involvement of government, with philanthropic and non-governmental organizations (NGOs) assuming primary responsibility for ECEC. India has relied on NGOs for services and funding (Munirathnam, 1995, Sharma, 1995), although governmental involvement at the policy level has increased over the last few decades (Jalongo et al., 1997). In 1985 India established the Department of Women and Child Development under the Ministry of Human Resource Development in order to coordinate governmental and NGO efforts within India (Sharma, 1995). In Kenya, a major ECEC effort was spearheaded by foundation funding, but "parents and local communities are the most important partners in the ECCE program. They have started and maintained 75% of the preschools in the country" (Swadener, Kabiru & Njenga, 1997, p. 287). In Zimbabwe, rural crèches were started by the main political party after they achieved independence in 1980. Although the government emphasizes the importance of a renewed educational system, no funds are provided (Cleghorn & Prochner, 1997).

In the Caribbean, countries such as Grenada, Guyana, and Cuba have the most developed programs due to high levels of government involvement. Although hindered by limited resources, other countries have seen increased involvement of the government, usually in the form of program regulation and monitoring (Davies 1997a, b). In Jamaica, the Early Childhood Education (ECE) Unit of the Ministry of Education oversees all types of ECE (Morrison & Milner, 1997). The Bahamian interest in quality and standards will be addressed via program licensing and monitoring, which has begun with legislation requiring all ECE organizations to register with the Ministry of Education (Johnson, 2000). Although the Netherlands is known for its comprehensive system of ECEC that emphases care for minority and at-risk children, an investigation of national policy highlights the minimal role of government in enforcing child-care quality (Singer, 1996b). The Japanese government has recently taken a more active role in ECEC through the establishment of standards for *hoiku mama* (family day care) (Newport, 2000). Although Canada does take responsibility for its children through various social support policies, it also has a rather fragmented approach to ECEC (Friendly & Oloman, 2000; Prochner & Howe, 2000). As a result of the Canadian government's diminishing role during the 1990s, responsibility for services is now at the local level (Friendly & Oloman, 2000). No public system of ECEC exists in Hong Kong as "government rhetoric about the importance of preschool has not been matched with financial support" (Mellor, 2000, p. 108).

Funding and Financing. Closely tied to the role of government is the source of funding for ECEC. Generally, if the role of government is minimal, so is the amount of public funding. For example, in 1998 only 15 government-run programs existed in the Bahamas, a country where government involvement in ECEC is quite recent (Johnson, 2000). A number of other Caribbean nations – Antigua-Barbuda, Belize, Dominica, Grenada, Jamaica, Montserrat, St. Lucia, St. Kitts, St. Vincent, and Trinidad and Tobago – allocate only 0.1 to 3% of their budget for ECEC, with the majority of these funds going toward preschool (ages three to six) (Davies, 1997a). In contrast, countries such as Barbados, Grenada, and Guyana, with high government involvement in ECEC, have higher amounts of public funding (Davies, 1997a). Overall, in the Caribbean nations, however, ECEC is primarily privately funded.

Many developing countries, including Azerbaijan, India, and numerous African nations, receive funds from international agencies and foundations (UNICEF, UNESCO, the World Bank, the Bernard Van Leer Foundation, etc.), in order to begin or maintain ECEC (Cleghorn & Prochner, 1997, Davies, 1997a, b). This can also result in a mix of publicly and privately-operated programs. For example, India, which had relied on NGOs, established a National Crèche Fund in 1994 that increased public funding of ECEC (Sharma, 1995).

Most higher income countries have systems supported by public funds. However, in a comparison of Denmark and the United States in 1997, Polakow (1997) characterizes the two countries as public and private respectively, given universal coverage through social policies in Denmark and the relative lack of social policies in the U.S. In Ireland, only junior classes for four year olds and kindergarten at age five are paid for by the government; all other programs are paid by voluntary organizations or run privately (Noirin, 1996). "Governmental support for education, particularly early childhood education, has been extremely minuscule in Korea" with only 1% of the education budget allocated to ECE (Choi, 1999, pp. 248–249). Significant funding cuts in countries such as Canada, Hong Kong, and New Zealand place the burden on families to cover the costs of ECEC (Friendly & Oloman, 2000; Meade, 2000; Mellor, 2000), although the picture in New Zealand is beginning to change. Prochner and Howe (2000) compiled a group of essays that provide historical background and an overview of current child-care programs and issues as a basis for understanding the current fragmented approach to ECEC in Canada. Prochner and Howe (2000) maintain that ambivalence regarding working mothers and child care, as well as public versus private responsibility for the care of children, has resulted in a reluctance to regulate child care in Canada, as well as a willingness to cut funding for kindergarten in recessionary periods.

Government funded programs are often targeted at children in poverty as they are in Trinidad and Tobago (Logie, 1997) and, in the United States, through the federal Head Start program and federal Block Grants to states. Korea also funded the Semaull Head Start preschool programs for low-income families, but this program no longer exists (Lee, 1997).

Access. Ensuring the availability of programs and enrollment of young children is of concern in many countries, especially amidst fluctuating birth rates and increased maternal employment. Government statistics and research are primary sources of data regarding availability and participation. These data reveal disparities within and among countries. India's 150 million children aged birth to six years are served by 130 programs, indicating a huge disparity between supply and demand (Sharma, 1995). Rayalaseema Seva Samithi (RASS), established in 1981 to address issues of child care in one district of India, currently serves 20,000 children in ECE centers and 9,000 children in day care centers/crèches (Munirathnam, 1995). Approximately 8,000 ECE centers exist in Zimbabwe, with the majority in rural villages (Cleghorn & Prochner, 1997). Jamaica has a number of programs – basic schools, infant schools, infant departments, independent/preparatory schools, and day care – serving 90% of four to six year old children, yet a lack of spaces prohibits the country from reaching all children under age six (Morrison & Milner, 1997). Out of the 14,000 four year olds in Trinidad and Tobago in 1995, 50% attended ECEC programs, indicating a decline that began in 1986 (Logie, 1997). As of 1997–1998, 1296 preschool programs (public and private) exist in the Bahamas, and 1999 data indicate that approximately 90% (12,476) of children in Grade 1 attended an organized early childhood education program (Johnson, 2000).

New Zealand estimates from the 1990s show that more than 50% of children from birth to age five, including almost all four year olds and more than 80% of three year olds, are enrolled in ECEC. Increasing enrollments, especially among Maori and Pacific Island children, have been occurring for the last decade (Meade, 2000). 1999 data from Japan indicate that 503,000 (14%) children under three and six million (80%) three to six year olds attend *yochien* (partial day kindergarten) (Newport, 2000). Similarly, 80% of three to five year olds in Hong Kong are enrolled in kindergarten (Mellor, 2000). In Korea, 15% of children from birth to three years were enrolled in day care centers, and 27% of three to five year olds attended kindergarten in 1999 (Choi, 1999). These figures reflect the lack of day care programs and the burden placed on families to pay for kindergarten (Choi, 1999). Although most mothers stay home with their children until age four or five in Korea, increased female employment and education necessitate the development of more programs for children under age three (Choi, 1999; Lee, 1997). Even established systems, such as those in Japan, need to respond to recent societal and economic changes that have increased the need for care and produced new challenges such as transportation, scheduling, and costs (Newport, 2000).

Research in the Netherlands indicates that child-care arrangements are used most by families with more educated parents (Singer, 1996b). In Greece in 1996, day care centers were primarily available in Athens with limited availability elsewhere in the country. Few care options existed for children under age four (Petrogiannis & Melhuish, 1996). Hayden (2000) argues that Australia's National Standards for Childcare emphasize quantity over quality in order to serve as many children as possible. This brief overview, although not representative, highlights the fact that more preschool children have access to services than children birth to age three.

Curriculum. Decisions about ECEC are being made within particular political, economic, and cultural contexts, resulting in a range and mix of emphases including academic readiness, socialization, culture, basic needs, and preparation for employment. Historically, most programs in Caribbean countries began with an academic focus (Davies, 1997b). A national standardized curriculum, The Readiness Programme, was developed by the Bahamian Ministry of Education's Preschool Unit and implemented in 1996 (Johnson, 2000). Jamaica has specified their national goals for ECEC as "the promotion of physical, social, emotional, cognitive, and creative competence through a child-centered approach geared to the Jamaican situation and experience" (Morrison & Milner, 1997, p. 52). Private programs in Trinidad and Tobago spend more time on academic skills, limiting time for free play and interaction with other children (Logie, 1997).

Ireland supports programs with a focus on readiness for four and five year olds. As in the U.S., publicly funded programs for children from birth to age three are for children who are economically disadvantaged or who have diagnosed disabilities. The primary goal of these programs is to ameliorate the effects of poverty (Noirin, 1996). In Japan, the focus during preschool is education and socialization, with a shift from "intellectual training to a more holistic approach" (Newport, 2000, p. 70). New Zealand's recent (1996) guidelines for a national curriculum also subscribe to a holistic approach that addresses the cognitive, social, cultural, physical, emotional, and spiritual aspects of children. It also is inclusive of Maori and other Pacific Island cultures and language and children with special needs (Guild, Lyons & Whiley, 1998). Notably, the *Te Whariki* curriculum is written on each page in both

English and Maori, and the one is not simply a translation of the other (Carr & May, 2000). Recent changes in the system of ECEC in India has led to an approach that is "child-centered, context-oriented, and responsive to the needs of the individuals and groups" (Jalongo et al., 1997, p. 180). In Korea, importance is placed on the role of mothers in their children's education with most children staying at home with their mothers until they go to *yoo chee won* (kindergarten) at age four or five. The focus is on preparing children "to adapt to society as an independent and responsible person" and to learn Korean culture and tradition (Lee, 1997, p. 49). According to the Early Child Education Development Law, the goal is to promote individual development so children can contribute to the future of the country (Choi, 1999). Specialized programs focusing on mathematics, language, art, and tae kwon do are available to children at age four (Lee, 1997). Hong Kong's two types of ECEC include kindergartens for children aged three to five and day nurseries for children aged two to six. Kindergartens emphasize academic skills for later school success and preparation of children for a technologically based economy, while day nurseries focus on the "development of the whole child" (Mellor, 2000, p. 102). Reflecting the focus on basic needs of children, the expectations of preschool programs in Kenya include health and nutrition, cultural transmission and maintenance, safety and custodial care (especially for children under age three), and school readiness (Swadener, 1997).

The recent shift from a planned to a free market economy in Lao PDR has resulted in a focus on education in order "to provide children with skills that will drive both economic change and prepare them to take part in the new economic structures" (Emblem, 1998, p. 36). There is a specific focus on socialization and skill acquisition, and children are viewed as active participants in their learning (Emblem, 1998). "Education in China has always been viewed as an avenue to train future leaders and citizens as well as to cultivate individual morality," but early childhood education is seen as helping children become ready for primary school (Jalongo et al., 1997, p. 175). These sources reflect varying degrees of government involvement, as well as broad and narrow goals for early education.

Staff Recruitment and Retention. The training, education, and salaries of teachers vary significantly across countries and closely follow the degree to which ECEC is developed in a particular country. ECEC programs in Azerbaijan and Fiji are not well developed nor is there a strong teacher base. In Azerbaijan teachers have minimal monthly salaries, of which they usually receive half (Starnes, 2000). Starnes (2000) claims that most teachers are not qualified or trained, and that they obtain their jobs through nepotism and bribes.

Most teachers in Fiji have no formal training, although a certificate program and training workshops have been implemented by the University of the South Pacific (Crane, 1998).

Finding trained teachers is problematic in other countries as well. A national survey of 175 teachers working in public and private programs in Trinidad and Tobago indicated that 54.5% have more than five years of experience; half are young (under age 20); one is male, and 57% in private programs have no ECEC training. Fully 89% of teachers' salaries were below the poverty level (Logie, 1997). Instruction is primarily teacher-initiated, with children sitting in a group listening to the teacher (Logie, 1997). The lack of qualified teachers in the Bahamas resulted in the development of an associate's degree and teacher certification program in 1990, with a preschool auxiliary teachers certificate program added in 1994, both through the College of the Bahamas (Johnson, 2000). The Bahamas Baptist Community College, along with other institutions, has also established short-term certificate programs for preschool teachers and owners (Johnson, 2000). In Jamaica, Basic School teachers have minimal qualifications, but on-site professional development and training are operated by the Early Childhood Education Unit of the Ministry of Education. Infant Schools and Departments have college-educated teachers who are paid according to an established scale (Morrison & Milner, 1997). A two-tiered system of a certificate and a diploma exists in New Zealand (Meade, 2000). Teachers in India undergo two-year training programs that prepares them for teaching in preschools or nursery schools, and recently established education colleges are also involved in preparing early childhood educators (Jalongo et al., 1997).

Countries with high incomes, such as Korea, tend to have strict requirements for teachers that can be fulfilled through established educational and training programs. Korean teachers must be certified and most attend two-year programs (Jalongo et al., 1997; Lee, 1997). Singer (1996a) found parents in the Netherlands to be less concerned with structural aspects of quality and more concerned with the qualities of the individuals caring for their children.

Parent Involvement
Involving parents in policy decisions and programming is an ongoing issue. For example, Dutch parents prefer informal arrangements, while the government focuses on center-based care, indicating the lack of parental input in policy decisions. This mismatch between parents' and policymakers' views suggests the need to incorporate parents into policy decisions (Singer, 1996a, b).

Challenges and Future Directions. Recent publications call attention to challenges faced by many countries. Low income countries face issues of funding, access, staffing, health, safety, nutrition, and lack of materials (Cleghorn & Prochner, 1997; Munirathnam, 1995; Sharma, 1995; Starnes, 2000; Swadener et al., 1997). Middle income countries such as Jamaica, Trinidad and Tobago are striving to establish national standards and an accreditation process, provide training for staff in rural areas and those working with children under age three, collaborate with social service agencies, and increase materials and resources (Logie, 1997). As Morrison and Milner (1997) state, "the challenge now is to both maintain them and extend their scope, particularly in the face of continuing economic decline" (p. 57). Seeing the benefit of regional collaboration, 19 Caribbean countries agreed, at the 1997 Regional Conference on ECE, to a Plan of Action aimed at program expansion and improvement (Davies, 1997b).

Higher income countries with more established systems of ECEC nonetheless confront issues of governance and regulation, funding, access, and staffing, among other challenges (e.g. OECD, 2001). For example, with no national policy, limited access, and the decentralization and privatization of services, Friendly and Oloman (2000) describe Canada in the throes of "a new childcare crisis" (p. 72). Mellor (2000) suggests that Hong Kong faces a different challenge. Strongly influenced by Western practices, Hong Kong yet needs to adapt these notions to a Chinese-influenced culture.

This discussion points to some of the current tensions in ECEC, tensions between centralized and decentralized approaches, between care and education, between broad and narrow goals regarding program content, and between increasing expectations and improving salary and working conditions for early childhood practitioners. Access is an issue within and across countries.

Comparative Studies Published After 1995
A second publication from the IEA Pre-primary Study (Olmstead & Weikart, 1995) appeared in 1995. While the first edited volume on the IEA Pre-primary Study was based on public records, the data reported in this volume derive from a household survey conducted within each participating nation with results compared across nations. Hungary, Kenya, and the Philippines are no longer represented, but some cross-national comparisons are included. Focal children for the survey were those $3\frac{1}{2}$–$4\frac{1}{2}$. The principal data-gathering technique involved 50-minute interviews eliciting information on parents' use of ECEC services during a typical week, a typical day in a child's life, families' household composition, parents' education and occupation, and information regarding specific country concerns.

A 1995 review by Kamerman and Kahn (1995) focused on infant and toddler care in the United States and selected European countries, namely Denmark,

England, Finland, France, Italy, and Sweden. The authors frame their examinations around the questions of "what are the programs like, what is their quality, how are they funded, and what kind of financial burden do the parents carry?" (p. 1287). They use the descriptions of European systems to suggest how similar policies could be utilized in the United States.

More recently, Penn (1999) considers what comprises good infant and toddler care in an international context. She contrasts the assumptions about children and childhood in which Anglo-American child development studies are based with the assumptions of European countries. In the Anglo-American context, Penn argues that children become commodities in market based systems. Children are both " a commodity which can be handled in a variety of ways to increase productivity" (p. 12) and part of the consumer market, as consumers themselves and as objects of consumerism, as seen in the extensive commerce around child-care material, equipment, etc. In European countries, by contrast, state provision for children is the norm with comprehensive care common for those three and over and limited provision for those under three. Penn calls attention to variations in state provision of service between and within the countries of the European Union and Eastern Europe. In 1999, she classified day care as social welfare based (Denmark, Sweden, Finland), health system based (France, Belgium), education based (France, Belgium), market based (U.K.), and communist (Eastern Europe). Emphasized is the connection between quality of care and public funding, which stands in contrast with the assumptions of the market system seen in the U.K. and the U.S. More detailed analysis is provided through a focus on five themes of practice: (1) the nature of relationships between adults and very young children, (2) children as learners, (3) health and well-being, (4) training and employment of those that work with young children, and (5) ecology and the environment.

Lubeck's (1995) cross-national comparison of the (former) German Democratic Republic, the United States, and France treats each nation as a unit of analysis that has integrity in its own right. Highlighting the influence of government, demographics, and views of children in regard to child care, she compares the three nations along dimensions of administrative and fiscal centralization and uniformity. While the GDR was high on both dimensions, the United States was low (i.e. more de-centralized and diverse). France, relative to the others, was illustrative of an approach that combined federal funding and local decision making, a uniform approach (the *école maternelle*) for children 4–6 and a more eclectic set of services for children 0–3. This representation is consistent with the more detailed descriptions of individual countries' policies and services regarding child care. She argues that, although efforts to change family support policies in the United States must be cognizant of the present ways in which ECEC is organized, as well as the historical circumstances that gave rise to them, cross-national comparison can suggest directions for change and alternatives for organizing, administering, and funding programs.

The International Child Care and Education Project (ICCE) includes data from the European Child Care and Education Study and the Cost, Quality, and Outcomes in Child Care Center Study in the United States. Two studies have investigated process quality in Austria (dropped from the later study), Germany, Portugal, Spain, and the United States using the Early Childhood Environment Rating Scale (ECERS) and Caregiver Interaction Scale (CIS) (Cryer, Tietze, Burchinal, Leal & Palacios, 1999; Tietze, Cryer, Bairrao, Palacios & Wetzel, 1996). The initial study found that countries differed on the ECERS, particularly on personalized care and the availability and use of space and play materials. Germany scored high on both dimensions, while Spain scored low on both, and the United States scored lower on personalized care and higher on space and play materials (Tietze et al., 1996). On the CIS Sensitivity scale, Austria and Spain scored better than the United States (Tietze et al., 1996). The authors include descriptions of the systems of care in each country, which help explain these differences. The second study added structural indicators of quality, which varied significantly across countries, yet only slight differences on the ECERS were found (Cryer et al., 1999). Using in-depth analyses, the authors found that differing emphases on structural aspects were related to process quality, although no single variable could be identified (Cryer et al., 1999).

The recent OECD *Thematic Review of Early Childhood Education and Care* is arguably the most in-depth cross-national examination of early childhood policies ever undertaken. Moe (2001) recently summarized some of the main findings of the OECD Comparative Report at a London conference. Across the 12 nations studied, the trend is toward expanding provision in order to achieve universal access, raising the quality of provision, promoting coherence and continuity, and ensuring adequate investment. Care and education have developed separately in most nations, with separate governing agencies, funding streams, and training schemes. A number of countries are now trying to consolidate care and education, as well as integrate services to better meet the needs of young children and their families. Cross-nationally the field is stymied by the low wages, limited career opportunities, and low status of the early childhood work force. In most nations, those working in day care centers and preschools make considerably less than those working in primary schools. Thus efforts are being made to improve training and working conditions. Many countries have developed pedagogical frameworks for young children. There is also broad interest in collaborations with parents, families, and communities. Overall, Moe stated that early childhood issues need to be addressed within broader frameworks aimed at reducing poverty, promoting equity, and valuing diversity.

Thus far, we have discussed some of the factors that are influencing early care and education policies and practices internationally, reviewed major studies of ECEC policies prior to 1995, and surveyed more recent studies (since 1995). In the next section, we look more closely at the policies of three nations that participated in the OECD Thematic Review: the Czech Republic, Sweden, and the United Kingdom.

CASE STUDIES OF THREE NATIONS

Case studies draw attention to the fact that globalization is not having a unitary effect on early education and care, as decisions are made within distinctive economic, political, and cultural environments. Table 1 displays development indicators for the United States and the three case study countries.

The United States, Sweden, and the United Kingdom are all considered high income countries by the World Bank; the Czech Republic is ranked as an upper

Table 1. Development Indicators for the United States and Case Study Countries.[1]

	Total population (thousands) 1998	Under 5 population (thousand) 1998	Annual no. of births (thousand) 1998	Under 5 mortality rate 1998	Per Capita GNP (U.S.$) 1997	Expenditure on Education (% of GNP)[2]	Pre-primary enrollment
United States	274028	19623	3788	8	29080	5.4	70%
Czech Republic	10282	503	88	6	5240	5.4	91%
Sweden	8875	508	86	4	26210	8.3	73%
United Kingdom	58649	3581	689	6	20870	5.4	30%[3]

[1] Except where noted, all statistics are from *The State of the World's Children* (UNICEF, 2000).
[2] Expenditure on education and pre-primary enrollment data are from the *1999 World Development Indicators* (World Bank, 1999).
[3] These data pre-date the Blair Government's initiative to establish universal pre-schooling. Universal provision for 4 year olds has been achieved, and provision for 3 year olds is expected to be achieved in the next few years.

middle income country (see Fig. 1 and Appendix A). Other countries included in the review of literature – Fiji, Jamaica, and Korea – are ranked as lower middle income countries, and Azerbaijan, India, Kenya, Lao PDR, and Zimbabwe as low-income countries.

As the table illustrates, the per capita gross national product (GNP) for the high income countries is substantially higher than that of the Czech Republic. Nonetheless, the U.S., the U.K., and the Czech Republic allocate the same percentage of GNP to education, and Sweden allocates considerably more. The U.S. figures include provision that is largely publicly funded (kindergarten) for fives in the year prior to first grade. As a residue of communism, the Czech Republic continues to provide considerable federal and municipal funding for early childhood education. With most funding coming from the federal and municipal governments and a smaller percentage from parents, Sweden has achieved full coverage for children 1–5 whose mothers work or are in school. In addition, a free preschool class is provided for children six years of age, prior to beginning primary school at age seven. Finally, since 1997, the U.K. has made a publicly-funded place available for all children four years of age, with universal provision for threes expected by 2002 in Scotland and 2004 in England. Children begin primary school at age five in England, Scotland, and Wales.

Each of these countries has responded to globalization in different ways. As Lee (2001) notes, in a discussion of educational reform in Japan, Korea, England and the U.S., regardless of a country's prior situation "globalization forces seem to destabilize" the status quo and each country must take into account its traditional cultural and institutional histories in their reaction to globalization. In the following case studies, we highlight how ECEC policies have changed and are changing in the Czech Republic, Sweden and the United Kingdom.

THE CZECH REPUBLIC

Background

The Czech Republic, formed in 1993 and formerly part of Czechoslovakia, has a population of 10.3 million (1994), with the majority (81%) of the population being Czech. Immigrants come primarily from Belarus, Germany, Poland, Russia, Slovakia, and the Ukraine.

The country has a strong history and commitment to early education and care dating back to 1631, when the first educational program was developed. However, the focus and intent of ECEC has changed over time. During the late

1700s, the educational function of preschool was emphasized, but in the next century social development was promoted. In the late 1800s, kindergartens, serving children ages three to six, were established with a preparatory education focus. Nurseries, for children three and older, and crèches for children birth to three focused on general child care and health. A reform effort, strongly influenced by the work of Maria Montessori, developed in the early 1900s and focused on children's individual rights and de-emphasized didactic instruction.

During the post-World War II era, an increase in maternal employment led to state-run child care with an educational emphasis. Kindergartens became a part of the school system, with compulsory schooling beginning at age six. As a result, there was an increase in the number of kindergartens, and all followed the same curriculum. The 1960 School Act included crèches as part of the school system. They were characterized by low parent involvement, large class sizes, and a focus on children's health. Enrollment in education-focused kindergartens reached its peak in the 1980s. This is largely attributed to high rates of female employment prior to the "Velvet Revolution" that led to independence from the Soviet Union. In the mid-1980s, almost 90% of females were working, and 97–99% of three to six year olds attended kindergarten. By 1989, nearly every child (99%) eligible for kindergarten was enrolled (OECD Background Report, 2000).

During this time, the country was undergoing massive political changes, culminating in the replacement of communism with democracy after the 1990 Revolution. A variety of options for early education and care developed as the government decentralized, parental influence on ECEC increased, and non-state (private and church-affiliated) operated preschools were introduced. The importance of an academically oriented education during kindergarten was widely embraced (OECD Background Report, 2000).

Along with these changes a shift in people's attitudes about the care of children has occurred. The OECD review team speaks of a "growing consensus within Czech society that it is better for young children to be brought up at home by their mothers than for them to attend preschool institutions" (OECD Country Note, 2000, p. 8). This attitudinal change has come at a time when female employment has dropped from an all-time high of 89% in the 1980s to 70% in 1998 due to radical economic and political shifts that have resulted in increased unemployment. Enrollment in higher education has increased, and retirement laws have also been altered. With fewer mothers working, the need for early care has lessened. Changes in family structure, reflecting general European trends, are also influencing the preschool system. Couples postponing marriage and childbearing, increasing proportions of children born to single mothers (18% in 1998), and unusually high divorce rates (24,000 per year since 1995) are some of these modern changes (OECD Background Report, 2000).

Current Provisions

Family Support
With these vast political and economic changes came significant modifications of early education and care. A number of policy decisions can help explain these shifts. New social support systems that focused on families have been established. Child allowances and social contributions, along with the Minimum Living Income Act, serve to assist families in caring for their children. Czech law mandates that family child-care costs cannot exceed 30% of a family's income. These policies have helped to keep child poverty to a minimum – 5.9% in 2000. Parental benefits were increased to cover lost wages during a child's first four years, thus enabling a parent (usually the mother) to stay home and care for her child. This benefit is extended to age seven if a child is chronically ill or has a disability (OECD Background Report, 2000).

Children 0–5
The new family support policies have significantly decreased the use of crèches for children under three. As of 2000, only 67 crèches serving 1,913 children were in operation. Three and four year olds in families receiving the parental leave benefit are limited to attending kindergarten three days per month, thus creating unequal access to preschool. In addition, the birth rate has decreased dramatically resulting in a decreased need for preschool programs. On average, 85.5% of children aged three to six are enrolled in preschool, with 98.4% of five-year-olds enrolled. However, this figure masks the high enrollments in the western and northern districts of Bohemia and the "historically low number of preschool aged children" (OECD Background Report, 2000, p. 9) in the central and eastern districts. These areas consist of many immigrant families who choose to care for their children at home until compulsory schooling at age six.

A number of government policies have been passed to ensure the education and care of children. One allows children up to age 10 to enroll in preschool services under special circumstances. An additional type of kindergarten, known as a special kindergarten, has been established for children with disabilities. Previously, these children were educated separately, but inclusion is now a priority. This is being accomplished by integrating children with disabilities into standard classrooms and by creating special classes as a part of standard kindergartens. Special education is delivered using a multidisciplinary approach, so children receive comprehensive services based on their individual needs. Support is provided by newly established advisory service centers. There are some problems with this approach, however, since class sizes are maintained when children with disabilities are included, and many staff members do not have the proper training to teach children with disabilities. In addition, programs are not as prevalent in rural areas, so children with disabilities must travel great distances to attend kindergarten (OECD Background Report, 2000).

Current Policies

Administration
Although kindergartens and crèches are both considered part of the preschool system, they are administered by two different agencies – the Ministry of Education, Youth, and Sports and the Ministry of Health and Care. This split of authority can be traced back to the roots of each type of setting and its primary focus. The State Authority and Self-Government in Education Act decentralized the government and created a new system of school management and financing, with programs managed by municipalities and primarily publicly funded through a combination of state and municipality support. Additional financial support comes from parents and private industry. It is anticipated that recent changes in district and regional boundaries will improve the financing of early education and care.

Regulation
Preschools are monitored regularly by state school inspectors, who rate each facility on its educational program, materials, and management. This information is used to rank preschools around the country. Family child care, including nannies or live-in caretakers, does not exist in this country, and, as a result, is not licensed or subsidized by the government.

Staffing
Staff qualifications are determined by the state. At a minimum, individuals must complete secondary schooling with a focus on preschool education. Higher vocational schools and university programs were recently added, but they are currently under-utilized. With the country's economic problems, already-low teacher salaries have been cut. Kindergarten teachers are limited to working 31 hours per week by law. Classes for children of socially and culturally diverse families (e.g. from Romany families) often include a teaching assistant who is familiar with the children's background. Given the feminization of the preschool teaching force, males are allowed to serve their military

duty as teaching assistants as one way to increase the number of male staff. In addition, voluntary professional development programs are being established, some of which are web-based (OECD Background Report, 2000).

Curriculum
Since independence, no national curriculum has existed. The government affirms each municipality's ability to decide how to educate and care for its young children. The range of programs, including "To Start Together," a Czech version of the international Step by Step program (see Coughlin, 1996) and "Healthy Kindergarten," a program begun in 1996 that focuses on children's health and education, has increased the variety from which families can choose. It has also led program staff to press for some national guidelines. The country is currently drafting a document which will highlight the important goals for preschool while still allowing the flexibility to create programs based on individually determined needs and increasing opportunities for parental involvement. Five aspects of the curriculum have been identified – biological, psychological, interpersonal, sociocultural, and environmental. The stated goal of preschools is to prepare children for primary school through stimulating environments that respect individual differences.

Parental Involvement
A national parent organization, the Union of Parents, exists to advocate for parental involvement in the educational system, and, as of 1995, School Boards have been established, and parents can be members. Since parents are now allowed to have a say in their children's education, some parents choose to postpone their children's enrollment. Reasons vary, but special preparatory classes for children five years of age were established in 1993 to address parents' concerns, along with the needs of children from socially and culturally diverse families. As of 2000, 106 such classes serving 1,358 children existed, and class size cannot exceed fifteen.

Research
The OECD Review Team reported that ECEC research has been virtually non-existent in the Czech Republic (OECD Country Note, 2000, p. 60). With the lack of a curriculum or generally agreed upon standards, it is difficult to conduct formal evaluations of the preschool system. The most recent and comprehensive study was conducted in 1999–2000 as part of the OECD project. Surveys were completed by 2,350 headmistresses, staff, and parents whose children attended 360 kindergartens throughout the country. Questions focused on topics such as parent-staff co-operation, access to kindergarten,

educational program development, staff qualifications, special education, policies and support, and quality of education and care.

SWEDEN

Background

Sweden, a country of 8.8 million people, has one of the highest living standards in the world. It also has had a relatively homogeneous population, consisting of the majority Swedish population and the minority populations of Finnish speakers and the Sami, who live in the far north of Sweden. However, in recent years, immigration has created a more heterogeneous population, thereby increasing cultural and linguistic diversity in society and in schools. People from more than 170 countries now live in Sweden, although two-thirds of those who are of non-Swedish origin are from other Nordic countries. In the 1990s, immigration, along with economic changes, placed added pressures on the Swedish system of early childhood education and care.

The current Swedish system of early childhood education and care must be considered within the context of the country's social welfare system, a system based on the "inviolability of human life, individual freedom and integrity, the equal value of all people, equality between the genders, and solidarity for the weak and vulnerable" (OECD Country Note, 1999, p. 7). From the 1930s, Sweden has increasingly addressed the living conditions of families through policies related to health, education, housing, child care, and parental leave. Although these policies are costly, there appears to be general agreement that this "system is a necessary prerequisite for the establishment of a society based on equality and a fair distribution of resources" (Gunnarsson, 1993, p. 499). This includes equalizing across families with and without children. The result of such policies, in a country with a low overall poverty rate, is that, after tax and transfers, only 3% of children are living in poverty (defined as less than half of the national median income) (UNICEF, 1996).

ECEC policies in Sweden can be traced to the establishment of infant crèches and kindergartens in the mid to late 1890s, but the primary influences on today's system stem from the late 1960s. In response to increasing numbers of women aspiring to work, labor needs, and demands for child care, recommendations were made for the expansion of the public child-care system. The guiding principles established at that time continue to guide today's system of care. These principles emphasize the role of society in providing high quality ECEC for all children in co-operation with parents. ECEC is to combine care and education, provide children with stimulating activities, allow parents to combine parenthood with work or studies and be available in neighborhoods in which families live.

By the mid-1980s the demand for public child care continued to be greater than municipalities could provide. Lengthy waiting lists persisted into the 1990s as increased provisions for care were accompanied by increasing demand due to a rising fertility rate that reached 2.1 children per woman in 1990–1992 (Gunnarsson, Korpi & Nordenstam, 1999, p. 12). Economic changes in the 1990s also affected ECEC policies. Unemployment, which had been as low as 2% in 1990, rose to 8% in the years between 1991 and 1993 (Gunnarsson et al., 1999, p. 14). Decreasing revenues resulted in cuts in funding. Responsibility for the provision of ECEC services was transferred to the municipalities. Lump sum subsidies from the state allowed municipalities to have increased local control and develop options in response to local needs. State and municipal governments cut spending for child care, health, and education and, in turn, parent fees for child care increased. The parental contribution to child care that had been 10%, with the municipalities and state each providing 45%, increased to 16.5% in the 1990s (Gunnarsson et al., 1999). As municipalities began setting parent fees, considerable variation and inequity also developed. In 1999, some families paid 2% of their disposable income for child care while others paid up to 20% (OECD Country Note, 1999, p. 37). A recent proposal that sets a maximum flat parental fee for preschool and school aged child care addresses these inequities in parental contributions.

Currently the child-care system, with its rapid expansion in the 1990s and a return to a low birth rate, is considered to have achieved "full coverage." Seventy-three percent of children ages 1 to 5 were enrolled in public child care in 1998, and 56% of those 6 to 9 were enrolled in school-age care (OECD Country Note, 1999, pp. 14–15). However, not all children are eligible for child-care services. Care is available only to those children whose parents work or study, or to those children in need of special care. Children, whose parents are looking for work, have immigrated and not yet found employment, or have taken parental leave due to the birth of another child are not eligible for services.

Two proposals are up for consideration to address these inequities of access. Universal provision of free half-day preschool services for all children ages four and five regardless of parents' employment or study status has been proposed. A second proposal aims to expand preschool activities to children whose parents are looking for work, thus increasing the services to those under four as well. These proposals would both alleviate disruptions experienced by children whose parents become unemployed or take parental leave and broaden access to many not currently eligible. Increased access would particularly assist the children of immigrant families who would benefit from exposure to the Swedish language and culture before entering compulsory schooling.

Current Provisions

Family Leave

Early childhood education and care policies are closely intertwined with family support policies. Of particular relevance is the parental leave program that was instituted in 1974 and has expanded since that time. To qualify the parent must have been employed for six months prior to the child's birth. In 1999 this leave was extended to 450 days with a financial benefit of 80% of salary for 360 days and a flat rate for the remaining 90 days. Of the now 450 days, 30 must be taken by the parent who is not the primary caregiver. Typically, this is the father. This policy was specifically, and successfully, instituted to encourage fathers to assist in the care of their children as care by both parents is considered important to children's development. Although parental leave can be taken anytime until the child reaches age eight, or finishes the first year of school, the majority of parents uses these days during the first 12 months following the child's birth. Consequently there is little demand for infant child care and only about 200 infants under age one are enrolled in non-parental care (OECD Country Note, 1999).

Children 1–6

Preschools

For those between the ages of 1 and 5 years there are three preschool options: preschools, family day care homes, and open pre-schools. Preschools offer the most complete care and education of these options as it is provided full time, year round. Municipalities set center hours based on the working hours of the parents and also determine the parental fee. In 1998, 338,000 children, 61% of children between the ages of 1 and 5, were enrolled in preschool (Gunnarsson et al., 1999).

Family Day-Care Homes

Family day care is supplied in the home of the provider. Children spend varying amounts of time in the family day care home, including evenings or weekends if the parents' schedules require this. The municipality again determines parent fees. In 1998 approximately 12% of children aged 1–5 and 4% of those aged 6–9 years were enrolled in day-care homes (Gunnarsson et al., 1999).

Open Preschool

For those not enrolled in a preschool or family day care home, there is the option of an open preschool. Since parents or a child minder must participate with the child, open preschools give parents the opportunity to interact with other parents while the children play. There is no fee or enrollment. The number of both family day-care homes and open preschools has decreased during the 1990s as utilization of preschools has increased.

Preschool Class

Changes implemented in 1998 have decreased the need for preschool care for 6 year olds. The compulsory school starting age in Sweden is 7 years of age. However, in 1998, municipalities were mandated to provide 6 year olds with a free preschool class for at least 525 hours per year in order to facilitate the transition from preschool to compulsory schooling. Although voluntary, virtually all children six years of age attend. This has meant that preschools no longer need to serve these children, but, in turn, that many more are enrolled in before and/or after school programs.

Leisure Time Centers

Rounding out the child-care options are the leisure time centers that are available for those aged 6–12. These centers provide before and after school care and often are located in school buildings.

Current Policies

Administration

A view of care and education as a single strand has guided developments in ECEC programs since the 1960s. In 1996, this principle was organizationally supported with the transfer of child-care responsibility to the Ministry of Education and Science from the Ministry of Health and Social Services.

In the 1990s the ECEC system was decentralized with primary responsibility for ECEC transferred to the local municipalities. The State gave increased decision making to the municipalities by providing them with lump sum subsidies. Within parameters set by the state, municipalities were to provide and monitor preschool and school age care and to provide services that met the needs of the parents in their locale. Specific national standards on such things as group size and adult-child ratios were eliminated. These standards, now set by municipalities, result in variation from one location to another.

Regulations

The municipalities originally had to follow state rules and regulations that detailed building design, space requirements, group size and so forth. However, these regulations also resulted in a system that was not responsive to the local needs of either the municipalities or families. Devolution of responsibility to the municipalities has increased the diversity of local programming but has also frequently increased group size and parental fees.

Staffing

Preschools are typically staffed by preschool teachers and childminders, who together form the center work-team. Preschool teachers, or early childhood pedagogues, have completed a three-year university program that includes theory and practice. The early childhood pedagogue is trained to work with children from birth to compulsory school age under the expectation that provision of care across this time period will be coherent and will combine education and care (Oberhuemer, 2000). The childminder program, also of three years duration, is part of the secondary school program and focuses on building basic child minding skills.

Curriculum

The first national curriculum for preschool was initiated in 1998. It provides the overall values, goals, and guidelines that municipalities are to implement, while leaving the means for achieving these goals to the preschool staff. Foundational values, such as respect for each person, individual freedom, equality, tolerance, and compassion, are clearly stated as the bases of the preschool. The child is considered "an active learner, searching for knowledge and developing this knowledge through play, social interaction, exploration and creativity, as well as through observations, discussions, and reflections" (Gunnarsson et al., 1999, p. 65). In addition to choosing between preschools or family day care homes, parents can also choose among different types of ECEC centers. Diverse programs are found in today's Swedish preschools, including curricula based on the environment, technology, and the principles of Reggio Emilia.

The preschool curriculum is linked with those for the after school centers, compulsory schools, and upper secondary schools. The curricula thus work together to provide a "common view of knowledge, development, and learning" (Ministry of Education and Science, 1998, p. 1) and facilitate the goal of life-long learning.

Parent Involvement

Although close co-operation is expected between the preschool and home, participation in day-to-day activities is not expected as parents are working or studying while their children are in care. However, parent involvement ranges from those who take no active role in the center to those who are involved in daily activities, as well as center planning and decision making.

Research

Considerable research on ECEC in Sweden has been completed over the last three decades. However, it has been suggested that there is currently a need for program evaluation at the local level and for work that addresses the application of research to policy and practice (OECD Country Note, 1999, p. 37).

THE UNITED KINGDOM

Background

The United Kingdom (U.K.) of Great Britain and Northern Ireland, with a population of 59 million, is comprised of England, Wales, Scotland, and Northern Ireland.[1] Approximately 6% of the population is considered ethnic minorities, including many immigrants from former Commonwealth countries, India (15%), the Caribbean (9.1%), Pakistan (8.7%), Africa (3.8%), Bangladesh (2.9%), Hong Kong, (2.8%), Ireland (15%), and the Mediterranean (Bertram & Pascal, 1999, p. 11). Britain also has the largest number of refugees and asylum seekers in Europe. Approximately 300,000 entered in the year 2000.

During the long period of Tory (conservative) Government in the 1980s and 90s, early childhood education and care in Britain was primarily purchased by parents. As in the U.S. public monies were expended only for children living in poverty or those with special needs. In 1997, the current (Blair) Government inherited what has been characterized as a diverse and fragmented system of care and education services (Oberhuemer & Ulich, 1997), in a context of increasing economic inequality. Indeed, the Treasury Inquiry of 1998 reported that approximately one out of every 3 children in the U.K. (14.3 million) was living in poverty, defined as a family income less than half the national median. However, UNICEF (1996) reports that, after taxes and transfers (Britain has national health care and child allowances), the child poverty rate of 30% was effectively reduced to 10%. This contrasts with the U.S., where, after taxes and transfers, the child poverty rate of 28% was reduced to 24% in 1996.

The new Prime Minister has pledged to eliminate child poverty in 20 years. He has also proposed reforms to address a wide array of social issues. In general, family patterns in Western nations have become less stable, but the U.K. currently has the highest percentage (20%) of single-parent families in Europe. The major growth in the labor force is projected to be the result of increasing female employment, but the U.K. has the second highest rate of part-time female employment (after the Netherlands) in the European Union. Moreover, the majority of single (lone) parents were not employed in 1999. The Government has thus seen the expansion and improvement of child care as key to addressing a range of social issues.

ECEC in the U.K. evolved as separate systems of "care" and "education" with often competing interpretations of the purpose of services. In 1870, publicly funded education became compulsory at age five, but children as young as two were admitted to state primary schools, especially when women were needed to work in local industries. The propriety of having very young children enrolled in formal educational settings has been an ongoing theme in ECEC discussion in the U.K. In 1873, the first publicly funded kindergarten was established in Salford. It provided nursery education, baths, meals, rest, and parent education (Bertram & Pascal, 1999). Others followed, but the dominant form of provision remained the state primary school, and a high percentage of three year olds in England and Wales were enrolled. These numbers were later reduced considerably, yet, to this day, the U.K. has among the earliest school starting ages in Europe. Primary schooling begins at age four in Ireland, and at age five in England, Scotland and Wales. Since widespread adoption of the Reception year, most children in England enter primary school at age four.

Current Provisions

Family Leave

In December 1999, major changes in maternity and parental leave policies took effect. Paid maternity leave was increased from 14 to 18 weeks, with the possibility of extending to a total of 40 weeks, and paternity leave was adopted for the first time. A parent now qualifies after one year of full-time employment rather than two. Mothers receive 90% of their salaries for the first six weeks and a flat rate for an additional 12 weeks. Parents are also entitled to 13 weeks *unpaid* leave from the time the child is born until age five, and they may take off work in times of family crisis or loss of child care. When parents in low-income families take parental leave, they are also entitled to claim additional funding to supplement their income during the leave period. These policies essentially bring the U.K. in line with the minimal standards on maternity and parental leave set by European Union Directives.

Children 0–3
According to the OECD Background Report, "current provision of education and care for the under 3s in the U.K. is uneven, of mixed quality and in short supply" (Bertram & Pascal, 1999, p. 21). With the exception of public provision for children "in need" (e.g. due to disability, family crisis, or abuse), care arrangements for young children must be purchased by parents after the maternity leave period (but see discussion of reforms below). Fully half of all mothers with children under three are now employed, so the provision of affordable, high-quality child care is a current and prominent concern of the Government.

Nonetheless, the child care and education scene in the United Kingdom is complicated. Different forms of care have evolved over time; there is public, private and voluntary involvement, and types of provision vary across regions. In England, providers of services include the local authorities (local authority day nurseries, before/after school clubs; state day nurseries), the voluntary sector (parent and toddler groups), and the private sector (private day nurseries, preschools, playgroups, before and after school clubs, childminders (family day care homes), nannies, and au pairs). There are also a small number of combined nursery/family centers, offering child care and early education, as well as other services (drop-in care, adult education, etc.). The Government currently sponsors a program of Early Excellence Centers, combined centers with extended care and education as well as other services, that serve as models of exemplary practice in the field.

According to a recent survey, the child-care workforce in England is predominantly comprised of childminders (24%) and nannies (26%) (HERA2 Report, 1999, p. 3), and most care for children this age is privately funded. Although costs vary by the age of the child, type of provision, and locale, care in England is generally costly. In 1999, the Daycare Trust reported that child care for two children, one school-age and one preschool-age would cost approximately £6000/year (roughly $9000/year) (Daycare Trust, 1999). In 1997, 93% of the cost was incurred by parents.

In 1998, the Blair Government launched the National Childcare Strategy, with plans to invest £8 billion to expand and improve child-care services for children birth to age three and out-of-school provision for children under 14 (under 16 for children with special needs) (Daycare Trust, 1999). Places for up to one million children are to be created by 2003. Due to the high rate of part-time maternal employment, each childcare place will be used by three children. The Government has also instituted a minimum wage and adopted a Childcare Tax Credit as part of the Childcare Strategy. These reforms are intended both to raise the wages of the lowest paid child-care workers and to refund some child-care expenses. Childcare Information Services (CIS) have been established in each local authority. Additionally, efforts are underway to improve the existing system through the development of an integrated training scheme and uniform regulatory practices (see the discussion below). The clear intent is to address issues of quality, affordability, and accessibility (Daycare Trust, 1999).

Children 3–5
The U.K. is in a period of transition with regard to preschool-age children. Since 1998, a free, part-time place has been guaranteed to all 4 year olds, and universal provision for 3s is expected to be achieved in Scotland by 2002 and in England by 2004. Four year olds in Scotland attend $4\frac{1}{2}$ hours per day, and, increasingly, children 4–5+ in England are attending a state-funded primary, first or infant school for $6\frac{1}{2}$ hours a day during the school year prior to compulsory schooling. In England, Wales, and Northern Ireland, this is referred to as the "Reception" class. Other types of early education provision include nursery schools and nursery classes (run by local education authorities), special schools and opportunity groups (run by local education authorities for children 3–5+ with special needs), preschools/playgroups (run by a community or voluntary group, by parents, or by a private, for-profit business), and private nursery schools, private day nurseries, and independent schools. Prior et al. (1999) report that, in 1999, the most common trajectory was for three year olds to attend a preschool/playgroup, younger fours to attend a state-funded nursery class, and older fours to attend a primary school Reception class.

Current Policies

In 1997, the Labor Party came into power in the United Kingdom (U.K.). The previous Tory Government had largely supported a neo-liberal agenda based on a "free market" approach to the expansion of services. The Blair Government has taken a much more active role, developing a wide-ranging plan of action to encourage employment, combat social exclusion, and decrease poverty. The reform of the "early years" system is thus one part of a general strategy to effect broad-based social change.

Administration

There has been a longstanding division between care and education in the United Kingdom, as evidenced by the fact that they have been overseen by different ministries, required different training, and operated on different salary scales. In 1999, the Government united care and education within the Department for Education and Employment (now called the Department for Education and Skills (DES)). Nonetheless, the child-care system for children birth to 14 remains largely privatized, while early education reform aims to provide universal provision for children three and four years of age prior to entry into compulsory schooling at age 5/5 +.

Under the former Government, care and education had been the province of local authorities, and considerable disparities developed over time. As a consequence, the Government has exerted strong centralized authority in effecting reform efforts, while, at the same time, working with new Early Years Development and Childcare Partnerships (EYDCP) and Local Education Authorities. The Partnerships assess the current provision of care and education in local areas, develop plans for future expansion, and address quality concerns. The EYDCP plans are linked to national targets for the provision of early education places for three and four year olds.

Regulation

In August, 1999, it was announced that the regulation of care and education would be consolidated under national standards in a new arm of the Office for Standards in Education (OFSTED), a non-ministerial government department set up in England as a regulatory body. The standards will include registration, inspection, investigation and enforcement.

Staffing

Although recent efforts have been made to consolidate care and education, staffing continues to be bifurcated by these categories. Training, staff-child ratios, and salaries differ across the education-care divide (Moss, 1999). A "qualified teacher" has either a three-year degree and a one year Postgraduate Certificate of Education (PGCE) or a four-year degree in higher education. Only about 20% of the early years workforce has acquired graduate-level qualification. A "nursery nurse" or qualified classroom assistant has two years post 16 specialist training. There are no training requirements for childminders, nannies, and au pairs, with the exception of 5–15 hours of training now required of childminders in some local authorities. Fully half of the staff in nurseries, playgroups, after-school clubs, breakfast clubs, and holiday play schemes have no training; classroom assistants in Reception classes may also be untrained.

Curriculum
Early Learning Goals (QCA, 1999a) for children and guidance for teachers (QCA, 2000) have been written to provide a curriculum framework for early years' practice in England. The Goals establish expectations for children to reach by age 5/5+. The Goals are subdivided into six areas of learning, with play considered to be quite important. However, some of the literacy and numeracy goals are extremely ambitious. Regarding literacy, for example, it is expected that most five year olds will be able to:

(1) Hear and say initial and final sounds in words, and short vowel sounds within words;
(2) Link sounds to letters, naming and sounding the letters of the alphabet;
(3) Read a range of familiar and common words and simple sentences independently;
(4) Write their own names and other things such as labels and captions and begin to form simple sentences sometimes using punctuation;
(5) Use their phonics knowledge to write simple regular words and make phonetically plausible attempt at more complex words. (p. 27)

A current concern is that the preschool experience will be considerably altered as teachers adapt to heightened expectations. Recent recommendations from the National Literacy Strategy state that the Reception year should include three hours of literacy instruction each day.

Parent Involvement
The Government has endeavored to address parents' needs by making parental leave policies statutory, implementing a Childcare Tax Credit, funding the National Childcare Strategy, and establishing Childcare Information Services, i.e. resource and referral services within each local authority. Early years providers are also encouraged to work closely with parents in support of young children (QCA, 1999).

Research
Funding for research and evaluation in early childhood in the United Kingdom has been limited until rather recently. Funding comes primarily from government departments, national research funding councils, charitable trusts and foundations, local governments, and institutions of higher learning (Bertram & Pascal, 1999, p. 76). The Government has recently funded a

number of policy and evaluation studies in relation to some of the ECEC policies outlined above.

SUMMARY AND CONCLUSIONS

This chapter has discussed features of globalization and reviewed case and comparative studies that examine how education and, in particular, early childhood education are changing internationally. In this period of global restructuring, early childhood education and care have become prominent policy issues. As more women have entered the labor force, the need for alternative forms of care has increased. Governments have also recognized that children benefit from early education, and attending to children's issues can be an important consideration as nations become more unequal (Moe, 2001). The review of case and comparative studies of ECEC policies and the more detailed case studies of policy and provision in three nations that participated in the recent OECD Thematic Review reveal considerable interest in how nations provide for young children. Some nations offer extensive parental leave and child allowances, in an effort to put a floor of support under all families. A few offer extensive public support for children birth to age five or six, and increasing numbers are striving for universal provision for three, four, and/or five year olds. Studies and reports describe family supports and forms of extra-familial provision and examine in more detail issues of governance, regulation, funding and financing, curriculum, staffing and retention, and family involvement and support.

Early education and care in every society is premised on the institutionalization of practices that derive from cultural/ideological beliefs about children, families and their place in society and from particular economic and political decisions that support these beliefs. Figure 2 illustrates how cultural, political, and economic issues relate to and influence the provision of early education and care in particular contexts.

Although globalization is exerting substantial influence internationally, the studies reviewed in this chapter reveal that its effects are neither unitary nor inevitable.

Culture/Ideology

Most European nations provide extensive supports to families in their child rearing role. When considering alternative reforms, it is important to recognize that people often talk about children – and children's issues – in ways that differ from the American discourse of individualism and "choice." The following questions and issues speak to cultural/ideological concerns:

What Conception of Children Drives ECEC Reform?

A useful perspective in addressing this question is the emerging paradigm of the new sociology of childhood, where childhood is seen as a social construction, a variable of social analysis, and a means of reconstructing childhood in society (e.g. Prout & James, 1997). Dahlberg, Moss and Pence (1999) suggest that the young child is primarily viewed as a future worker in countries such as the U.S., the U.K., and Australia where school readiness reforms are on the ascendancy. They contrast this image with that of the "rich" and capable child of Reggio Emilia, Italy, a child capable of developing theories about the world and representing complex ideas in various media and forms. Woodhead (1997, 2000) also cautions that undue focus on children's "needs" constructs children as weak and vulnerable and obscures the complexities of their lives.

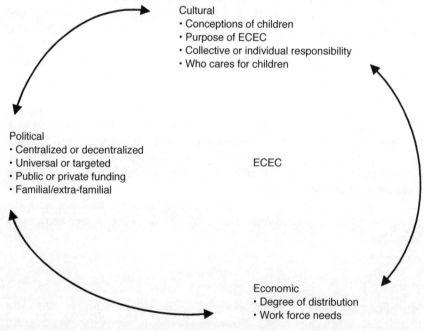

Fig. 2. Economic, political, and cultural issues related to an influencing ECEC.

What is the Purpose of ECEC?
Do services primarily exist to care for children while their parents work? To get children ready for formal schooling? To provide a broad range of supports to families? An espoused goal of ECEC in many nations is to create conditions so that all children can thrive. Policies that decrease the number of children living in poverty are thus crucial. Dahlberg and her colleagues (1999) also suggest that early childhood institutions might be viewed as "public forums" for democratic decision making.

Should there be Collective or Individual Responsibility for Children?
The United States has reconciled the tension between public and private responsibility very differently than have the nations of Europe. The American construction of public responsibility to families has primarily taken two forms. The first, premised on the English tenet of *parens patriae*, has been to justify public action only when there is evidence that parents cannot fulfill their child rearing roles without substantial assistance. When this is the case, public monies are used to provide services for children living in poverty or to those who have diagnosed disabilities. Tax credits now subsidize child care expenses, but provision is general market driven.

Who Cares for Children?
The question of whether young children should be primarily cared for by mothers and fathers or by others has become less salient as more and more women work outside the home. Nonetheless, it is striking how other Western nations provide substantial supports so that parents can both work and care for children. It is also noteworthy how ideological supports for maternal employment are withdrawn when women are no longer needed in the labor force. This occurred at the end of World War II in the U.S. and the U.K. and is now occurring in Eastern Europe, as research there suddenly shows that extra-familial care is "not good" for young children birth to age three.

Economics

Economic decisions play an important role in how children are cared for within a society. Economic issues related to ECEC include:

How Much Personal Income Should be Redistributed on Order to Support Families with Children?
Although there is a trend toward market-based economies internationally, nations redistribute income through various means and for various purposes

(e.g. national health care, public schooling, universal preschooling, targeted programs such as Head Start, etc.). Europeans generally pay higher taxes to support policies such as universal health care, as well as policies that guarantee paid parental leave, child allowances, and public preschools, to citizens who qualify.

How Does Support for ECEC Address Workforce Needs?
Interest in and support for ECEC can increase when women are needed in the labor force. Currently, much of the growth in the work force in the United States and the United Kingdom is due to the increased participation of women and minorities.

Politics

Finally, policy makers face a number of issues when considering early childhood education and care:

Should Governance be Centralized or Decentralized?
Tensions exist between centralized and decentralized authority, but governance need not be an either-or proposition. Many Western nations provide funding and guidance at the federal level while devolving administrative responsibility to states or municipalities. In the example of Sweden above, the government decentralized to provide municipalities greater flexibility in addressing local needs, but doing so increased inequity (i.e. some parents were paying 2% of their salaries while others were paying 20%). The recent proposal to cap parental contributions to child care is an interesting and important example of how federal-local tensions might be reconciled.

Should ECEC Programs be Universal or Targeted?
Universal provision, as the name implies, makes ECEC services available to all children, while targeted programs serve discrete populations of children. Head Start and state-funded pre-K programs are examples of targeted programs. Many European nations have achieved universal provision for preschoolers, but, the age of children served varies. Universal provision has advantages. For example, poor children or children with risk factors would be less likely to be segregated and labeled. However, Head Start currently sponsors many full-day programs, and children receive additional health, social, nutritional, and other benefits that could be lost if part-day programs were made available to all children.

Should Programs be Publicly or Privately Funded?

This question again addresses the extent to which individuals should be responsible for young children's care and education. Nations address this question in different ways. Policy makers can also provide public funds to build infrastructure and to improve and expand the system.

Should Governments Provide Support for Familial and/or Extra-Familial Provision?

As the case studies illustrate, policies can support familial and extra-familial care. In the U.S. parental leave is unpaid and only pertains to those who work in companies that have 50 employees or more. With the exception of targeted programs and tax credits, ECEC is privately funded. By contrast, European nations generally provide public support for maternity leave, paternity leave, and typically for extended leave (at a reduced rate or unpaid), sick leave, and child allowances. Child care for children 1–3 is often, although not always, privately funded, while, increasingly, preschool programs are publicly funded for at least some portion of the day.

Despite the fact that these issues address common concerns, there is substantial variability in what nations actually do. Distinctive histories create the possibilities for certain kinds of policies and make others less likely, although change can also be both abrupt and transformative. Of necessity, the examples cited depict snapshots within films. The issues discussed above work in tandem and cannot easily be separated from the context in which they arise. This complexity is best captured in actual cases.

Czech Republic

Although the Czech Republic has adopted a market economy, government support of collective responsibility for children and child care retain features of the former (communist) approach. The recent overhaul of government in the Czech Republic has had an enormous impact on their economy, policies, and views about children, all of which influence ECEC. The country's economy was robust in 2000, and many mothers were working full-time, thus necessitating early care and education for children aged birth to six. At the same time, however, there is a return to traditional values that support maternal child rearing for young (birth to three) children. Social policies such as parental benefits, child allowances, and social contributions have been adopted to ensure parental leave for the first three years and assist families in financial need through subsidies. Legislation has also been passed to minimize the out-of-pocket costs of ECEC for families with young children, and efforts are made to keep child poverty rates quite low. In sum, these strategies highlight how young children and families are valued in the Czech Republic despite a new system of rule.

Along with the adoption of a market economy, the government has shifted from a communist to democratic political system. Decentralization has occurred, and ECEC is managed and financed by a combination of state and municipal funds. This change to local control has allowed for flexibility, choice, and increased parent involvement. One example of the effort to balance the two levels of government involvement is the discussion of a national curriculum framework, which will provide some general recommendations the profession is calling for, while still allowing local authorities to adapt them as appropriate (OECD Background Report, 2000). Issues of access, affordability, and research remain, and are dependent on economic, political, and ideological factors.

Sweden

Sweden's comprehensive early childhood education and care system, "embedded in an enviable range of supportive policies" (Brennan, 1994, p. 3) is premised on longstanding adherence to socialist values. It serves as a tangible reminder of Sweden's collective commitment to children and families. Sweden has a long tradition of equity-based policies through which traditional values of equality, solidarity, and fairness are enacted. Subsidized preschool services and parental leave policies that allow a parent to be at home for the child's first year (with 80% of salary) are included in the extensive social welfare policies provided to Sweden's citizens. Consequently most children from 0–1 are cared for at home. Children from 1–6 have access to publicly funded ECEC services that focus on the development of the child as an active and life-long learner. These services have been limited to those children whose parents are studying or working, or to children in need of special care. Current proposals would extend access to children whose parents are looking for work and to all four and five year olds. Open preschools, attended by both parents and children, are available to all, regardless of employment status.

Maintaining these services and adhering to traditional values has been put to the test with changing market conditions. Sweden has not been immune to the effects of globalization and economic change and these have had an impact on the maintenance of social welfare policies. In the 1990s the Swedish government responded to recessionary conditions and an economic picture "clouded by budgetary difficulties, inflation, high unemployment and a gradual loss of competitiveness in international markets" (Coutsoukis, 1999, p. 1). In relation to ECEC the response was to decentralize, with more control and responsibility given to the municipalities. This was accompanied by a decrease in spending for child care, health, and education. As a result of these changes, greater disparities in access and cost developed. However, a range of diverse services that were more attuned to parental needs and local situations also developed as decision making devolved to the local level. Maintaining this diversity of services while also providing equitable care is a challenge currently faced by the Swedish ECEC system.

Kallós and Tallberg Broman (1997) contend that effects on the ECEC system in the 1990s were particularly severe due to growing "marketization tendencies" based on the "neo-liberal" economic policies of the conservative coalition government. Parents were viewed as consumers, and the values emphasized were "freedom to choose, marketization, increased efficiency, creative competitiveness, enhanced quality" (Kallós & Broman, 1997, p. 280). Set aside were the concepts of equality, solidarity, justice, and a fair distribution of resources that had been hallmarks of Swedish life. Although the Social Democrats returned to power in 1994, Kallós and Tallberg Broman (1997) suggest that the Social Democrats also accept this economic approach. They note that, rather than seeing high spending on social programs as positive, they primarily see them as "an indicator of possibilities of further reductions" (p. 281).

However, recent proposals instituting universal preschool for four and five year olds, eligibility for those whose parents are looking for work, and a maximum flat fee for care will improve access and promote equity. These proposals are another step in the evolution of the Swedish ECEC system.

The United Kingdom (England)
Policies such as national health insurance, parental leave, and child allowances have been adopted for some time in the United Kingdom, and recent reforms extend maternity leave, make it statutory, and make paternity leave available for the first time, effectively bringing the U.K. in line with the European Social Contract.

Child care for children birth to three and after-school care will continue to be privately funded, although a new minimum wage and Child Care Tax Credit are intended to help fund child care indirectly. Through the National Childcare Strategy, the Government has pledged £8 billion to improve and expand child care.

Children enter formal school at age four in Northern Ireland and at age five in England, Scotland, and Wales. Universal provision for 4 year olds has now been achieved. In Scotland, children are guaranteed $4\frac{1}{2}$ hours per day; in England, most fours are now in Reception classes in state-funded primary schools for $6\frac{1}{2}$ hours per day.

Other reform efforts in England include uniting care and education in the Department for Education and Employment (now the Department for Education and Skills); establishing curriculum goals (QCA, 1999a) and guidance for teachers (QCA, 2000), developing of a national "climbing frame" (QCA, 1999b), and designating the Office for Standards in Education (OFSTED) as the regulatory body for both care and education. Decentralization had led to fragmentation and inequity. For example, some local authorities had provided substantial public funding for child care, while others had provided very little; some had supported care by childminders, while others had encouraged the development of center-based services. As a consequence, the Blair Government has taken a strong role in guiding reform efforts and fostering "joined up" thinking. The Early Years Development and Childcare Partnerships assess needs and set goals locally. Universal provision for three year olds is expected to be achieved by 2004.

Overall, a number of conclusions might be drawn from this review. First, globalization is often construed as a unitary phenomenon driving particular types of reforms, yet this review highlights the fact that globalization's effects are strongly mediated by historical precedent and cultural factors. For example, although the Czech Republic has adopted a market economy, the provision of early education is viewed as a collective responsibility. Secondly, the vast majority of the world's children do not have access to a well-developed system of early education and care, and the responsibility of Minority World nations to these children must be acknowledged. The *World Declaration on Education for All* (UNESCO, 1990) reports that there are over 100 million children, the majority girls, who have no access to primary school; for many more early education and care are beyond their reach. Thirdly, and particularly within the Minority World, recent ECEC policy concerns and initiatives are evident in the areas of governance and regulation, funding, access, curriculum, staff recruitment and retention, and parent involvement. Although these features take shape differently in different countries, some patterns and trends can be discerned cross-nationally. For example, increasing numbers of countries offer statutory paid parental leave, and the trend in Europe is toward universal access to preschool programs for children 3–6 (OECD, 2001). It is also increasingly common for countries to house early education and care in a single ministry, to develop curricular frameworks for early childhood education, and to increase training and credentialing requirements. There is noticeably less discussion about raising salaries. Finally, although a substantial amount of work is now being done on ECEC policy issues both nationally and internationally, studies and reports are under-theorized. Explaining different approaches, within the context of global change, and exploring their long-term effects would be a starting point for developing new forms of inquiry into ECEC policies and programs.

NOTE

1. Since devolution, there are differences in the ways in which each country has responded to early education reform initiatives. The discussion here focuses primarily on England.

REFERENCES

Adamson, P. (1996). *Commentary: Beyond basics* [On-line]. Available: www.unicef.org/pon96/ inbasic.htm

Apple, M. W. (2000). Between neoliberalism and neoconservatism: Education and conservatism in a global context. In: N. C. Burbules & C. A. Torres (Eds), *Globalization and Education: Critical Perspectives* (pp. 57–77). New York: Routledge.

Bertram, T., & Pascal, C. (1999). *The OECD thematic review of early childhood education and care: Background report for the United Kingdom.* Worcester: Centre for Research in Early Childhood.

Bredekamp, S. (1987). *Developmentally appropriate practice in early childhood programs serving children from birth through age 8.* Exp. Ed. Washington, D.C.: NAEYC.

Brennan, D. (1994). *The politics of Australian child care: From philanthropy to feminism.* Cambridge, U.K.: Cambridge University Press.

Burbules, N. C., & Torres, C. A. (2000). Globalization and education: An introduction. In: N. C. Burbules & C. A. Torres (Eds), *Globalization and Education: Critical Perspectives* (pp. 1–26). New York: Routledge.

Carnoy, M. (1999). *Globalization and educational reform: What planners need to know* (Vol. 63). Paris: UNESCO: International Institute of Educational Planning.

Carr, M., & May, H. (2000). *Te Whariki:* Curriculum voices. In: H. Penn (Ed.). *Early childhood services: Theory, policy and practice* (pp. 53–73). Buckingham, U.K.: Open University Press.

Choi, S.-H. (1999). The notion of the child and early childhood education in Korea. In: G. Brougere & S. Rayna (Eds), *Culture, Childhood and Preschool Education* (pp. 241–251). Paris: UNESCO.

Cleghorn, A., & Prochner, L. (1997). Early childhood education in Zimbabwe: Recent trends and prospects. *Early Education & Development, 8*(2), 339–352.

Clyde, M., Parmenter, G., Rodd, J., Rolfe, S., Tinworth, S., & Waniganayake, M. (1994). Child care from the perspective of parents, caregivers and children: Australian research. *Advances in early education and day care* (Vol. 6, pp. 189–234). Greenwich, CT: Jai Press, Inc.

Cochran, M. (Ed.) (1993). *International handbook of child-care policies and programs.* Westport, CN: Greenwood Press.

Convention on the Rights of the Child (1990).

Copley, J. V. (2000). *The young child and mathematics.* Washington, D.C.: NAEYC.

Coughlin, P. (1996). Child-centered early childhood education in Eastern Europe: The Step by Step approach. *Children & Society, 12*(3), 223–227.

Coutsoukis, P. (1999). Sweden: Economy [On-line]. Available: http://www.photius.com/wfb1999/ sweden/sweden_economy.html

Crane, J. (1998). The pacific world of early childhood education. *International Journal of Early Childhood*, *30*(2), 47–50.

Cryer, D., Tietze, W., Burchinal, M., Leal, T., & Palacios, J. (1999). Predicting process quality from structural quality in preschool programs: A cross-country comparison. *Early Childhood Research Quarterly*, *14*(3), 339–361.

Dahlberg, G., Moss, P., & Pence, A. (1999). *Beyond quality in early childhood education and care: Postmodern perspectives*. London: Falmer Press.

Davies, R. (1997a, April 1–5, 1997). *A historical review of the evolution of early childhood care and education in the Caribbean*. Paper presented at the Second Caribbean Conference on Early Childhood Education, Barbados.

Davies, R. (1997b). *Striving for quality in early childhood development programmes: The Caribbean experience* (ERIC Document ED 413077). Kingston, Jamaica: University of the West Indies.

Daycare Trust (1999). *Making the most of the National Childcare Strategy*. London: Daycare Trust.

Emblem, V. (1998). Providers and families: Do they have the same views of early childhood programmes? Some questions raised by working on early childhood education in the Lao People's Democratic Republic. *International Journal of Early Childhood*, *30*(2), 31–37.

Feeney, S. (1992). *Early childhood education in Asia and the Pacific: A source book*. New York: Garland Publishing, Inc.

Friendly, M., & Oloman, M. (2000). Early childhood education on the Canadian policy landscape. In: J. Hayden (Ed.), *Landscapes in Early Childhood Education* (pp. 69-82). New York: Peter Lang Publishing.

Guild, D. E., Lyons, L., & Whiley, J. (1998). *Te Whaariki*: New Zealand's national early childhood education curriculum guideline. *International Journal of Early Childhood*, *30*(1), 65–70.

Gunnarsson, L. (1993). Sweden. In: M. Cochran (Ed.), *International Handbook of Child-care Policies and Programs* (pp. 491–514). Westport, CN: Greenwood Press.

Gunnarsson, L., Korpi, B. M., & Nordenstam, U. (1999). Early childhood education and care policy in Sweden: Background report prepared for the OECD thematic review [On-line]. http://www.oecd.org/els/education/ecec/

Hayden, J. (2000). Policy development and change on the Australian landscape: A historical perspective. In: J. Hayden (Ed.), *Landscapes in Early Childhood Education* (pp. 49–68). New York: Peter Lang Publishing.

HERA 2. (1999). *Final report: Childcare training in the U.K.*. Suffolk: Suffolk County Council.

Jalongo, M. R., Hoot, J. L., with Pattnaik, J., Cai, W., & Park, S. (1997). Early childhood programs: International perspectives. In: J. P. Isenberg & M. R. Jalongo (Eds), *Major trends and issues in early childhood education: challenges, controversies, and insights* (pp. 172–187). New York: Teachers College Press.

Johnson, P. (2000). *The EFA 2000 Assessment: Bahamas Country Report* [On-line]. Available: http://www2.unesco.org/wef/countryreports/bahamas/contents.html

Kallós, D., & Broman, I. T. (1997). Swedish child care and early childhood education in transition. *Early Education and Development*, *8*(3), 265–284.

Kamerman, S. B. (1991). *Child care, parental leave, and the unders 3s: Policy innovations in Europe*. New York: Auburn House.

Kamerman, S. B., & Kahn, A. J. (1981). *Child care, family benefits, and working parents: A study in comparative policy*. New York: Columbia University Press.

Kamerman, S. B., & Kahn, A. J. (1995). Innovations in toddler day care and family support services: An international overview. *Child Welfare*, *74*(6), 1281–1300.

Katz, L. (1989). Afterword: Young children in international perspective. In: P. Olmstead & D. Weikart (Eds), *How nations serve young children: profiles of child care and education in 14 countries* (pp. 401–406). Ypsilanti, MI: High/Scope Press.
Lamb, M. E., Sternberg, K. J., Hwang, C.-P., & Broberg, A. G. (Eds). (1992). *Child care in context: Cross-cultural perspectives*. Hillsdale, NJ: Lawrence Erlbaum Associates.
Lee, G. (1997). The characteristics of early childhood education in Korea. *International Journal of Early Childhood*, 29(2), 44–50.
Lee, J. (2001). School reform initiatives as balancing acts: Policy variation and educational convergence among Japan, Korea, England, and the United States. *Education Policy Analysis Archives* [On-line serial], 9(13). Available: http://epaa.asu.edu/epaa/v9n13.html
Levin, B. (2001). Conceptualizing the process of education reform from an international perspective. *Education Policy Analysis Archives* [On-line serial], 9(14). Available: http://epaa.asu.edu/epaa/v9n14.html
LeVine, R., Dixon, S., LeVine, S., Richman, A., Leiderman, P. H., Keefer, C., & Brazelton, T. B. (1994). *Child care and culture: Lessons from Africa*. New York: Cambridge University Press.
LeVine, R., & White, M. (1986). *Human conditions: The cultural basis of educational development*. New York: Routledge & Kegan Paul.
Logie, C. (1997, April 1–5, 1997). *The status of ECCE provision in Trinidad and Tobago*. Paper presented at the Caribbean Conference on Early Childhood Education, Barbados.
Lubeck, S. (1989). A world of difference: American child-care policy in a cross-national perspective. *Educational Policy*, 3(4), 331–354.
Lubeck, S. (1995). Nation as context: Comparing child-care systems across nations. *Teachers College Record*, 96(3), 467–491.
Lubeck, S., & Jessup, P. (in press). Globalisation and its impact on early years funding and curriculum: Reform initiatives in England and the United States. In: T. David (Ed.), *Applied Research in Early Childhood Education*. Stamford, CT: Jai Publishing Ltd.
Meade, A. (2000). The early childhood landscape in New Zealand. In: J. Hayden (Ed.), *Landscapes in Early Childhood Education* (pp. 83–92). New York: Peter Lang Publishing.
Melhuish, E. C. (1993). Preschool care and education: Lessons from the 20th for the 21st century. *International Journal of Early Years Education*, 1(3), 19–32.
Melhuish, E., & Moss, P. (Eds) (1991). *Day care for young children: International perspectives*. London: Tavistock/Routledge.
Meljeteig, P. (1994). *Children's involvement in the implementation of their own rights: Present and future perspectives*. Paper presented at the International Society for the Study of Behavioral Development, Amesterdam.
Mellor, E. J. (2000). Hong Kong's early childhood landscape: Division, diversity, and dilemmas. In: J. Hayden (Ed.), *Landscapes in Early Childhood Education*. New York: Peter Lang Publishing.
Melton, G. B. (1993). Is there a place for children in the new world order? *Journal of Law, Ethics & Public Policy*, 7(2), 491–532.
Ministry of Education and Science (1998). *Curriculum for pre-school*. Stockholm, SW: The Ministry of Education and Science.
Moe, T. (2001, February). *An international perspective: Early findings of the OECD Thematic Review of Early Childhood Education and Care*. Paper presented at the conference on Children, Families and the Community: The U.K. and International Experience. London, U.K.

Morrison, J. W., & Milner, V. (1997). Early education and care in Jamaica: A grassroots effort. *International Journal of Early Childhood, 29*(2), 51–17.
Moss, P. (1999). Renewed hopes and lost opportunities: Early childhood in the early years of the Labour Government. *Cambridge Journal of Education, 29*(2), 229–238.
Munirathnam, G. (1995, November 3–4). *Child care – The RASS experience.* Paper presented at the Consultation on Government and Non-Governmental Organisations Partnership in Child Care, Madras, India.
National Association for the Education of Young Children (1996). *Guidelines for the preparation of early childhood professionals.* Washington, D.C.: NAEYC.
National Association for the Education of Young Children (2000). NCATE program standards: Initial and advanced programs in early childhood education. Washington, D.C.
National Research Council (1998a). *Preventing reading difficulties in young children.* Washington, D.C.: National Academy Press.
National Research Council (1998b). *Starting out right: A guide to promoting children's reading success.* Washington, D.C.: National Academy Press.
National Research Council (2001). Eager to learn. Washington, D.C.: National Academy Press.
Neuman, S., Copple, C., & Bredekamp, S. (1999). *Learning to read and write: Developmentally appropriate practices for young children.* Washington, D.C.: NAEYC.
Newport, S. F. (2000). Early childhood care, work, and family in Japan: Trends in a society of smaller families. *Childhood Education, 77*(2), 68–75.
Noirin, H. (1996). *Early education in Ireland – Towards collision or collaboration?* (ERIC Document ED 403067). Dublin, Ireland: Dublin Institute of Technology.
Oberhuemer, P. (2000). Conceptualizing the professional role in early childhood centers: Emerging profiles in four European countries. *Early Childhood Research & Practice* [On-line serial], *2*(2). Available: http://ecrp.uiuc.edu/v2n2/oberhuemer.html
Oberhuemer, P., & Ulich, M. (1997). *Working with young children in Europe: Provision and staff training.* London: Paul Chapman.
OECD (2001). *Starting strong: Early childhood education and care.* Paris: OECD.
OECD Background Report (2000). *OECD thematic review of early childhood education and care policy: Background report for the Czech Republic* [On-line]. Available: http://www.oecd.org/els/education/ecec/
OECD Country Note (1999). *OECD country note: Early childhood education and care policy in Sweden* [On-line]. Available: http://www.oecd.org/els/education/ecec/
OECD Country Note (2000). *OECD country note: Early childhood education and care policy in the Czech Republic* [On-line]. Available: http://www.oecd.org/els/education/ecec/
Olmstead, P., & Weikart, D. (Eds) (1989). *How nations serve young children: Profiles of child care and education in 14 countries.* Ypsilanti, MI: High/Scope Press.
Olmstead, P., & Weikart, D. (Eds.) (1995). *The IEA preprimary study: Early childhood care and education in 11 countries.* Oxford, U.K.: Pergamon.
Penn, H. (1999). *How should we care for babies and toddlers? An analysis of practice in out-of-home care for children under three.* Toronto: Centre for Urban & Community Studies.
Penn, H. (2000). How do children learn? Early childhood services in a global context. In: H. Penn (Ed.), *Early Childhood Services: Theory, Policy and Practice* (pp. 7–14). Buckingham, U.K.: Open University Press.
Petrogiannis, K., & Melhuish, E. C. (1996). Aspects of quality in Greek day care centers. *European Journal of Psychology of Education, XI*(2), 177–191.

Piscitelli, B., McLean, V., & Halliwell, G. (1992). Early childhood education in Australia. In: S. Feeney (Ed.), *Early Childhood Education in Asia and The Pacific: A Source Book* (pp. 197–236). New York: Garland Publishing, Inc.

Prochner, L., & Howe, N. (Eds). (2000). *Early childhood care and education in Canada.* Vancouver: UBC Press.

Qualifications and Curriculum Authority (QCA) (1999a). *Early learning goals.* London: Department of Education and Employment.

Qualifications and Curriculum Authority (QCA) (1999b). *Early years education, childcare and playwork: A framework of nationally accredited qualifications.* London: QCA.

Qualifications and Curriculum Authority (QCA) (2000). *Curriculum guidance for the foundation stage.* London: QCA.

Robinson, N. M., Robinson, H. B., Darling, M. A., & Holm, G. (1979). *A world of children: Daycare and preschool institutions.* Monterey, CA: Brooks/Cole Publishing.

Sharma, A. (1995, November 3–4). *Critical issues in partnership.* Paper presented at the Consultation on Government and Non-Governmental Organisations Partnership in Child Care, Madras, India.

Singer, E. (1996a). Children, parents and caregivers: Three views of care and education. In: H. Eeva (Ed.), *Childhood education: International perspectives* (pp. 159–170). Oulu, Finland: University of Oulu.

Singer, E. (1996b). Dutch parents, experts and policymakers: Conflicting views of day care. *Childhood Education, 72*(6), 341–344.

Starnes, L. (2000). Early childhood education in Azerbaijan. *Childhood Education, 77*(1), 6–12.

Swadener, E. B., Kabiru, M., & Njenga, A. (1997). Does the village still raise the child? A collaborative study of changing child-rearing and community mobilization in Kenya. *Early Education & Development, 8*(2), 285–306.

Tietze, W., Cryer, D., Bairrao, J., Palacios, J., & Wetzel, G. (1996). Comparisons of observed process quality in early child care and education in five countries. *Early Childhood Research Quarterly, 11*, 447–475.

UNESCO (1990). *World declaration on education for all.* Available: http://www2.unesco.org/wef/en-leadup/inter_efa.htm

UNICEF (1989). *The convention on the rights of the child.* Available: www.unicef.org/crc/fulltext.htm

UNICEF (1996). *The progress of nations 1996.* UNICEF. Available: www.unicef.org/pon96/indust4.htm.

UNICEF (2000). *The state of the world's children 2000.* New York: UNICEF.

Webster, F. (1995). *Theories of the information society.* London: Routledge.

Woodhead, M. (1997). Psychology and the cultural construction of children's needs. In: A. James & A. Prout (Eds), *Constructing and reconstructing childhood: Contemporary issues in the sociological study of childhood* (pp. 63–84). London: Falmer Press.

Woodhead, M. (2000). Towards a global paradigm for research into early childhood. In: H. Penn (Ed.). *Early childhood services: Theory, policy and practice* (pp. 15–35). Buckingham, U.K.: Open University Press.

Woodill, G. (1992). International early childhood core and education: Historical perspectives. In: G. Woodill, J. Bernhard & L. Prochner (Eds), *International handbook of early childhood education* (pp. 3–19). New York: Garland Publishing.

Woodill, G., Bernhard, J., & Prochner, L. (Eds). (1992). *International handbook of early childhood education.* New York: Garland Publishing.

World Bank (1999). *World development indicators.* Washington, D.C.: The World Bank.

APPENDIX A

The World by Income (World Bank, 1999)

Low Income

Afghanistan	Ethiopia	Nicaragua
Albania	Gambia, The	Niger
Angola	Ghana	Nigeria
Armenia	Guinea-Bissau	Pakistan
Azerbaijan	Haiti	Rwanda
Bangladesh	Honduras	São Tomê and
Benin	India	Principe
Bhutan	Kenya	Senegal
Bosnia and	Kyrgyz Republic	Sierra Leone
Herzegovina	Lao PDR	Somalia
Burkina Faso	Lesotho	Sudan
Burundi	Liberia	Tajikistan
Cambodia	Madagascar	Tanzania
Cameroon	Malawi	Togo
Central African	Mali	Turkmenistan
Republic	Mauritania	Uganda
Chad	Moldova	Vietnam
Comoros	Mongolia	Yemen Republic
Congo, Dem. Rep.	Mozambique	Zambia
Côte d'Ivoire	Myanmar	Zimbabwe
Eritrea	Nepal	

Lower Middle Income

Algeria	Dominican Republic	Jamaica
Belarus	Ecuador	Jordan
Belize	Egypt, Arab Rep.	Kazakhstan
Bolivia	El Salvador	Kiribati
Bulgaria	Equatorial Guinea	Korea, Dem. Rep.
Cape Verde	Fiji	Latvia
China	Georgia	Lithuania
Columbia	Guatemala	Macedonia, FYR
Costa Rica	Guyana	Maldives
Cuba	Indonesia	Marshall Islands
Djibouti	Iran, Islamic Rep.	Micronesia, Fed. Sts.
Dominica	Iraq	Morocco

Namibia
Panama
Papua New Guinea
Paraguay
Peru
Phillippines
Romania
Russian Federation

Samoa
Solomon Islands
Sri Lanka
St. Vincent and the Grenadines
Suriname
Swaziland
Syrian Arab Republic

Thailand
Tonga
Tunisia
Ukraine
Uzbekistan
Vanuatu
West Bank and Gaza
Yugoslavia, FR

Upper Middle Income
American Samoa
Antigua and Barbuda
Barbados
Botswana
Brazil
Chile
Croatia
Czech Republic
Estonia
Gabon
Grenada
Guadeloupe

Hungary
Isle of Man
Lebanon
Libya
Malaysia
Malta
Mauritius
Mayotte
Mexico
Oman
Palau
Poland

Puerto Rico
Saudi Arabia
Seychelles
Slovak Republic
South Africa
St. Kitts and Nevis
St. Lucia
Trinidad and Tobago
Turkey
Uruguay
Venezuela

High Income
Andorra
Aruba
Australia
Austria
Bahamas, The
Belgium
Bermuda
Brunei
Canada
Cayman Islands
Channel Islands
Cyprus
Denmark
Faeroe Islands
Finland
France
French Guiana

French Polynesia
Germany
Greece
Greenland
Guam
Hong Kong, China
Iceland
Ireland
Israel
Italy
Japan
Korea, Rep.
Kuwait
Liechtenstein
Luxembourg
Macao
Netherlands Antilles

New Caledonia
New Zealand
Northern Mariana Islands
Norway
Portugal
Qatar
Reunion
Singapore
Slovenia
Spain
Sweden
Switzerland
United Arab Emirates
United Kingdom
United States
Virgin Islands (U.S.)

CHILD CARE QUALITY: A MODEL FOR EXAMINING RELEVANT VARIABLES

Eva L. Essa and Melissa M. Burnham

ABSTRACT

Child care research has progressed over the past several decades to a level of sophistication and depth that begins to give some answers to the question, "what is the impact of child care on young children?" This paper provides a model within which this complex body of literature can be viewed and presents a comprehensive review of the literature. The Child Care Quality Model, based on ecological theory, helps to organize and conceptualize the relationship among the salient components of the child care research literature. Central to the model is the relationship of child care quality to child outcomes. In addition, both proximal and distal influencing variables are considered.

The examination of the literature expands on these aspects of the model by first reviewing the elements of structural and process quality, and how these are measured. It then considers studies that report the impact of child care on child outcomes in social, behavioral/emotional, and cognitive/language development. Research that focuses on additional influencing factors, which interact with child care to impact child outcomes, are also reviewed. These include proximal variables such as family characteristics, child characteristics, and program characteristics, and more distal community and societal variables, including child care licensing standards. A summary synthesizes the literature in the context of the Child Care Quality Model, and points out some of the gaps in the current level of understanding of how child care influences young children.

Child care research, it has been said, has progressed in stages or "waves." The earliest wave, lasting into the late 1970s, asked the straightforward question, "Is child care good or bad for children?" (Belsky, 1984). As Howes and Hamilton (1994) tersely summarized this wave, "The answer to that question was a resounding 'maybe'" (p. 325). The second wave broadened its scope by acknowledging that the element of quality needs to be part of the equation; thus the question focused on how quality of child care influences children's development. Much current child care research still reflects an exploration of this second question, but a new, third wave has amplified the scope of inquiry even further by considering the effects of child care in relationship to other contexts, such as the family and community. Thus child care research can be depicted as representing an expanding set of systems: the first phase focused on the child, the second added the impact of aspects of the child care environment, and the third considered interactions with other contexts within which children are socialized. The ecological systems approach (Bronfenbrenner, 1979) suggested by the evolution of child care research offers an excellent framework within which to consider quality of child care in relation to child outcomes. It also allows us to examine the interaction of these variables in relation to different contexts such as child, family, child care, community, and societal characteristics. This theoretical approach has, of course, been suggested as relevant in conceptualizing child care in context (e.g. Kontos & Peters, 1988).

In recent years, child care research has become progressively more sophisticated and complex as researchers attempt to tease out the potential impact of numerous variables on the development of young children in nonmaternal care. A number of review articles have explored this growing body of research over the past two decades, most often focusing on the question of what impact child care really has on young children (e.g. Caldwell, 1993; Cryer, 1999; Lamb & Sternberg, 1990; McGurk, Caplan, Hennessy & Moss, 1991; Melhuish, 2001; Scarr & Eisenberg, 1991; Vandell & Wolfe, 2000; Zaslow, 1991). Other papers have focused only on specific aspects of quality, for instance, child-to-adult ratio and group size (Dunn, 1993b) or the contribution of child care to the development of social competence (Howes, 1987). These reviews provide important insights and syntheses of the research. This paper aims to add to existing literature by providing a model into which the research can be organized. In addition to building on the research, the

Model also incorporates a modification of Dunn's (1993a) concept of proximal and distal aspects of child care quality in relation to children's development. A brief discussion of this Model will be followed by a thorough review of the literature.

THE "MODEL"

In order to give structure to the extensive literature on quality child care, the Child Care Quality Model (Fig. 1) has been developed to help organize and conceptualize the relationship among the salient components. The central components of the model, particularly seen in "wave two" research, have included *elements of quality* and *child outcomes*. Quality factors have included such variables as ratio, group size, teacher education and training, and child-teacher interaction. Outcome measures have typically examined such factors as child socialization, behavior, language, cognition, and academic achievement. The "third wave" of research includes a number of additional, *influencing variables*, that contribute to and have an impact on the relationship of quality factors to child outcomes. Some of these variables are proximal and include characteristics of the family such as socioeconomic status, maternal education, marital status, and ethnicity; characteristics of the child, such as temperament and gender; and characteristics of the child care setting, such as center philosophy, hiring practices, and auspice (profit or not-for-profit status). Other variables are more distal, including characteristics of the larger community, such as licensing standards and monitoring, and the growth of the knowledge base in the field of early childhood education. These components are by no means exhaustive, but represent the major variables examined in the extant literature.

The significance of these variables is expressed in the Child Care Quality Model that delineates their relationships (see Fig. 1). Central to the model is the child, as affected by the child care experience. Elements of child care quality, measured in static *structural* or dynamic *process* terms, generally assess quality and serve as independent variables in research. The impact of such variables on the child is measured in terms of developmental outcomes, the dependent variables. These two core elements – child care quality and child outcomes – are the critical microsystems of this model. Yet, the relationship between child care quality and child outcomes is not straightforward or linear; other, influencing, variables, both proximal and distal, exert an impact.

The model shows three groups of proximal influencing variables that are found in the literature, related to characteristics of the family, child, and child care center. The child's family is powerful in shaping his or her development.

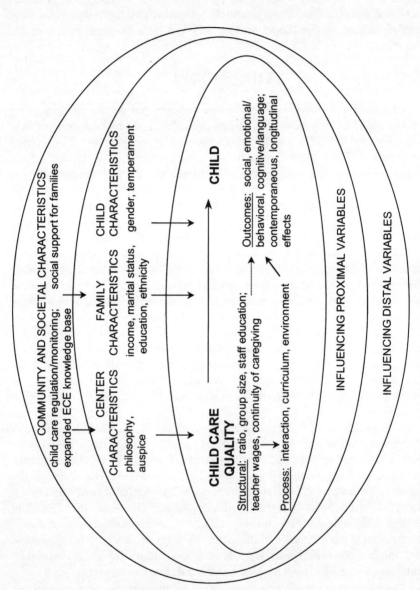

Fig. 1. The Child Care Quality Model.

Thus, one must consider that the effect of family certainly mitigates the impact of child care. A positive family life may make up for a negative child care experience, just as high quality child care may ameliorate a poor home environment (e.g. Caughy, DiPietro & Strobino, 1994). The model shows the family as influencing the child care quality-child outcomes relationship. In addition, there is evidence that characteristics of the child also can influence the relationship between child care quality and the child's development. Boys and children with difficult temperaments appear to be more vulnerable and may require especially sensitive care (e.g. Desai, Chase-Lansdale & Michael, 1989). A third proximal variable involves characteristics of the child care center. Such characteristics are differentiated in the model from the elements of quality because they are, in essence, a step removed from those structural and process variables. Furthermore, the model shows characteristics of child care centers as exerting an impact in a somewhat different way from family and child characteristics, which are depicted as intervening between the relationship of child care quality and child outcomes. Characteristics of child care centers influence this relationship in a more linear way. Attributes such as program philosophy and auspice have a direct effect on child care quality which, in turn, impacts the child (e.g. Helburn, 1995; Whitebook, Howes & Phillips, 1990).

The model shows the community and larger society exerting more distal influences on the child care quality-child outcomes relationship. As Phillips (1996) reminds us, those often ignored variables are important, though more difficult to pinpoint and measure. The model shows community and societal impact through three paths. Two exert influence on child care center characteristics, one through regulations and monitoring that establish minimum standards for centers and the other through the expanding knowledge base in the field of early childhood education. The final distal path is through family characteristics. The support provided to families by the community and the larger society affect the options families have related to child care and the decisions they make to provide non-familial care for their child. As an example, some research has shown that low family income means lower quality child care for many families (e.g. Burchinal et al., 1996; Kontos et al., 1997); however, the availability of subsidized programs can offer high quality care to some more vulnerable children. International comparisons (e.g. Hofferth & Deich, 1994), such as with studies done in Sweden, point to the positive effect of society's support for the nurture and care of children in high quality centers.

One noticeable feature of the Model is that most of the arrows designating relationships among components are unidirectional. This reflects the state of the research rather than the many potential bidirectional relationships that can exist among the variables. For example, teachers who work with young children often comment on how a particular child or group of children can impact the tenor-hence the quality-of their program. Yet the existing research has not explored this aspect of child care quality, rather it has focused on how quality components impact children's development.

This Model presents a way to organize and conceptualize the sizeable literature that examines the effects of child care quality. This literature will now be reviewed in three sections, which reflect the broad categories of variables that are factored into the various studies and which are incorporated into the Child Care Quality Model: (1) elements of quality, (2) child outcomes, and (3) additional, influencing variables.

ELEMENTS OF QUALITY

The specification of different elements of quality has been of interest to researchers since the advent of the second wave of child care research. Defining the essential elements underlying quality child care would presumably add valuable insight into designating which elements correspond to positive outcomes for children and which may be superfluous. However, as noted by Phillips and Howes (1987) well over ten years ago, "quality, by its nature, is a fuzzy concept" (p. 3). While the quality concept has come into clearer focus in recent years, the number and nature of its elements can be somewhat staggering. Many criteria for quality have been delineated (e.g. NAEYC's statement on Developmentally Appropriate Practice, Bredekamp & Copple, 1997; Health, Education, and Welfare Day Care Requirements, Federal Register, 1980) and numerous research investigations into the important variables related to quality child care have been conducted.

In general, as reflected in the Child Care Quality Model, two main categories of quality variables have resulted from these investigations: *structural* and *process*. Structural variables are those elements of quality that are essentially descriptive in nature and relatively easy to measure, particularly when a large sample of programs is studied. Indeed, some researchers use the terms "structural" and "regulatable" synonymously. Structural elements include such aspects of child care programs as adult-to-child ratio, group size, teacher education and training, and teacher salaries. Process elements, on the other hand, are more dynamic, involving such variables as the quality and warmth of teacher-child interactions and the activities and learning opportunities available for the children (Phillipsen, Cryer & Howes, 1995). These variables are both more challenging and more time consuming to measure. Both structural and process variables have been used in research investigations of child care quality.

The two types of quality have unique methodological concerns. Specifically, investigators studying structural quality often have focused on the relationship between structural and process elements to determine the relative importance of each. In addition, researchers have examined the feasibility of using the more easily-measured structural elements as proxies for overall program quality. In contrast, the main research issue facing investigations of process quality is measurement. Process quality is a much more difficult construct to measure than are structural elements, and numerous and sometimes overlapping instruments have been developed to assess this set of variables. Thus, there have been some conflicting results on the relative importance of specific process variables. Each element of quality will be thoroughly examined in turn.

Structural Elements

As mentioned briefly above, structural aspects of child care quality are those easily quantified elements of care which focus on describing the framework of a child care program. In the Child Care Quality Model, these appear as a subgroup under child care quality. Regulatable structural elements include such variables as teacher-child ratio and group size. Other structural variables which have been found to impact on child outcomes but are not necessarily regulatable include teacher tenure in the program, hours in child care, and the continuity of care. Structural variables can be measured through a number of means, including questionnaires, observation, interviews, or any combination of the three. To insure concurrent validity, a combination of these techniques is often used in any assessment of structural characteristics.

In general, structural elements of quality are thought to provide the minimal base for a high quality child care program. It is commonly believed that structural variables provide the base from which other, more dynamic elements of quality reflecting children's actual experience can be realized (Phillipsen, Burchinal, Howes & Cryer, 1997; Phillips & Howes, 1987). For instance, it is thought that a caregiver would be more able to be responsive to the emotional needs of an individual child in a program with a small group size and ratio (Doherty-Derkowski, 1995). Thus, structural elements are often used to predict process quality. In a few investigations, they also have been directly correlated with child outcomes (e.g. Scarr, Eisenberg & Deater-Deckard, 1994). However, it is typically assumed that structural variables are useful predictors of child outcomes only through the mediating influence of process variables (Phillips &

Howes, 1987; File & Kontos, 1993). In other words, the relationship between the two elements of quality and child outcomes is asserted to flow linearly from structural to process quality to outcomes. These relationships are depicted by arrows in the Model and will be examined in turn. Occasionally, researchers have attempted to use structural elements as the sole method of measuring child care quality, particularly in large investigations where program observation is prohibitive (e.g. Kisker, Hofferth, Phillips & Farquhar, 1991). More commonly, structural elements are included in studies as an important but not singular indicator of quality. Studies that have included structural characteristics of care to examine quality will be reviewed first.

Ratio and Group Size
By far, the most recognized and researched elements of structural quality are ratio and group size (e.g. Doherty-Derkowski, 1995; Helburn & Howes, 1996), both shown in the Child Care Quality Model. The recommended numbers of caregivers per child and the number of children in a group vary according to the age of children, with smaller group sizes and higher ratios recommended for the youngest children in care (Bredekamp & Copple, 1997). With few exceptions, ratio and group size have been included as variables in both large, multi-site investigations (Helburn, 1995; Howes, 1997; Kisker et al., 1991; NICHD Early Childhood Research Network, 1996, 1999; Phillipsen et al., 1997; Scarr, Eisenberg & Deater-Deckard, 1994) and smaller investigations (Burchinal, Roberts, Nabors & Bryant, 1996; Deater-Deckard, Pinkerton & Scarr, 1996; Field, 1980; Field, 1991; Howes, 1988; Howes, 1990; Howes & Olenick, 1986; Phillips, McCartney & Scarr, 1987; Vandell, Henderson & Wilson, 1988; Vandell & Powers, 1983). Favorable ratios and group sizes have generally been related to positive child outcomes. These relationships will be examined in the section on Child Outcomes.

Higher ratios of adults to children have been related to measures of process quality. Using data from the *Cost, Quality, and Child Outcomes Study*, Howes (1997) has reported that teachers in classrooms with ratios in compliance with professionally-recommended standards were rated as more sensitive and responsive than were teachers in classrooms which were out of compliance. Although significant, however, the differences found were not striking; the lowest mean "sensitive" rating for teachers in classrooms with poor ratios were 2.2, on a 4-point scale, while the highest rating given to teachers in classrooms in compliance with ratio standards were 2.9 ($F = 6.25$, $p < 0.01$). Indeed, in the same study, Howes also reported an analysis of the *Florida Quality Improvement Study* data, in which a relationship between ratio and teacher behavior was not found. While Howes' analyses leave questionable the relationship between ratio and teacher behavior, other investigations have found relatively robust relationships between ratio and teacher characteristics. For example, in the formal report of the *Florida Quality Improvement Study*, Howes, Smith, and Galisky (1995) found that classrooms which were in compliance with the professional ratio standards set by the National Academy of Sciences rather than the less stringent state standards had more sensitive, less harsh, and less detached teachers. In addition, as would be expected, those classrooms in compliance with more stringent professional ratio standards were also given higher process quality ratings than either those programs which met Florida standards or were completely out of compliance. What seems apparent from these various analyses is that programs with higher ratios tend to have more responsive teachers, although this relationship is by no means unequivocal.

Similarly, in their investigation, Phillipsen and colleagues (1997) found that higher ratios were significantly associated with three measures of process quality in both infant/toddler and preschool classrooms. Interestingly, these researchers did not ubiquitously find a significant relationship between group size and process quality measures. Rather, the only association between group size and process quality was a small negative correlation between group size and preschool teachers' responsiveness ($r = -0.12$, $p < 0.05$). This finding is congruent with Dunn's (1993b) conclusion that ratio is a more important indicator of quality than group size. The NICHD Study Team (1999) also found that ratio, though not group size, was associated with positive outcomes for young children. Indeed, the bulk of research substantiates this conclusion. For example, Scarr and colleagues (1994) report moderate correlations between ratio and two measures of process quality (r's $= 0.36$ and 0.31, significance level not reported) but much smaller correlations between group size and process measures (r's $= -0.10$ and -0.17). File and Kontos (1993) also report significant correlations between ratio and teacher involvement, but no significant correlations between group size and teacher involvement. Additionally, Howes and colleagues (1992) report a significant association between ratio and appropriate caregiving as well as developmentally appropriate activities but no association between group size and appropriate caregiving.

In a more general investigation of the relationship between ratio and program process quality using hierarchical regression analysis, Mocan and colleagues (1995) report that "the adult-child ratio is the single most important factor in determining child care quality" (p. 294). In other words, these researchers found that ratio was the single best predictor of process quality in their national data set. McCartney and colleagues (1997) also found a significant positive relationship between ratio and teacher-child interaction in their multi-state sample of center care for infants, toddlers, and preschoolers. However, ratio was not found to distinguish between child-centered, intermediate, or didactic preschool and kindergarten programs in one study (Stipek, Daniels, Galluzzo & Milburn, 1992). Further, at least one investigation has reported significant correlations between group size and process quality measures, but not between ratio and these measures (Holloway & Reichart-Erickson, 1988). Nevertheless, while there appear to be some confusing results regarding the relative contributions of ratio and group size to quality care, for the most part ratio, and to some extent, group size, appear to be vital characteristics of a high quality child care program.

Teacher Education and Training
A second structural element of quality that has been examined in great depth is teacher education and/or specialized training. Education refers to attendance at formal institutions which issue a degree not necessarily related to child development while training refers to hours spent in workshops or other methods of instruction related to child development and care. Often, education is further partitioned into general education and that specifically related to child development. A number of investigations have used teacher education and/or training to categorize quality (Arnett, 1989; Burchinal et al., 1996; Deater-Deckard et al., 1996; Field, 1991; Howes, 1988; Howes, 1990; Howes & Olenick, 1986; Kisker et al., 1991; Phillips et al., 1987; Vandell et al., 1988; Vandell & Powers, 1983). While some discrepancy exists in the literature regarding the relative importance of general education, education specific to child development, or training in contributing to the quality of a child care program, it is generally recognized that higher levels of education and more training are optimal. In the multi-site study reported by Phillipsen and colleagues (1997) and in other investigations (e.g. Dunn, 1993a; Howes, 1997; Scarr et al., 1994), formal education and training were highly correlated. Further, Phillipsen's group (1997) found significant correlations between the lead teachers' having a Bachelor's degree and the global quality of both infant/toddler and preschool classrooms. Having a Bachelor's degree was also correlated with preschool teachers' sensitivity and responsiveness. On the other hand, some college education but no degree was not correlated with indicators of process quality. Similar results were reported by Howes (1997) in her analysis of two large national data sets. Specifically, she found that teacher education related to early childhood was associated with teacher sensitivity and responsiveness. Relatedly, Scarr and colleagues (1994) report a moderate association between teacher education and training and two measures of process quality.

Findings from the Florida Quality Improvement Study indicate that higher levels of teacher education specific to early childhood are related to improved teacher responsiveness (Howes et al., 1996). In this study, caregivers who earned a Child Development Associate's (CDA) credential generally scored better on measures of sensitivity, responsiveness, and overall classroom quality than did caregivers with less education; caregivers with a more advanced education in early childhood generally scored even higher on these measures. In a smaller investigation of the influence of training/education on teacher attitudes and behavior, Arnett (1989) found that caregivers with more training/ education had less authoritarian child rearing attitudes and were observed to have more positive interactions, were less punitive, and were less detached than caregivers with less training. Caregivers with a four-year college degree in early childhood were significantly less authoritarian than any of the other three groups, including the group which had attended extensive training workshops in early childhood education. In a prospective investigation of the effects of child-related training on the quality of classrooms and teacher beliefs, Cassidy and colleagues (1995) found that attending 12–20 hours of community college coursework related to child development and early childhood education significantly improved the developmental appropriateness of teacher practices and beliefs. Although the changes reported were moderate, they are an indication that training can impact practice. Such moderate changes as a result of some coursework would be expected, given other findings that teachers with higher levels of education behave more appropriately.

Other investigators also have found a significant association between teachers' formal education and/or specialized training and both the global quality of a program (Howes & Smith, 1995; Mocan et al., 1995) and teacher-child interaction (McCartney et al., 1997). Further, in a qualitative study examining the discipline strategies that preschool teacher's use, Essa (1998) found that teachers with less formal education reported using less developmentally-appropriate discipline strategies than their more educated counterparts. However, at least one investigation has reported that teacher education was not related to the developmental appropriateness of the program in a sample of preschool and kindergarten classrooms. Stipek and colleagues (1992) distinguished between child-centered, intermediate, and didactic programs. While measures of process quality were able to differentiate these types of programs, teacher education proved to have little or no association with type of program. In another investigation, File and Kontos (1993) reported that specific training in early childhood education, but not overall education level, was related to positive teacher involvement. Further, although significant, Ghazvini and Readdick (1994) report a very small relationship between specialized training and process quality ($r=0.11$, $p<0.05$). Clearly, however, most investigations have found that teacher education and training do impact on the quality of care given to children.

Other Structural Characteristics

Some other structural characteristics of child care programs include square footage per child, equipment availability, structure of the environment, teacher characteristics (wages, experience, tenure in the program, continuity), and hours of care. While some of these are regulatable (e.g. square footage and teacher experience), a few are program characteristics that can easily be quantified and appear to be important elements of quality but cannot be regulated by licensing institutions (e.g. teacher tenure, wages, hours of care, and continuity of caregivers). The latter group may be important for research purposes to help designate quality programs, but is less helpful for formal regulation. Regulatable structural elements, such as square footage per child, partitioning of the environment, and toy availability have been used as indicators of quality in some investigations (Field, 1980; Howes, 1988; Vandell et al., 1988; Vandell & Powers, 1983). In general, investigators agree that more space, more toys and equipment, and a partitioned environment rather than a large open space per child are indicators of higher quality programs.

Researchers have found that other, non-regulatable aspects of structural quality relate to program quality to varying degrees. Teacher wages have been robust indicators of quality in a number of investigations to date. For example, Phillipsen and colleagues (1997) found that lead teacher wages were positively correlated with all three measures of process quality that they used (r's ranged from 0.13 to 0.54, p's <0.05). Similarly, Scarr and colleagues (1994) found that the best structural predictor of process quality was highest wage paid to a teacher. Results of the *Child Care Staffing Study* indicate that, of all work environment variables including teacher education and training, teacher wages were the most important predictor of process quality (Whitebook et al., 1989). Teacher income also has been found to relate significantly to the quality of teacher-child interactions (McCartney et al., 1997). In addition, using data from the *Cost, Quality, and Child Outcomes Study*, Mocan and colleagues (1995) report that teacher wages were significant predictors of process quality using two different types of statistical analyses, even when education was controlled. The authors conclude that "the consistent association between teaching staff wages and process quality in correlational, regression, and discriminant analyses suggest that wages index other factors that are not measured by teaching staff training, education, experience, or by the center and classroom structure" (p. 287). What these factors may be remains to be determined.

Nonetheless, it appears that teacher wages are important structural indicators of quality.

The continuity of caregivers also has been implicated as important in studies examining quality child care. Continuity relates to teacher tenure in the program, turnover rates, and program structure. The literature consistently reports higher quality and outcomes among programs with teachers who remain constant (Howes, 1987; Howes & Hamilton, 1993; Howes, Hamilton & Phillipsen, 1998; Howes, Matheson & Hamilton, 1994), while change in teachers has been associated with negative outcomes (Field, Vega-Lahr & Jagadish, 1984; McCartney et al., 1997). While a six-year longitudinal study of the child care workforce (Whitebook, Sakai, Gerber & Howes, 2001), found the same relationship between continuity of caregiving and program quality, it nonetheless concluded that "the teaching staff workforce is alarmingly unstable, even among ... teachers in relatively high-quality programs" (p. v). Better wages were directly associated with better retention of good teachers in this study. However, some studies have not reported consistent relationships between teacher turnover and measures of process quality (Phillipsen et al., 1997; Scarr et al., 1994). It is possible that this structural characteristic has a direct relationship to child outcomes, as will be discussed in a later section, rather than one mediated through process quality or that turnover is simply a correlate of other structural characteristics. The possibility of a direct influence of structural features of child care on child outcomes has been suggested by other researchers (e.g. Dunn, 1993a; Kontos, 1991; Vandell & Powers, 1983). This relationship is depicted in the Child Care Quality Model by a dashed arrow connecting structural quality to outcomes to convey that it may well exist in the context of some variables but not others.

Combining Structural Variables

Although the structural characteristics discussed above have been categorized as separate elements of care, it is well-recognized in the literature that many structural variables co-occur in any given program. It is unusual, for example, to see a program which has professionally-recommended group sizes, but non-compliant ratios. This fact led Phillips and Howes (1987) to conclude that, when it comes to elements of quality in child care, "good things go together" (p. 3). In the *Cost, Quality, and Child Outcomes Study*, for example, teacher wages were correlated with education, experience, and tenure in the program (Helburn, 1995). However, this study did find that wages and education independently contributed to process quality in a regression analysis, indicating the importance of measuring both in an analysis of quality. Scarr and colleagues (1994) report similar correlations between teacher training and education and wages. It is important to note that Dunn (1993a) did not find robust intercorrelations between structural variables or between process and structural variables. Additionally, Clarke-Stewart (1987) has cautioned against the assumption that elements of quality necessarily coexist. Interestingly, the NICHD Study reported an additive effect for structural variables in that child outcomes became increasingly more favorable as more such variables met higher standards. Despite the usual high degree of intercorrelation between structural variables and given that some studies report small intercorrelations among quality variables, it is generally recognized as best to include more elements rather than fewer in any investigation of child care quality.

At least two research instruments have been developed to give a composite indication of the structural quality of a program. The Day Care Environment Inventory (Prescott, Kritchevsky & Jones, 1972) is an interview which focuses on a number of descriptive aspects of the child care environment. A second instrument, the Child Development Program Evaluation-Indicator Checklist (CDPE-IC), is a 15-item scale with items such as ratio, staff qualifications, safety, supervision, and food preparation rated as pass/fail (Fiene & Nixon, 1985). According to Kontos (1991), the CDPE-IC can be considered as a measure of the bare minimum standard for child care programs. Interestingly, Kontos (1991) found that the CDPE-IC score, but not the score on a global measure of process quality, was related to children's social development, presumably indicating the importance of structural variables in impacting children directly. Kontos also reported a moderate correlation ($r=0.36$, significance level not reported) between the CDPE-IC and a common measure of process quality (the Early Childhood Environmental Rating Scale; Harms & Clifford, 1980). Similarly, McCartney (1984) reported moderate correlations between items on the Day Care Environment Interview and children's language development scores as well as an overall measure of process quality. Combining a number of structural elements into a composite index appears to be a fruitful method to assess quality.

Summary of Structural Quality
Using multiple methods to assess structural elements is a good way to assess the validity of the measures. Indeed, most studies of child care quality have included multiple variables and ways of measuring quality (e.g. Helburn, 1995; NICHD Early Child Care Research Network, 1996; Vandell & Powers, 1983). It is also important to keep in mind that, despite the relatively robust relationships reported above between structural and process elements, measuring only structural variables to assess quality is generally not accepted. The body of empirical knowledge on the strength of the relationship between

structural and process variables is not strong enough to conclude that structural indicators can be used alone. In her review of the literature on ratio and group size, for example, Dunn (1993b) concludes that, although these variables were found to be decent proxies of overall quality, ratio and group size "may be necessary but not sufficient conditions for optimal child development" (p. 222). Future research into the empirically vital question of whether structural characteristics can be used in lieu of process measures to characterize quality is needed. Overall, it is apparent that a number of structural characteristics of quality exist and that many of them appear to be at least moderate indicators of global aspects of quality and, to some extent, related to child outcomes. Elements of process quality will now be reviewed.

Process Elements and Global Quality

Process variables are considered to be more sensitive indicators of quality child care than are their structural counterparts. Indeed, process variables seem to encompass the most important elements of a high quality program. Process elements of quality generally reflect the more dynamic, relationship-oriented processes which occur within child care programs. As discussed above, process quality refers to specific elements of the caregiving environment which are thought to impact directly on children's development. For this reason the relationship between process quality and child outcomes is shown in the Child Care Quality Model as connected by a solid arrow. A later section of this paper will focus on research examining the impact of specific elements of process quality on children's development. This section will focus on what particular variables have been used and which appear to be solid indicators of high quality care. A distinction has been made in the literature between process and global quality. While the exact factors which separate the two remain elusive, in general, process quality focuses on more specific, individual aspects of the child care environment, such as teacher-child interaction or teacher beliefs while global quality focuses on multiple process (and sometimes structural) elements combined. After describing the various ways that process quality has been operationalized, the most common process elements considered in the literature will be discussed, followed by an account of studies using global measures.

Measurement of Process Elements
The most commonly assessed process variables fall under two broad categories: the quality of interactions between caregiver and child, and the quality of activities available for the children. Other process elements include

caregiver attitudes and beliefs, emotional climate of the program, and curriculum characteristics. Measurement of the process elements of quality has proven more difficult than measurement of structural variables. Indeed, process quality is a more "slippery" construct, with varying definitions and methods of assessment. Methods devised to assess process quality include a great number of checklists, questionnaires, and observations (see Table 1).

Indeed, there have been at least ten different ways to assess caregiver characteristics and caregiver-child interactions presented in the literature. Among these measures are both formal instruments (Belsky & Walker Checklist; ORCE; Appropriate Caregiving subscale of the ECERS; ECOI subscale; Adult Involvement Scale; Arnett Scale of Provider Sensitivity; COFAS; Teacher Didactic Beliefs Scale; Teacher Questionnaire; subset of items from the Assessment Profile) and structured observations developed

Table 1. Instruments Used to Measure Process Quality.

Instrument	Reference
Caregiver Characteristics/Interactions	
ECERS/ITERS App. Caregiving Subscale	ECERS: Harms & Clifford, 1980 ITERS: Harms, Cryer & Clifford, 1990
Caregiver Interaction Scale	Arnett, 1989
Adult Involvement Scale	Howes & Stewart, 1987
Early Childhood Classroom Observation Scale/Instrument: Interaction Subscale	Bredekamp, 1986; Modified: Holloway & Reichart-Erickson, 1988
Belsky & Walker Checklist	Belsky & Walker, 1980
Observational Rec. of the Caregiving Env.	NICHD Early Ch. Res. Network, 1996
Caregiver Observation Form and Scale	Fiene, 1984
Assessment Profile	Abbott-Shim & Sibley, 1992
Educational Attitudes Scale	Rescorla et al., 1990
Teacher-Child Interaction Index	McCartney et al., 1997
Teacher Didactic Beliefs Scale Teacher Questionnaire: Teacher Beliefs & Instructional Activities Subscales	Stipek et al., 1992 Burts et al., 1992
Curriculum/Activities	
Rubenstein & Howes Object Play Scale	Rubenstein & Howes, 1979
ECERS/ITERS Dev. Appr.Activities Subscale	ECERS: Harms & Clifford, 1980 ITERS: Harms, Cryer & Clifford, 1990

specifically for a given study (e.g. those used by McCartney, 1984; File & Kontos, 1993). Similarly, numerous ways of characterizing the quality of activities and curriculum for child care programs have been presented in the literature. Again, researchers have either developed their own methods of assessing activities and curriculum (e.g. Howes & Smith, 1995; Kontos, Howes, Shinn & Galinsky, 1997) and/or have used formal instruments to characterize this element of process quality (e.g. Howes et al., 1992; Kontos et al., 1997; Stipek et al., 1992). The diversity of instruments used in the literature proves to be both a blessing and curse for quality research. On the one hand, instrument diversity provides a check on the validity of the process quality construct. On the other, it precludes clear comparisons between studies which attempt to measure the same underlying construct using different instruments. Studies which have used the two main elements of process quality will be examined in turn, with specific focus on which measures were used by each investigation.

Caregiver-Child Interactions
Perhaps the most widely accepted element of process quality is the type and quality of interactions between caregivers and children. While the measurement of interactions has proven to be a particularly challenging methodological endeavor, a number of studies have attempted to assess them nonetheless (e.g. Bredekamp, 1986; Deater-Deckard et al., 1996; Dunn, 1993a; File & Kontos, 1993; Hestenes, Kontos & Bryan, 1993; Howes & Smith, 1995; McCartney, 1984; McCartney et al., 1997; NICHD Early Child Care Research Network, 1996). A number of instruments have been developed which attempt to assess these interactions (see Table 1). However, many of these measures actually reflect caregiver behaviors toward children and would seem better categorized more generally as caregiver characteristics rather than interactions. Since the two often reflect similar elements of quality in the literature, the distinction made here should be considered more of an organizational than conceptual discrimination. Interaction refers to teachers' behaviors toward children, specifically engaged, whereas other characteristics may or may not be specific to interactions with children. In general, studies which have attempted to assess caregiver-child interactions using either formal instruments or measures derived for a specific study have found this element of process quality to relate to other aspects of quality as well as to child outcomes.

Formal Instruments
One instrument designed in part to tap the quality of caregiver-child interactions is a subscale of the Early Childhood Observation Instrument

(ECOI; Bredekamp, 1986). Representative items from the six-item "interaction" subcomponent of the ECOI, revised by Holloway and Reichart-Erickson (1988), include: "staff interact frequently with children, showing affection and support" and "staff speak with children in a friendly, courteous manner." Following a detailed program observation, observers rate each item on a three (original) or four (revised) point scale. The "interaction" subcomponent of the ECOI has been found to relate significantly to structural aspects of quality as well as to child outcomes (see below; Holloway & Reichart-Erickson, 1988).

Similarly, in their analysis of caregiver behaviors using Howes and Stewart's (1987) Adult Involvement Scale, Hestenes and colleagues (1993) report significant associations between caregivers' level of engagement and child outcomes. Howes and Stewart (1987) also report a significant correlation between scores on the Adult Involvement Scale and an overall measure of global quality for both boys and girls. This scale requires a detailed observation of caregiver behaviors. Specifically, caregiver behaviors toward a specified target child are coded on a six-category scale ranging from "ignores" to "intense" following the observation period. While mean levels of engagement derived from this scale were used by Howes and Stewart, Hestenes and colleagues chose to dichotomize the scale into "low level" and "high level" engagement for their analyses. Both uses of the scale appear to have derived similar results.

Specifically-Developed Measures
In another study designed to examine caregiver-child interaction as an element of child care quality, McCartney and colleagues (1997) used a composite index of teacher-child interaction derived from interaction items from the ECERS/ ITERS (Harms & Clifford, 1980; Harms, Cryer & Clifford, 1990) and Assessment Profile (Abbott-Shim & Sibley, 1992) in their large investigation of child care centers in three states. For preschool classrooms, eight ECERS items and 35 Profile items load into the Interaction Index. For infant-toddler classrooms, four ITERS items and 39 Profile items load into the Index. Items were chosen which specifically reflected the quality of interactions between caregivers and children. While other investigations have reported a significant association between caregiver-child interactions and child outcomes, McCartney and colleagues (1997) found a significant association between interaction and structural indicators of quality care, but little association between teacher-child interaction and child outcomes. The authors note that this finding may have been due to moderate reliability on the interaction index.

In another investigation using caregiver behavior with children as an indicator of quality, File and Kontos (1993) developed an observation scheme for coding teacher involvement. Specifically, during time-sampled free-play intervals, interactions between caregivers and specified target children were observed and teacher behaviors were coded into five categories ranging from "no involvement" to "support social play." Using this scheme, teacher involvement was found to be significantly related to the ECERS composite score of global quality (File & Kontos, 1993). McCartney (1984) found a similar relationship between her measure of the quality of verbal interactions between caregivers and children and an overall measure of global quality. Specifically, McCartney's (1984) observation scheme focuses on the function and quantity of verbal interactions during a time-sampled observation period. Function is coded along four dimensions: control, expressive, representational, and social. As well as the reported relationship between verbal interaction and global quality, both McCartney (1984) and Phillips and colleagues (1987) found that the verbal interaction index was a robust predictor of children's social development as assessed by both caregivers and parents. Thus, while there is some degree of disagreement in the literature as to the strength of the relationship between caregiver behaviors and child outcomes, the majority of investigations have found a positive relationship between the two. Further, caregiver-child interaction has also been found to be correlated with other measures of quality child care.

Other Teacher Characteristics
In addition to teacher-child interactions, other specific characteristics of caregivers are often considered valuable indicators of quality. The most common characteristics assessed include teacher attitudes and beliefs and teacher sensitivity or warmth. In general, high quality programs have caregivers who are characterized as more child-centered and sensitive. The ECOI further considers the level of affection, responsiveness, courteousness, use of positive discipline techniques, appropriate expectations, and encouraging independence as important characteristics for teachers in high quality programs (Bredekamp, 1986; Holloway & Reichart-Erickson, 1988). Indeed, as mentioned above, Holloway and Reichart-Erickson (1988) found a significant relationship between a composite index of caregiver characteristics derived from the ECOI and other elements of quality as well as child social outcomes.

Caregiver Attitudes and Beliefs
In addition to these characteristics, a number of investigations have pointed to the importance of caregiver beliefs in distinguishing quality care. For example,

Stipek and colleagues (1992) found that scores on a scale designed to assess teachers' didactic beliefs were able to differentiate between types of preschool and kindergarten programs (e.g. child-centered versus didactic). Of course, it is not surprising that more didactic programs have more didactic teachers. Perhaps more importantly, this study also found that teacher warmth, measured by a direct full-day program observation, loaded significantly into a factor which distinguished between types of programs. In a similar type of investigation, scores on the Teacher Beliefs Scale, a subscale of the Teacher Questionnaire designed to assess the developmental appropriateness of kindergarten teachers' beliefs, were found to be significantly related to the developmental appropriateness of the teachers' classrooms as assessed by a structured observation (Burts, Hart, Charlesworth, Fleege, Mosley & Thomasson, 1992).

Caregiver Warmth
In another assessment of quality using caregiver characteristics, Lamb, Hwang, Broberg, and Bookstein (1988) used the Belsky and Walker checklist to categorize quality. This instrument provides a list of positive and negative events which are checked off during 3-minute observation sampling intervals. It specifically focuses on a number of characteristics reflecting the degree of warmth and responsiveness of caregivers, such as "positive regard," "verbal elaboration," and "care providers in non-child conversation." A pilot study reported in Lamb and colleagues (1988) indicated substantial consistency in the ratings across a six month interval. One benefit of using the Belsky and Walker checklist is that it can be used across settings, so that the care provided at home, in family child care, and center care can be compared. Interestingly, Lamb and colleagues (1988) failed to find a relationship between the quality of alternate care as assessed by the Belsky and Walker Checklist and children's later personality or social skills.

Another instrument which focuses on caregiver characteristics and can be used across settings was specifically developed for the NICHD's study of early child care in the United States. The Observational Record of the Caregiving Environment (ORCE) assesses caregivers' behaviors with a specific target child, including such items as "sensitivity/responsiveness to infant distress" and "flat affect" (NICHD Early Childhood Research Network, 1996). This group has validated the ORCE with the Child Care HOME inventory (Caldwell & Bradley, 1984) in family child care settings. A composite variable derived from the ORCE which focuses on positive caregiving has been reported by this group to significantly relate to caregivers' nonauthoritarian child rearing beliefs. Further, positive caregiving ratings also were found to be related to other structural aspects of quality as well as to a subset of items from the Assessment Profile which attempt to tap the appropriateness of the physical environment. Although the ORCE is a new instrument, its specificity and validity are encouraging. It may prove to be a good method for observing specific caregiving aspects of the child care environment for future investigations.

Other instruments which attempt to tap specific caregiver characteristics also have been found to correlate with both global quality and child outcomes. For example, the Caregiver Observation Form and Scale (COFAS), has been found to be moderately correlated with a measure of global quality ($r = 0.38$; Kontos, 1991). This instrument requires a 20-minute observation of caregivers during which the frequency of 29 behaviors is coded. The behaviors are multiplied by their respective frequencies and summed to obtain an overall score which reflects the appropriateness of caregiver behavior. In a more recent investigation, Kontos and colleagues (1997) used the Arnett Scale of Provider Sensitivity to characterize provider behavior in their study of the quality of family child care programs. Interestingly, they found that provider sensitivity differed as a function of the income level of parents, indicating the interaction between family factors and the provision of quality care. File and Kontos (1993), using their own observational scheme discussed above, also report a significant relationship between quality of teacher-child interactions and global quality as well as with children's social play.

Another element of process quality which presumably depends upon the characteristics of caregivers and has been included in at least one investigation of child care quality is the emotional climate of a program. In their investigation of Swedish preschools, Hagekull and Bohlin (1995) used a composite measure of quality which included an assessment of the "emotional tone between adults and the child" (p. 511). Unfortunately, the authors did not use a standardized instrument nor do they discuss how this construct was measured. However, they did find a significant relationship between their composite index of quality and children's socioemotional outcomes. Clearly, more research into this construct as well as a more detailed definition and method of measurement are necessary before the implications of "emotional climate" as an important aspect of quality are fully determined.

In her detailed investigation of child care quality, Dunn (1993a) used caregivers' curriculum goals, strategies, and guidance of social-emotional development to characterize process quality. Curriculum goals were measured using the Educational Attitudes Scale (Rescorla et al., 1990), while caregiver strategies and guidance were measured using a behavioral observation scheme developed for this investigation which relied on coding audio transcripts of caregiver-child verbal interactions. Not only did Dunn find few intercorrelations among these variables, but there were even fewer correlations between these variables and other structural and global indicators of quality. This curious finding presumably indicates the importance of measuring multiple aspects of care in order to get a reliable indication of the overall quality of a program. A similar finding of a lack of a robust relationship between global quality and positive social interactions has been reported by Howes and Smith (1995) using a modified version the Adult Involvement Scale. Thus, it appears that while caregiver characteristics and interactions with children have often been associated with global measures of overall quality, this relationship is by no means resolved.

By far, the most widely used measure of global quality which includes an assessment of caregiver characteristics is the Early Childhood Environment Rating Scale (ECERS; Harms & Clifford, 1980) or one of its derivatives, the ITERS or FDCRS. The ECERS measures seven dimensions of quality, one of which is the nature of interactions between caregivers and children. The entire ECERS scale comprises 37 items, rated subjectively on a Likert-type scale from 1 to 7 after a thorough observation of the program. One of the subscores derived by factor analysis from this instrument reflects appropriate caregiving practices (Whitebook et al., 1989). Twelve items which focus on the nature of teacher-child interactions load into this factor (Stipek et al., 1992). The appropriate caregiving subscale has been related to child positive affect (Hestenes et al., 1993) and to security of attachment with caregivers (Howes, Phillips & Whitebook, 1992). Additionally, Stipek and colleagues (1992) report that scores on the appropriate caregiving subscale were able to distinguish between child centered, intermediate, and didactic child care programs ($F = 22.45$, $p < 0.001$). Thus, the ECERS caregiving subscale appears to reflect what studies using other instruments have found. The quality of interactions between children and caregivers appears to be an important aspect of quality child care.

Activities/Curriculum
The second major element of process quality which has been examined in depth is the quality and nature of the curriculum and activities available for the children in a given program. It is presumed that the type of activities available and the level of engagement by children in these activities reflects a classroom's level of quality. A number of investigators have included a measure of the activities available and/or curriculum to assess quality. For example, in their investigation of family child care, Kontos and colleagues (1997) assessed the types of activities that children were engaged in during a time-sampled

observation period. Scores were derived which reflected amount of time spent watching television, engaged in learning activities, engaged in gross-motor activities, and proportion of time engaged in no activities. These researchers found that activity ratings differed by income level and ethnicity of the children such that children from low-income families and Latino children were engaged in no activities more often than high-income children and European-American children. Latino children were also more likely to be engaged in watching television and less likely to be engaged in learning activities than were European-American children. Howes and Smith (1995) also measured children's engagement in play activities using their own observation scheme in which children were observed and activities were classified into one of 11 categories. Five clusters were educed from these categories: creative, language arts, didactic teaching, gross motor, and manipulatives. Howes and Smith (1995) found significant correlations between amount of creative play and a measure of global quality in infant/toddler classrooms. In contrast, few significant correlations were found between activities and global quality in preschool classrooms.

A second subscale derived from the ECERS is a measure of the developmental appropriateness of activities available for children. The appropriateness subscale is a composite of 10 items from the ECERS which focus on the nature and quality of activities available. While seemingly less robust than the caregiver interaction subscale, the appropriateness scale has been used in a few investigations of quality. For example, Howes and colleagues (1992) found a significant relationship between the developmental appropriateness score and children's social orientation. Further, Stipek and colleagues (1992) report that score on the appropriateness of activities available was able to distinguish between child-centered, intermediate, and didactic programs. However, it should also be noted that the two ECERS subscales have been reported to be highly correlated (Hestenes et al., 1993). Indeed, some investigators have been unable to derive two separate factors from the ECERS (e.g. Scarr et al., 1994), indicating that perhaps the ECERS would be better used to indicate one overall quality score. Studies which have used the ECERS as an indicator of overall quality as well as the methodological limitations of the ECERS will be discussed in the next section. It seems clear that, at least to some extent, the availability and quality of activities in a child care program do relate to the overall quality of the program.

Combining Process Elements: Global Quality

While there is some inconsistency in the literature, global quality is generally used to refer to an aggregate of multiple elements of process quality, often

Table 2. Global Measures of Child Care Quality.

Measure	Reference
ECERS/ITERS: Total Score	ECERS: Harms & Clifford, 1980 ITERS: Harms, Cryer & Clifford, 1990
FDCRS: Total Score	Harms & Clifford, 1989
Early Child. Classroom Obs. Scale/Inventory	Bredekamp, 1986
Classroom Practices Inventory	Hyson, Hirsh-Pasek & Rescorla, 1990
Assessmt Profile for Early Childhood Programs	Abbott-Shim & Sibley, 1987; Research Version: Abbott-Shim & Sibley, 1992
UCLA Early Childhood Observation Form	Stipek et al., 1992
Checklist for Rating Developmentally Appropriate Practice in Kindergarten Classrooms	Burts et al., 1992
HOME Inventory	Caldwell & Bradley, 1984

combined with some elements of structural quality. A number of measures have been developed which attempt to tap this construct (see Table 2). For example, using the total score on the ECERS/ITERS/FDCRS would be one way to measure global quality. Similarly, the CPI and Assessment Profile provide another way to measure the global quality of a program. While the majority of instruments to measure global quality are formal, at least two investigators have used their own composite indices to tap global quality. While using a global index may prove to be a more valid way to assess the true quality of a program, global indices are not as useful for researchers interested in delineating the specific elements which appear to be the most important indicators of quality and predictors of positive child outcomes. Investigations using formal and self-derived global measures of quality will be examined in turn.

Formal Instruments

A number of researchers have used the ECERS/ITERS as the instrument of choice to derive an overall measure of global quality. The ECERS measures seven dimensions of quality, including personal care routines, creative activities, language/reasoning experiences, fine and gross motor activities, social development, furnishings and display for children, and adult needs (Harms & Clifford, 1980). While seven subscales can be derived from this measure, the ratings are typically summed to create an overall quality score. The total scores, then, can be divided into overall quality categories, such as "poor," "mediocre," and "high." In fact, a number of investigations have used the instrument in this way, either alone or in combination with other measures (Burchinal et al., 1996; Deater-Deckard et al., 1996; Dunn, 1993a; File & Kontos, 1993; Ghazvini & Readdick, 1994; Hausfather, Toharia, LaRoche & Englesmann, 1997; Howes & Smith, 1995; Kontos, 1991; McCartney, 1984; Phillips et al., 1987). When used to assess overall global quality, the ECERS has generally been positively associated with other measures of quality and good child outcomes.

Indeed, Scarr and colleagues (1994) assessed multiple aspects of the child care environment using both structural and process measures and found a high degree of concordance between the ECERS/ITERS scores and scores derived from the Assessment Profile. The concordance between items within each instrument and between instruments was such that Scarr and her associates report the acceptability of using smaller subsets of items on both scales to assess program quality. In a paper presented at the 1998 conference of the National Association for the Education of Young Children, however, DeVries and Zan furnished evidence that classroom quality ratings using the ECERS were moderately related to the quality ratings given by two other commonly-used instruments, the C-DAP (Checklist for Rating Developmentally Appropriate Practice in Kindergarten Programs; Charlesworth et al., 1991) and CPI (Classroom Practices Inventory; Hyson, Hirsch-Pasek & Rescorla, 1990). Interestingly, and contrary to the results reported by Scarr, the ECERS ratings were largely discordant with those derived from the Assessment Profile (Abbott-Shim & Sibley, 1992). Indeed, in 11 out of 20 preschool and kindergarten classrooms assessed, the ECERS gave the classroom a "high" rating while the Assessment Profile gave a "mediocre" or "poor" rating. This large degree of variability in quality ratings between instruments indicates the need for further research into the reliability and validity of these measures. As the differences between the results reported by Scarr and DeVries suggest, there is some disagreement in the literature as to the empirical validity of using the ECERS, as well as other instruments, to assess process quality. The differences found in these two studies may have been due to the saqmples each used. Scarr's investigation included 40 preschool classrooms across three states, while DeVries' preliminary report included only 10 preschool and 10 kindergarten classrooms. Nonetheless, the fact that DeVries found such discordant quality ratings using multiple measures despite high inter-rater reliability, does raise questions about the validity of these instruments. Using multiple measures then, appears warranted given this disagreement.

While there appears to be some question in the literature as to the concordance between ratings derived from the ECERS and those derived from other instruments, other researchers have questioned the use of the ECERS in predicting child outcomes. In particular, Bjorkman Poteat, and Snow (1986) have questioned the validity of quality ratings assessed by the ECERS given their finding of a lack of clear association between children's social behavior and ECERS quality rating. These researchers note the importance of using multiple measures to gain a stronger indication of the quality of a child care program. Similarly, other researchers also have failed to find a relationship between ECERS scores and a variety of other child outcomes (Hestenes et al., 1993; Kontos, 1991). While it is possible that elements of quality assessed by the ECERS actually do not impact on certain aspects of children's development, it is equally possible that the measure itself is lacking. Further research is necessary to disentangle these possibilities. Given the degree of uncertainty provided by research into the use of the ECERS, then, using multiple methods and measures to assess quality would seem most appropriate.

Specifically-Derived Measures
As mentioned above, there have been at least two major investigations of quality which have developed their own global composite index by combining elements of process quality as well as instruments designed to assess process or global quality. Those studies, which have standardized and combined the scores from a number of different process/global instruments, may better be categorized as "meta-global" investigations. In one such investigation using the data from the *Cost, Quality, and Child Outcomes Study*, Peisner-Feinberg and Burchinal (1997) used a global composite index combining scores from the ECERS, CIS, ECOF, and AIS to categorize quality. These researchers found moderate to high correlations between these measures (r's ranged from 0.26 to 0.91). A principal component analysis concluded that one factor could be derived from the four measures which accounted for 68% of the variance in child care quality. Thus, these individual mesures of process/global quality appeared to be tapping the same underlying construct.

In their study of Swedish preschools, Hagekull and Bohlin (1995) used a composite index comprising a number of structural and global elements including ratio, group size, a measure of the stimulation provided by the facilities, and emotional tone between caregivers and children, among others. These researchers used both direct observation and interviews to make the quality ratings and found a relationship between overall quality and children's socioemotional development. Using a composite index such as those developed by Hagekull and Bohlin (1995) and Peisner-Feinberg and Burchinal (1997) appears to be a good way to validly assess the overall quality of a program.

However, as mentioned above, composite indices are less useful for research interested in tapping the most important specific elements of quality.

Summary of Process and Global Quality
Clearly, the two most researched elements of process quality are teacher characteristics or teacher-child interactions, and the quality of activities available for the children. There have been a large number of measures developed which attempt to assess these elements of process quality. While most of these measures have been found to correlate positively with other elements of quality as well as child outcomes, some investigations have failed to find these clear relationships. Unfortunately, the differences in methods of measurement and instrumentation, compounded with varying sample sizes, different sampling techniques, and methods of data aggregation and analysis preclude the elaboration of a clear conclusion regarding the relative importance of these variables in characterizing quality. Given that the number of studies which have concluded that caregiver characteristics and activities are important outweigh the number which have failed to find a relationship between these variables and child outcomes, it is probably safe to presume that caregiver characteristics and the quality of activities available are important aspects of overall quality which impact on at least some child outcomes. It is also important to note that the relationships reported between process and global measures may be equivocal. Depending upon which process factors are compared to which global factors, it is likely that the two are measuring either the same or similar underlying construct(s). The meaning of a correlation between a process and a global measure may not be very useful. In other words, showing that a composite index which includes scores derived from the ECERS is indeed correlated with the ECERS does not have much empirical utility, except possibly to show that a smaller subset of items can be used to assess quality. Such correlations should certainly not be interpreted as evidence for the reliability of a measure. While the measurement of process and global quality suffers a number of problems, it does seem clear, as will be shown below, that this type of quality has a direct impact on children's development.

Summary of Quality

Analyzing the quality of a program using either individual process elements or combining elements into a global evaluation appears to be a more valid method of program assessment than using structural elements alone. Indeed, there seems to be little doubt that the interactions between caregivers and children as well as the activities and curriculum provided afford a better index of the true quality of a program than do simple structural characteristics such as ratio and teacher training. However, the reliability of assessments using process and/or global measures of quality is somewhat less stable than assessments of structural elements of quality. In other words, while two observers will likely have little disagreement as to the number of caregivers per child in a given classroom at a given time, an assessment of the sufficiency of learning activities rated on a seven point scale after a two-hour observation will likely be more open to observer bias. Thus, in order to achieve an optimal balance between reliability and validity, the quality of a program would appear best evaluated using both structural and process elements. In addition, the use of overall impressions of quality by observers should weigh less in an assessment of quality than measures of specific structural and process aspects of the caregiving environment. Measuring specific aspects instead of relying on subjective assessments would elevate the reliability of global measures and aid research interested in tapping the most important elements of quality. While there have been great gains in our knowledge of the aspects of quality child care in the past twenty years, it is clear that much work remains to be done before an unclouded interpretation of "quality" can be realized.

CHILD OUTCOMES

The second major variable found in the center of the Child Care Quality Model is child outcomes. The elements of quality, whether measured in structural or process terms, provide insight into what constitutes and differentiates high from mediocre or low quality child care. The question that must be answered, however, is whether variation in quality matters. In other words, does high quality care enhance children's development and, conversely, does poor quality care put children at risk (Blau, 2000; Wilcox-Herzog, Fortner-Wood & Kontos, 1998)? This section considers the relationship of quality measures to children's social, behavioral/emotional, and cognitive/language development. It ties together the two central components of the Child Care Quality Model.

Social Development

One outcome variable, as can be seen on the Model, and, in fact, the most widely studied one, is children's emerging socialization as it is impacted by quality of child care. Many studies have examined this relationship in a relatively straightforward way. Several early studies, for instance those done in Bermuda (McCartney, Scarr, Phillips, Grajek & Schwartz, 1982; Phillips et al., 1987) and in Chicago (Clarke-Stewart, 1987) found a clear link. Helburn

(1995), in the large-scale *Cost, Quality, and Child Outcomes Study* also found that higher quality programs produced children with more advanced social abilities. Dunn (1993a) found a moderate impact of quality factors on children's social development, as did Kontos (1991). Similarly, Holloway and Reichhart-Erickson (1988) concluded that positive teaching style, one indicator of process quality, resulted in children who were more prosocial. In one of her earlier studies, Howes, and colleague Olenick (1986), concluded that children enrolled in high quality child care were more socially mature than children not enrolled in child care.

Several longitudinal studies also examined the more pervasive effect of child care quality on children's social development. A follow-up to the large national *Cost, Quality, and Child Outcomes Study* (Peisner-Feinberg et al., 1999) found that, at second grade, the children who had been in better quality child care programs when they were younger were still more socially competent, even when taking the quality of their intervening school experiences into account. As a follow-up to an earlier study (Vandell & Powers, 1983) in which four-year-olds in higher quality child care were found to be more socially competent than those in poorer quality care, Vandell and co-researchers (1988) measured the same children four years later. They found that the difference continued to be evident. Children who, at age four, experienced positive interactions with their caregivers, remained more socially competent, more accepted by peers, more skillful in conflict negotiation, and more empathic. In Sweden, Andersson followed children from their early child care experience to age 8 (Andersson, 1989) and age 13 (Andersson, 1992). At both later ages, children who had experienced high quality early care were more socially adept. Another Swedish study followed a group of children from age 1 to age 15 (Campbell, Lamb & Hwang, 2000). The researchers concluded that quality of early care seemed most important during the earliest years, up to age 3, and that patterns of social development remained more stable after that. One reason why children in high quality care fare better socially may be because they experience more stable caregiving, which appears to have long-lasting social effects (Howes & Hamilton, 1992b; McCartney et al., 1997); children's perceptions of their relationship with peers and with their elementary school teacher at age 9 was best predicted by their relationship with their first teacher, when they were toddlers (Howes et al., 1998). On the other hand, in Sweden, researchers (Lamb et al., 1988) found little impact of changes in child care arrangement on the children's socialization, perhaps a residual effect of the generally high quality of care in that country.

Other studies that examined the relationship of child care quality and socialization considered additional dimensions such as age of entry into child care, length of time in child care, and number of changes in child care arrangement. In the above cited studies by Andersson (1989, 1992), children who entered the relatively uniformly high quality child care of Sweden before the age of 1 had the highest scores on social measures, compared to children who entered child care at later ages. The opposite was found by Vandell and Corasaniti (1990) in their follow-up of children who had been in child care in a state with minimal quality standards; the lowest scores in social as well as other outcome measures were among children who had been in full-time child care since infancy. On the other hand, another study (Bates, Marvinney, Kelly, Dodge, Bennett & Pettit, 1994) found that it was not so much age of entry but length of time in child care that made an impact on children's development. Children who experienced overall more time in care had more negative adjustment scores at later ages, even when compared to children who had been in early infant care but had less cumulative child care experience. Greater amounts of time in a high quality center, however, resulted in children who were more socially mature (Schindler, Moely & Frank, 1987).

Attachment
Children's emerging social abilities also are related to attachment. Young children with secure attachment have a base from which to form new relationships, both with adults and with peers. A number of studies have explored the effect of child care on attachment; Wave 1 research, in fact, was often centered on this question (e.g. Belsky, 1988; Clarke-Stewart, 1989). A number of more recent studies have considered the relationship of child care quality and attachment, finding that children in high quality programs with caring, stable caregivers showed more secure attachment to their caregivers than children in low quality care (Anderson, Nagle, Roberts & Smith, 1981; Howes & Hamilton, 1992b; Howes et al., 1996). Interestingly, Howes and Hamilton (1992a) found that if children are more securely attached to their mothers, their teachers are more responsive to them and the children are more likely to develop a secure attachment to their caregivers. Yet, Howes and colleagues (1988) concluded that "one secure attachment relationship [with caregivers] may at least partly compensate for an insecure attachment relationship [with the mother]" (Howes, Rodning, Galluzzo & Myers, 1988, p. 413).

One of the intriguing issues related to attachment security and early entry into child care has revolved around how many hours infants spend in child care. Several studies conducted in the 1980s found that when babies were in care more than 20 hours per week, they exhibited a relatively high incidence of insecure avoidant attachment (Belsky & Rovine, 1988; Clarke-Stewart, 1989;

Lamb & Sternberg, 1990). Two more recent studies have failed to replicate this finding (NICHC, 1997b; Roggman, Langlois, Hubbs-Tait & Rieser-Danner, 1994). The NICHD researchers suggest that, perhaps, the differences in findings are a function of the cohort of mothers and caregivers involved in these studies; technical and popular writings have raised awareness about the importance of the quality of maternal and nonmaternal care in relation to the formation of attachment. More important, the NICHD team also made the link between child care and family variables; children whose mothers are less sensitive and responsive and who are in low quality child care are most vulnerable (NICHD, 1997b).

Play
Several studies have examined the effect of child care quality on children's play, as one manifestation of social development. Schindler and colleagues (1987), using Parten's categories of play, found a "positive relation between frequency of associative play and time in child care, and strong negative relations between the frequencies of unoccupied and onlooker behavior and time in child care" (p. 259), particularly among children in the more developmentally appropriate program of the two programs observed. Similarly, Vandell and Powers (1983) concluded that the quality of a child care center was clearly linked to children's free play behaviors. Children in high quality programs were engaged in significantly more positive social interaction and play than children in moderate and low quality care who spent more time in solitary or unoccupied behaviors. Elements of quality, for instance better trained teachers, also resulted in children with more sophisticated and complex play. The *Florida Child Care Improvement Study* (Howes et al., 1995) found that children in higher quality care were engaged in more complex social play with peers. File and Kontos (1993) found that positive teacher interaction was related to higher levels of children's play, although this result did not hold true for teacher involvement in play; children engaged in higher levels of play when teachers were less involved, perhaps because their involvement interfered with the play or because the children were perceived as having appropriate skills, thus not requiring intervention. Centers with developmentally appropriate spacial arrangements facilitated more focused play and fewer antisocial responses (Holloway & Reichhart-Erickson, 1988).

Almost uniformly, research has found a positive relationship between child care quality and children's social development. A few exceptions exist, however. Bjorkman, and associates (1986) found no relationship between ratings on the ECERS and children's social ratings; however, they question the validity of the ECERS. In a more recent and more complex study, McCartney and colleagues (1997) found a surprising lack of association between one measure, teacher-child interaction, and children's social development. One reason posited by the authors for this finding is the "moderate reliability" of the measure.

Behavioral/Emotional Development

Researchers also have examined the effect of child care quality on a variety of children's emotional and behavioral outcomes, another variable shown in the Child Care Quality Model. Howes and Olenick (1986) concluded that toddlers in high quality child care programs were more compliant and showed better self-regulation than children in lesser quality care. Howes (1988) also found that high quality child care was predictive of fewer behavior problems in first grade. Another study (Andersson, 1989) established that children in higher quality child care were more persistent, more independent, and less anxious than peers in lower quality programs. In her large-scale national study, Helburn (1995) found that children in higher quality child care had more positive self-perceptions, a finding that was particularly evident for children of mothers with low levels of education. In an interesting pre-post research opportunity in Florida, Howes and her colleagues (Howes et al.,1996; Howes et al., 1998) were able to measure children's development before and after the state implemented more stringent child care licensing regulations. They found that once quality, in terms of ratio and teacher training, was improved, children exhibited fewer behavior problems, in particular, were less aggressive, less anxious, and less hyperactive. In a secondary analysis of data from three larger studies, Burchinal, Peisner-Feinberg, Bryant, and Clifford (2000) concluded that children in low quality child care displayed more behavior problems than those in higher quality care.

A more specific aspect of quality, adult-child verbal interaction, was identified as the link to children's emotional adjustment in one study (McCartney et al., 1982). In the same study, caregivers in classrooms with less adult-child verbal interaction, hence lower quality, rated the children in their care as "more anxious, hyperactive and aggressive" (pp. 147–148). Hestenes and colleagues (1993) also concluded that engagement by teachers was a significant factor. Children in lower quality care, with less engaged teachers, showed less positive and more negative affect. The closeness of the teacher-child relationship was also a factor in behavioral ratings given by teachers (Peisner-Feinberg & Burchinal, 1997). Teachers' scores on the Child Behavior Index were higher in classroom of higher quality.

Age of entry into child care also has been considered in some studies as a variable related to behavioral outcomes. Children who began care at an early age in a low quality center exhibited higher levels of anger and defiance as preschoolers (Hausfather, 1997). In their retrospective longitudinal study, Vandell and Corasaniti (1990) found that children who had been in full-time child care since infancy in a state with minimal child care licensing standards were rated lower in compliance by both parents and teachers. Using data from the National Longitudinal Survey of Youth, Baydar and Brooks-Gunn (1991) concluded that children who had entered child care during the first year of life later had significantly more negative scores on the Behavioral Problems Index. While neither of these studies controlled for quality variables in child care, their results pose an interesting contrast to data from Sweden (Hagekul & Bohlin, 1995) which found that early entry into child care had a positive effect four years later. Children were less fearful, were less unhappy, and had greater ego-strength if they were in high quality as compared to low quality care from their first year of life.

Another factor considered as part of child care quality is the consistency of caregiving. Unfortunately, recent reports have indicated considerable turnover of caregivers during the first year of life (NICHD Early Childhood Research Network, 1997) and beyond (Kisker et al., 1991; Whitebook et al., 1989; Whitebook et al., 2001). There is growing indication that children whose caregivers follow them across age fare better than those in programs where children transition to different caregivers at different ages (Essa, Favre, Thweatt & Waugh, 1999; Howes & Hamilton 1993; Whitebook et al., 1989). McCartney and associates (1997) report that preschool-aged children with a history of multiple child care are more dependent and experience more behavior problems. Indeed, change in caregiver has been associated with greater amounts of fussing, longer latency to sleep, and more affectionate and aggressive physical contact (Field, Vega-Lahr & Jagadish, 1984). Howes and Hamilton (1992b) also found that when toddlers are put in the care of a new teacher, this change is "disturbing" to them (p. 871). In fact, children who experience change in primary caregiver at age 1 were more aggressive at age 4 (Howes & Hamilton, 1993).

Two interesting studies considered the developmental appropriateness of the overall child care program in relation to children's anxiety. Although the two studies used different instruments to measure developmental appropriateness, their findings are similar. In inappropriate classrooms, children displayed more stress behaviors than in settings that followed developmentally appropriate practice (Burts et al., 1992). This finding was particularly evident during inappropriate activities and was significantly higher for boys than for girls.

Similarly, Hyson, Hirsh-Pasek, and Rescorla (1990) found lower anxiety scores among children in developmentally appropriate programs than among youngsters in more highly structured, academic classrooms. In another study, children in highly didactic programs were also more anxious, as well as more dependent and less motivated, than children in child-centered programs (Stipek, Feiler, Daniels & Milburn, 1995). An interesting physiological study lends support to such findings. Children in low quality child care programs had significantly elevated levels of cortisol, a hormone associated with stress, than children in higher quality child care (Dettling, Parker, Lane, Sebanc & Gunnar, 2000).

Researchers have found a relationship between emotional and behavioral development and child care quality, although the traits that have been measured vary widely among these studies. In one longitudinal study, however, little effect on children's behavioral adjustment as a function of child care quality was found after four years (Deater-Deckard et al., 1996). The findings of this study are in contrast to those of other researchers (e.g. Vandell & Corasaniti, 1990). Deater-Deckard and fellow researchers found children's emotional and behavioral ratings were related much more significantly to parental stress and disciplinary practices than to variations in child care quality. A partial corroboration is provided by the large-scale NICHD study (NICHD Early Child Care Research Network, 1998), which concluded that mothering was a stronger predictor of child behavior outcomes than child care; however, these researchers also found that among child care variables, child care quality was the single most consistent predictor of young children's self-control, compliance, and problem behaviors. Many studies have found that variables other than child care quality, especially ones related to the family, need to be considered in the equation, as will be discussed later in the section on other influencing variables.

Cognitive/Language Development

The linkage between quality in child care and children's cognition, language development, and later school adjustment has generally been demonstrated by the research. High quality child care produced children with higher scores on a variety of cognitive and language instruments (Burchinal et al., 1996; Helburn, 1995; Howes et al., 1996; McCartney, 1984; NICHD Early Child Care Research Network, 2000; Peisner-Feinberg & Burchinal, 1997).

In the Bermuda study, McCartney (1984) concluded from results of a variety of language measures that children enrolled in higher quality centers were clearly better communicators. "Quality of the day care environment appears to have a profound effect upon language development. Indeed, quality of the day care environment appears to be as predictive of language skills as the family background variables" (p. 251). Similarly, Peisner-Feinberg and Burchinal (1997) also found higher language scores among children in higher quality centers, even when family variables were controlled. In the Florida study, Howes and her associates (1995, 1998) also found a clear link between improved child care quality and children's cognitive and language development. "Children engage in more cognitively complex play with objects . . . and [play] with other children in more complex ways. In addition, children's adaptive language scores increase, meaning that their narrative and discourse skills are higher" (p. 14). The NICHD Study Team (2000) also found quality of child care to be a predictor of cognitive and language competence. In particular, the frequency of language stimulation by caregivers during infancy was associated with later linguistic ability.

There is also longitudinal evidence of enhanced cognitive and language ability for children in high quality care. Second grade follow-up of the children in the *Cost Quality, and Child Outcomes Study* (Peisner-Feinberg et al., 1999), found that children who had been in classrooms of high quality and had experienced close teacher-child relationships as preschoolers had better language and math skills in second grade. These findings held, regardless of the children's intervening school experiences. In another study, high quality child care showed a consistent relationship to better cognitive and language skills, as examined longitudinally between 6 and 36 months in one study (Burchinal, Roberts, Riggins, Zeisel, Neebe & Bryant, 2000). Adherence to professionally recommended adult-to-child ratios was especially related to better receptive and expressive language scores, while higher teacher education seemed to impact girls' cognitive and receptive language abilities. In his Swedish studies, Andersson's (1989, 1992) longitudinal age 8 and age 13 follow-ups of children who had been in child care since infancy found superior results in cognitive tests and higher teacher ratings in relation to school achievement at both ages. School performance was rated highest for those children who had entered child care before the age of 1. The investigator in another longitudinal study (Field, 1991) discovered a significant correlation between amount of time children spent in a quality infant program and inclusion in a gifted program in elementary school. Howes (1988) found that high quality child care predicted higher school skill ratings from first grade teachers. In a later study, Howes (1990) concluded that children who had been in poor quality child care from an early age were the least task-oriented and most distractible in kindergarten. Quality and stability of child care were important predictors of later school adjustment for children.

The findings related to cognitive and language development are particularly evident for children from low income backgrounds. Burchinal, Lee, and Ramey (1989), for instance, found that disadvantaged children benefitted from a high quality child care program. When children had at least one year of a high quality child care experience, their intellectual development and IQ were enhanced. In another study (Burchinal et al., 1996), infants with the highest gains in cognitive and language development scores were those enrolled in high quality programs. They found that both lower adult-child ratios and teacher education were associated with more advanced receptive and expressive language skills. Quality child care participation during the first three years was also positively related to later reading skills of low-income children and, less strongly, to later math skills (Caughy et al., 1994). Finally, in their secondary analysis of data from three large studies, Burchinal and colleagues (2000) concluded that quality child care was particularly critical in enhancing the language development of low-income children of color.

Summary

As has been shown, a number of studies demonstrate a positive relationship between quality in child care and children's development. Yet, one must question the strength of that relationship, examining whether it is, indeed, a robust reflection of how elements of quality impact aspects of children's development or whether it is merely a spurious reflection of statistical rather than substantive meaning. Wilcox-Herzog and her colleagues (1998) examined this concern by considering effect size rather than simple statistical significance as a way to explore the magnitude of results from studies looking at the quality-outcomes relationship. Using this approach, these authors reviewed a sub-sample of research for which effect size is reported. They found that quality, measured in structural, process, and global terms, moderately accounts for variations in children's social-emotional development. Specifically, quality accounted for a median of 17% to 20.5% of the variance in social-emotional development, based on eight studies. Process quality proved to be moderately related to language development (a median of 13% based on 5 studies), while structural quality was most closely related to variation in cognitive development (a median of 12.5% based on 5 studies). This analysis of the data of a number of studies provides evidence of robustness and generalizability. "Because similar results were obtained with different samples across developmental domains using different ways of measuring children's development, replicability is enhanced" (Wilcox-Herzog et al., 1998, p. 95). However, this analysis also reveals that elements of quality account for a relatively small proportion of the variance in child outcomes and highlights the importance of examining additional factors that exert an impact on children. These factors are reflected in the outer circle of the Child Care Quality Model, as discussed next.

ADDITIONAL INFLUENCING FACTORS

The core of child care quality research has focused on the relationship between the elements of quality, both structural and process, and children's development. Wave 3 research, however, has incorporated consideration of other variables that might impact this relationship. As shown in the outer layers of the Child Care Quality Model, these include variables related to the family, the child, the child care center, the community, and the larger society. Additional influencing factors that either impact child care quality directly, or impact the relationship between quality and child outcomes, will be examined in this section.

Family Characteristics

Few would argue that the family exerts the most potent influence on the development of young children. It is one of the influencing proximal variables shown in the Child Care Quality Model. More recent research has acknowledged that different characteristics of families – for instance, socioeconomic standing, maternal education, marital status, and ethnicity – exerts a differential impact on how child care quality affects children. Howes and Stewart (1987) have, in fact, commented that "the development of children in child care cannot be studied without examining concurrent family influences" (p. 429). They conclude that by considering both child care and the family, research explains more of the variance in children's behavior. But the results of studies that have included family impact have not been uniform, with some studies resulting in clear indication that quality effects disappear when family factors are controlled, others showing that both variables are influential, and still others finding child care quality to be the most salient variable (Kontos, 1991). Scarr (1997), in fact, considers that the confounding effects of family characteristics can lead to an overestimation of the effects of child care. While many studies control for family characteristics in their measures of child care, "it is impossible to covary out all family effects, because one has only a limited set of measures of the families" (Scarr, 1997, p. 103).

Family Socioeconomic Status
Family income has been shown to exert a potent influence in relation to child care quality in one recent study. In a study of family child care, Kontos and co-researchers (1997) found that income and inadequate child care were strongly related; 73% of children from very low income families were in poor care compared to 43% of low income and 13% of moderate income children. The researchers found that children from very low income families were with caregivers who were less sensitive in their interactions, making it less likely that the children would be able to establish a secure relationship. The NICHD research team (1997a) also found that less advantaged families tend to rely more on multiple caregivers, one indicator of poorer quality; however, they found that children from the lowest and the highest income levels received higher quality care than those from middle income families (NICHD, 1997b), a finding consistent with some previous research (Helburn, 1995; Phillips, Voran, Kisker, Howes & Whitebook, 1994). Another recent study, however, reported a linear relationship, with low income children more frequently enrolled in low quality centers (Hausfather et al., 1997). Some research has found a link between socioeconomic status and problem behaviors in children (Hagekul & Bohlin, 1995; Vandell & Corasaniti, 1990). Other studies, however, have not found as clear a relationship between quality, child outcomes, and income (Desai et al., 1989; Goelman & Pence, 1987; Kontos & Fiene, 1987), although some researchers consider that high quality child care can mitigate the effects of impoverished home environments (e.g. Caughy et al., 1994).

One of the conclusions of research that has controlled for family impact in relation to quality variables is that families tend to make different choices about child care depending on socio-economic level; families from higher social classes place their children in better quality child care (Burchinal et al., 1996; Kontos, 1991; Vandell et al., 1988). Families who choose low quality child care, according to Howes and Olenick's research (1986) had "more complex lives, and at home were less involved and invested in their children's compliance than were parents who placed their children in a high-quality day-care center" (p. 202). In a replication of Howes and Olenick's study in family child care settings, Howes and Stewart (1987) also found that stressed families chose low quality child care. Thus, the relationship between child care quality, child outcomes, and family variables are confounded even further because of the choices parents make. Yet, it is not necessarily a matter of high quality care costing more, since the *Cost, Quality and Child Outcomes Study* (Helburn, 1995) found only a small cost difference between higher and lower quality group care. Accessibility and availability of child care as well as type of care may also be relevant factors to low income families.

Maternal Education

Parental, especially maternal, level of education is also a significant family variable that is related to child care quality. Mothers with higher education chose higher quality child care and the effects of that choice are still evident several years later (Peisner-Feinberg, 1998). In both the original and subsequent studies of the *Cost, Quality, and Child Outcomes* data, maternal education proved to be related to child care choice. Peisner-Feinberg and Burchinal (1997) concluded that children whose preschool teachers reported a closer relationship with them also had mothers with higher levels of education. When these same children were in kindergarten and two years later when they were in second grade, maternal education was still a moderating variable in the relationship of child care quality to later measures of problem behaviors and sociability (Peisner-Feinberg et al., 2000). Another study concluded that mothers with less education tended to place their children in the care of adults with poorer teacher-child interaction skills; these children had higher problem behavior ratings (McCartney et al., 1997). In a similar vein, an earlier study found that parents with higher levels of education tended to choose child care arrangements where caregivers are more responsive to the children and demand less compliance than do parents with lower levels of education (Clarke-Stewart, Gruber & Fitzgerald, 1994). Maternal education is also an indicator of child's age of entry into child care. Children of mothers with higher levels of education tend to enter child care at earlier ages (Howes, 1988; NICHD, 1997).

Parental Marital Status

A few studies have considered other family variables in their relationship to quality and child outcomes. In an examination of child care quality and the impact of various influences on children's development, Dunn (1993a) found that children from intact families, who were enrolled in child care programs, were rated as better adjusted on the Parent Behavior Questionnaire than were children from divorced families. Vandell and Corasaniti (1990) concluded from their data that children from divorced families were more likely to be in full time care. However, in another study, Vandell and her associates (1988) found little difference in choice of child care in relation to quality as a function of marital status.

Family Ethnicity

Ethnicity as a family variable is more difficult to measure because of unequal numbers within various groupings. In the previously discussed study of children's experiences in family child care (Kontos et al., 1997), a clear relationship was found between poor quality care and very low income families. The researchers also found a relationship to ethnicity in their measurement of African-American, Latino, and European American children. Latino children were the least likely to be in high quality care, a finding that was especially compounded by low income. Yet, since the majority of children from very low income families in this study were also minorities, the authors acknowledge the difficulty in disentangling the effects of income and ethnicity. One study, for instance, found that although child care quality could be predicted as a function of social class in both African American and European American families, the relationship was seen through different paths (Howes, Sakai, Shinn, Phillips, Galinsky & Whitebook, 1995). In European American families, the association between class and quality was direct; in African American families this relationship was mitigated by parents' work demands. Another issue that factors into the differential use of care in relation to quality has to do with family values. Latino families, in particular, often disagree with the values espoused in American center-based care, values such as the promotion of independence. The NICHD researchers (1997c) also found children from ethnic minority groups, with the exception of those in center care, to be in lower quality settings. In other words, variables determined by researchers to measure high quality may not be viewed in the same way by low income and ethnic minority parents.

In the discussion of the effects of quality variations on social development outcomes, the findings related to the age at which the child entered child care were considered. The interaction among age of entry, child care quality, and child outcomes also has been examined in the context of family variables. Howes (1990) considered the comparative influence of family and child care socialization practices on adjustment in kindergarten in relation to age of entry. She found that "family socialization was a less powerful influence for those children enrolled in child care as infants . . . . Child care quality, as represented by teacher socialization practices, was a more powerful predictor for these children" (p. 300). This finding is congruent with the conclusion that a good relationship with a caregiver can compensate for insecure attachment with the mother (Howes et al., 1988). Early entry into child care can, however, be detrimental if the quality of care is poor (Vandell & Corasaniti, 1990). An interesting consideration of the differential influence of family variables in relation to child outcomes is suggested by Kontos (1991). She proposes that family background may be a more significant predictor of language and cognitive development while child care quality is a more potent precursor of social development. It is, after all, in the peer milieu of group care that children most readily develop social skills.

In numerous ways, family characteristics impact child outcomes, as indicated in the Child Care Quality Model. Some of the variables discussed above – socioeconomic status, maternal education, family structure, and ethnicity – contribute to an understanding of how children's development is affected by child care in the context of family characteristics. There are numerous other family factors which have been examined in only a few studies or have not been addressed at all.

Child Characteristics

Another influencing proximal variable that affects the child care quality-child outcomes relationship is characteristics of the child, as shown in the Model. Some studies consider attributes of the child as potential influences on developmental outcomes in relation to child care quality. Several studies suggest that boys are more vulnerable than girls to variations in child care. Desai and colleagues (1989) found that boys from higher income families experienced more negative effects from child care, especially if they entered care during the first two years of life. Similarly, boys with extensive infant care histories had more behavioral problems at age 8 (Vandell & Corasaniti, 1990) and were less able to establish self-regulation as toddlers if they were in poor quality care (Howes & Olenick, 1986). On the other hand, Hagekul and Bohlin (1995) found no gender differences in their Swedish study. In fact, they found that boys with more child care experience had fewer concentration problems and greater ego strength, attesting to the potentially positive effects of high quality child care on children who may be more vulnerable. These authors also considered temperament in their examination of quality and outcomes, finding a weak relationship between child care quality and child temperament measures. On the other hand, temperament did not seem to impact the relationship between quality of child care and children's display of affect (Hestenes et al., 1993), although the authors expressed concern about methodological problems in relation to the measurement of temperament.

Child Care Program Characteristics

Some characteristics of the child care program, apart from the earlier discussed measures of quality, can have a less direct, though important, effect on children's development. Thus, while child care quality is difficult to disentangle from child care program characteristics, there are features of the program that should be considered separately as influencing variables, including program philosophy and auspice of program, seen as a third influencing proximal variable in the Model.

Program Philosophy
Certainly a supportive philosophy as established by administrative policy can make a difference for children in an indirect way. When caregivers are stressed and isolated, they are "less able to engage in harmonious interactions with the children in their care" (Howes & Rubenstein, 1985). Similarly, the *Child Care Staffing Study* (Whitebook et al., 1988) found that better quality centers provided higher wages and better work environments for staff, which resulted in more positive child outcomes. One follow-up to this study (Whitebook et al., 2001) also found that centers paying higher wages were better able to retain qualified staff, although, overall, this study concluded that staff and director turnover are alarmingly high (p. vi). In addition, centers that hire teachers who have child-specific education and training are more likely to be programs of higher quality (Whitebook et al., 1988). Nonetheless, in the Florida study, Howes concluded that teacher education, while making significant differences in terms of child outcomes (Howes et al., 1995), is not sufficient to ensure high quality; a well trained teacher in a setting with appropriate child-adult ratios is most effective. There may well be other, as yet unmeasured, variables that contribute to program quality. Blau (2000), for instance, suggests that the leadership style of the director, though difficult to measure, may be more potent in establishing quality than measures such as ratio and group size. Overall, the philosophy of the center contributes to the mix of variables that impact child care quality.

Auspice
The auspice of a program, whether it is a for-profit or a not-for profit operation, is another characteristic that is related to quality of child care. *The Cost, Quality and Child Outcomes Study* concluded that not-for-profit centers provided higher than average quality care (Helburn, 1995). McCartney and her colleagues (1997) found the highest teacher-child interaction scores in not-for-profit programs. This finding is supported by the *Child Care Staffing Study* (Whitebook et al., 1988), which discovered that not-for-profit centers had better educated and trained staff; the relationship between staff education and quality of care has been discussed earlier in this paper and support the link between better educated teachers and positive teacher-child interaction (e.g. Arnett, 1989; Howes, 1997). The *Child Care Staffing Study* also found that not-for-profit centers offered higher wages and better work environments for staff and had lower adult-to-child ratios. In addition and probably as a result of such attributes, these centers had lower staff turnover (Whitebook et al., 1988). More frequent parent-caregiver communication was also found in not-for-profit centers (Ghazvini & Readdick, 1994). In the Florida study, Howes and her

co-researchers (1998) found that not-for-profit centers were more likely to comply with licensing regulations than for-profit centers. One other study linked not-for-profit centers with quality. Kontos (1991) found that better educated parents were more likely to choose not-for-profit centers because these offered higher quality care for their children.

Community and Societal Characteristics

A final set of variables that influence child care quality and, in turn, child outcomes, are embedded in the community and the larger society, as shown in the outer circle of influencing distal variables on the Child Care Quality Model. From an ecological systems perspective, community and societal characteristics, as all the other variables previously discussed in this paper, exert an influence on young children. Most of the child care quality research, however, does not address these more distal community and societal influences. Phillips (1996) comments that our definitions of quality "remain somewhat parochial" (p. 44) because they focus primarily on features within the early childhood setting. But factors beyond child care programs, such as the opportunities provided by the community for teacher training, resource and referral, appropriate health care for all children, policies that affect children and families, regulation and monitoring of child care, and "broader financing mechanisms that either support or fail to support quality care and education" (Phillips, 1996, p. 44), all contribute to the total picture. Phillips' words were echoed by Valora Washington, one of the speakers at the 1997 White House Conference on Child Care. Child care also becomes an issue of social equity when findings, such as those discussed earlier, indicate that children of lower income families often receive lower quality in care. Other writers lament the lack of social policy support for working families, for instance, paid maternity leave, parental leave, job sharing, part-time employment arrangements, and universal affordable child care (McCartney et al., 1997). Gallagher and Clifford (2000) reflect that, at a more fundamental level, the United States lacks a comprehensive infrastructure under which services for children under age 5 can be coordinated and suggest steps for development of such a support structure. The differences in outcomes of child care children in the United States and Sweden underscore the national context for examining the impact of child care (Andersson, 1989).

Child Care Licensing Standards

One of the indicators of how society cares about its young children is revealed in child care licensing regulations and standards. The standards of most states reflect "a philosophy of preventing harm to children in child care, rather than of providing a high-quality, developmental environment" (Phillips, Lande & Goldberg, 1990, p. 159). A recent in-depth analysis of the licensing standards of the four states included in the *Cost, Quality, and Child Outcomes Study* (Gallagher, Rooney & Campbell, 2000) also found that their regulations set much higher standards for child protection than for the enhancement of children's development. Such requirements, the authors posit, tend to perpetrate stereotypes about the nature of child care and child care workers. In a state-by-state assessment of how licensing standards have changed between 1981 and 1995, some improvements were found (Snow, Teleki & Reguiero-de-Atiles, 1996); by 1995, increasing numbers of states improved ratio and group size requirements for infants and toddlers, although standards for four-year-olds declined during those years. The impact of the stringency of state regulations was seen in the *Child Care Staffing Study* (Whitebook et al., 1990), with children in states with higher regulations faring better on developmental measures. Vandell and Corasaniti (1990) found that children who had attended child care in a state with minimal standards were later impacted negatively by the extent of their child care experiences. The NICHD Study Team (1999) examined child outcomes in relation to professionally recommended quality standard on ratio, group size, caregiver training, and caregiver education. Such standards generally are considerably more stringent than ones set by state licensing. Not surprisingly, given the research reviewed in this paper, they found a clear relationship between meeting standards and child outcomes. In fact, they discovered a linear relationship; the more standards met, the better the results for children.

One recent large-scale study (Scarr, Phillips, McCartney & Abbott-Shim, 1993) specifically examined 120 child care centers in three states, one with high, one with medium, and one with low licensing standards. There was a clear relationship among state standards, quality, and child outcomes. They also found considerable lack of compliance with existing state regulations, a finding echoed by the Florida research team (Howes et al., 1998), especially in terms of infant/toddler ratios. Non-compliance was highest in the state with the lowest standards. Scarr and fellow researchers (1993) conclude, "the importance of enforcement of state regulations cannot be exaggerated, given these data on actual ratios that occur in the ordinary practices of centers. We also observed violations of health and safety regulations, inappropriate programming, and punitive interactions between caregivers and children.... More ... enforcement of existing regulations seem[s] to be essential to improving the quality of child care" (p. 187). The authors also acknowledge that bettering child care in this country will take a "massive infusion of public money" (p. 187).

Expanded Early Childhood Knowledge Base

Another distal impact on quality child care comes from the field of early childhood education itself. In the past few decades, there have been substantial changes in the field, in terms of the breadth of its knowledge base and of the sources where that knowledge can be accessed (Reifel, 1997). The very conception of the field of early childhood education has changed as those with a strong commitment to young children have broadened their sphere of impact through expanded research, education, and policy avenues. Such new information, often conveyed through journals, texts, other books, and on-line sources, can have an impact on the quality of early childhood programs. New findings may be incorporated into the education of pre-service and in-service teachers and potentially can also be included into their practice by those who already work with children. Impact on child care regulations and policies through lobbying efforts also have an effect of the field. Thus, the early childhood profession itself is one of those distal influences that is impacting child care in various ways.

Summary

The review of studies which examine some of the distal and proximal influencing variables that interact with or intervene in the relationship between child care quality and child outcomes begins to illustrate the complexity of these relationships and interrelationships, as illustrated in the Child Care Quality Model. From an ecological systems perspective, children can be viewed as interacting with a wide range of factors in both near and more distant systems. The child herself or himself, the family, and the child care center all have characteristics that impact such interactions. More removed community and societal characteristics also influence children's outcomes in more indirect ways. Undoubtedly, there are many other potential ways in which the child care-child outcomes relationship is impacted by influencing variables beyond that relationship.

SUMMARY AND CONCLUSIONS

The research on child care quality covers a wide range of topics, which provide a composite picture of how child care affects young children. Elements of quality, their relationship to child outcomes, and the interaction of these two in relation to other influencing variables combine to provide insights into the critical question of the impact of child care on children. The relationships among these variables is depicted in the Child Care Quality Model, which shows how many of the variables that have been explored in the research on child care relate to each other. One aspect needing exploration is the lack of bidirectionality of the relationships among the variables shown in the Model; clearly many of these components impact each other rather than the one-way effects explored in the extant literature. Nor do the variables discussed above and shown in the Model represent a complete catalog. In each of the major elements of the model, other variables could and should be explored. In addition, relationships and interactions among the various elements need to continue to be delineated, as Wave 3 research has begun. Following are a few examples of under- or unexplored topics. For instance, little is known about how similarity or disparity between child care staff and parental child-rearing practices impacts children. In addition, little research sheds light on how child care providers make decisions about guiding and disciplining children. The child care center director is an important force in determining quality, yet little research has provided insight into the impact of directors with varying characteristics; in addition, it would be useful to discern how much autonomy directors have in centers of different auspices.

The most challenging gaps in the research are, perhaps, at the exo- and macrosystems levels. Interesting research from Florida (Howes et al., 1995; Howes et al., 1998) gives insight into how changes in licensing standards can impact children's development, just as earlier research (e.g. Scarr et al., 1993; Vandell & Corasaniti, 1990) made the link between stringency of state standards and child outcomes. But little research has examined how less formal facets of community support can improve child care quality and, in turn, child outcomes. For example, it would be relevant to learn how the presence of a strong and active professional early childhood organization and the availability of higher education in the field or other appropriate training options affect the quality of care within a community. Similarly, more needs to be learned about how different types of employer support impact families and children, or how government policies related to parental leave, sick-child care, job assurance, and flex time impact quality of care. These represent only a few areas where exploration can lend further insight into how child care impacts young children's development.

While more research is needed to fill in gaps in our knowledge of the effects of child care, the more important question about the relevance of the research also needs to be addressed. Blau (2000) questions the true effect of often studied structural variables such as ratio, group size, and teacher education, suggesting instead that other, more difficult to measure variables may be the ones that make an actual difference. Wilcox-Herzog and colleagues (1998) raise the question of how much variance in child outcomes is accounted for by quality; basically, how important are quality variables in impacting children's lives? They cite Scarr's (1992) perspective on "good enough" environments as a point of argument that, perhaps, quality is not as relevant in assessing child outcomes, particularly mediocre quality. Indeed, the small percentage of variance in child outcomes accounted for by quality that was reported by these investigators illustrates this view.

Scarr's position is that, from an evolutionary perspective, environments within a normal range do not much impact variations in development; rather it is the child's genetic endowment in the context of any good enough environment that contributes to child outcomes. In more recent writings, Scarr (1995, 1997, 1998) expanded on her argument that a good enough environment is acceptable in child care. In this view, the link between child care quality and child outcomes in the Child Care Quality Model is more tenuous, and might well be depicted as a broken rather than as a solid line. She argues that mediocre (differentiated from unsafe or abusive care) does not permanently impact children and that professionals and law makers should seriously consider less stringent standards that would make care more affordable to a wider range of parents. Scarr does not argue the bulk of literature showing that poor quality child care can impact children negatively, or that high quality child care can be beneficial, especially for certain populations of children. Rather, she stresses that the distinctions between mediocre and good quality care are less important in determining child outcomes. However, a delineation of the "normal range" of child care within which no detrimental impact on children will occur has yet to be suggested. Indeed, such a normal range may differ according to the child, as illustrated by the research cited above suggesting that child care affects different children differently. Nonetheless, Scarr concludes that most studies show little impact on children's development as a function of variation in child care quality and that even this small amount of variance in outcomes may be more a function of unmeasured family effect rather than of child care. Thus, from Scarr's perspective, one might conclude that high quality child care is not an essential ingredient for the welfare and future success of children. This assumption, however, can be questioned as a result of some more recent findings from large scale studies, such as those of the NICHD Team, of significant cumulative impacts of child care quality on children's development, effects that exceed concurrent gains (Vandell & Wolfe, 2000).

Questions such as those raised by Scarr's provocative view need to be carefully considered and addressed, however. One study sought to identify "thresholds" of quality by comparing children's engagement in activities and how much attention they received from teachers under different ratio and group size conditions (Howes et al., 1992); higher quality clearly was more beneficial to children. But the question of whether higher thresholds of quality are critical to the current and future well-being of young children needs to be asked (Love, Schochet & Meckstroth, 1996).

Continuing research into the effects of child care is needed. The majority of *past* studies which have found significant relationships between child outcomes and quality in child care were contemporaneous rather than longitudinal in design, thus not addressing the concern about whether less than optimal child care has an impact on children's future development. More recent large-scale studies include longitudinal components. This should not imply, however, that concurrent studies are not valuable; they certainly provide insight into the ongoing effect of child care on young children. In addition, the question of significance of findings needs attention. As discussed, earlier, one approach to gauging the impact of quality dimensions more carefully has been taken by Wilcox-Herzog and colleagues (1998) by examining data in terms of effect size, not just statistical significance. McCartney and Rosenthal (2000) support such a strategy, particularly if results of research are to be used in more practical ways, to design social policy for children. Furthermore, methodological issues related to reliability, validity, and consistency among the various instruments used to measure quality also need to be resolved (DeVries, 1998). Love and colleagues (1996) discuss some of the methodological shortcomings of much of the child care research; they express concern that statistical methods are not used which "correct and test for systematic unobservable differences between the characteristics of children who enroll in centers of different quality" (p. 45). They also recommend that more pre-post designs be utilized, similar to the Florida study (Howes et al., 1995; Howes et al., 1998), but with more rigorous design components.

Child care is a reality in the lives of the majority of young American children today. The question of its impact on children's current and future development and success is a critical one that needs to be addressed. The Child Care Quality Model helps to place the salient variables under study into perspective. It can help researchers continue to refine and examine just what role child care plays in impacting the future of our children.

REFERENCES

Abbott-Shim, M., & Sibley, A. (1987). *Assessment profile for childhood programs.* Atlanta, GA: Quality Assist, Inc.

Abbott-Shim, M., & Sibley, A. (1992). *Research version of the assessment profile for childhood programs.* Atlanta, GA: Quality Assist, Inc.
Anderson, C. W., Nagle, R. J., Roberts, W. A., & Smith, J. W. (1981). Function of center quality and caregiver involvement. *Child Development, 52,* 53–61.
Andersson, B-E. (1989). Effects of public day-care: A longitudinal study. *Child Development, 60,* 857–866.
Andersson, B-E. (1992). Effects of day-care on cognitive and socioemotional competence of thirteen-year-old Swedish school children. *Child Development, 63,* 20–36.
Arnett, J. (1989). Caregivers in day-care centers: Does training matter? *Journal of Applied Developmental Psychology, 10,* 541–552.
Bates, J. E., Marvinney, D., Kelly, T., Dodge, K. A., Bennett, D. S., & Pettit, G. S. (1994). Child care history and kindergarten adjustment. *Developmental Psychology, 30,* 690–700.
Baydar, N., & Brooks-Gunn, J. (1991). Effects of maternal employment and child-care arrangements on preschoolers' cognitive and behavioral outcomes: Evidence from the children of the National Longitudinal Survey of Youth. *Developmental Psychology, 27,* 932–945.
Belsky, J. (1984). Two waves of day care research: Developmental effects and conditions of quality. In: R. C. Ainslie (Ed.), *The Child and the Day Care Setting: Qualitative Variations and Development* (pp. 1–34). New York: Prager.
Belsky, J. (1988). The effects of infant day care reconsidered. *Early Childhood Research Quarterly, 3,* 235–272.
Belsky, J., & Rovine, M. (1988). Nonmaternal care in the first year of life and the security of infant-parent attachment. *Child Development, 59,* 157–167.
Belsky, J., & Walker, A. (1980). *Infant-toddler spot observation system.* Unpublished manuscript, Pennsylvania State University, Department of Individual and Family Studies, University Park.
Bjorkman, S., Poteat, G. M., & Snow, C. W. (1986). Environmental ratings and children's social behavior: Implications for the assessment of day care quality. *American Journal of Orthopsychiatry, 56,* 271–277.
Blau, D. M. (1996). The production of quality in child care centers. *The Journal of Human Resources, 32,* 354–387.
Blau, D. M. (200). The production of quality in child care centers: Another look. *Applied Developmental Science, 4,* 136–148.
Bredekamp, S. (1986). The reliability and validity of the early childhood classroom observation scale for accrediting early childhood programs. *Early Childhood Research Quarterly, 1,* 103–118.
Bredekamp, S., & Copple, C. (1996). *Developmentally appropriate practice for programs serving children ages birth through 8.* Washington, D.C.: National Association for the Education of Young Children.
Bronfenbrenner, U. (1979). *The ecology of human development.* Cambridge, MA: Harvard University Press.
Burchinal, M., Lee, M., & Ramey, C. (1989). Type of day-care and preschool intellectual development in disadvantaged children. *Child Development, 60,* 128–137.
Burchinal, M. R., Peisner-Feinberg, E., Bryant, D. M., & Clifford, R. (2000). Children's social and cognitive development and child care quality: Testing for differential associations related to poverty, gender, or ethnicity. *Applied Developmental Science, 4,* 149–165.
Burchinal, M. R., Roberts, J. E., Nabors, L. A., & Bryant, D. M. (1996). Quality of center child care and infant cognitive and language development. *Child Development, 67,* 606–620.

Burchinal, M. R., Roberts, J. E., Riggins, R., Zeisel, S. A., Neebe, E., & Bryant, D. (2000). Relating quality of center-based child care to early cognitive and language development longitudinally. *Child Development, 71,* 339–357.

Burts, D. C., Hart, C. H., Charlesworth, R., Fleege, P. O., Mosley, J., & Thomasson, R. H. (1992). Observed activities and stress behaviors of children in developmentally appropriate and inappropriate kindergarten classrooms. *Early Childhood Research Quarterly, 7,* 297–318.

Caldwell, B. M. (1993). Impact of day care on the child. *Pediatrics, 91,* 225–228.

Caldwell, B. M., & Bradley, R. H. (1984). *Home observation for measurement of the environment.* Little Rock: University of Arkansas at Little Rock.

Campbell, J. J., Lamb, M. E., & Hwang, C. P. (2000). Early child-care experiences and children's social competence between $1\frac{1}{2}$ and 15 years of age. *Applied Developmental Science, 4,* 155–175.

Cassidy, D. J., Buell, M. J., Pugh-Hoese, S., & Russell, S. (1995). The effect of education on child care teachers' beliefs and classroom quality: Year one evaluation of the TEACH Early Childhood Associate Degree scholarship program. *Early Childhood Research Quarterly, 10,* 171–183.

Caughy, M. O., DiPietro, J. A., & Strobino, D. M. (1994). Day-care participation as a protective factor in the cognitive development of low-income children. *Child Development, 65,* 457–471.

Charlesworth, R., Hart, C. H., Burts, D. C., & Hernandez, S. (1991). Kindergarten teachers' beliefs and practices. *Early Child Development and Care, 70,* 17–35.

Clarke-Stewart, K. A. (1989). Infant day care: Maligned or malignant? *American Psychologist, 44,* 266–273.

Clarke-Stewart, K. A. (1987). Predicting child development from child care forms and features: The Chicago study. In: D. A. Phillips (Ed.), *Quality Child Care: What Does the Research Tell Us?* (pp. 21–42). Washington, D.C.: National Association for the Education of Young Children.

Clarke-Stewart, K. A., Gruber, C. P., & Fitzgerald, L. M. (1994). *Children at home and in day care.* Hillsdale, NH: Lawrence Erlbaum.

Cryer, D. (1999). Defining and assessing early childhood program quality. In: S. W. Helburn (Ed.), *The Silent Crisis In U.S. Child Care* (pp. 39–55). Thousand Oaks, CA: Sage.

Deater-Deckard, K., Pinkerton, R., & Scarr, S. (1996). Child care quality and children's behavioral admustment: A four-year longitudinal study. *Journal of Child Psychology and Psychiatry, 37,* 937–948.

Desai, S., Chase-Lansdale, P. L., & Michael, R. T. (1989). Mother or market? Effects of maternal employment on the intellectual ability of 4-year-old children. *Demography, 26,* 545–561.

Dettling, A. C., Parker, S. W., Lane, S., Sebanc, A., & Gunnar, M. R. (2000). Quality of care and temperament determine changes in cortisol concentrations over the day for young children in childcare. *Psychoneuroendocrinology, 25,* 819–836.

DeVries, R., & Zan, B. (1998, November). *Research regarding developmentally appropriate practices: Findings and methodological considerations.* Paper presented at the meeting of the National Association for the Education of Young Children, Toronto, Canada.

Doherty-Derkowski, G. (1995). *Quality matters: Excellence in early childhood programs.* Reading, MA: Addison-Wesley.

Dunn, L. (1993a). Proximal and distal features of day care quality and children's development. *Early Childhood Research Quarterly, 8,* 167–192.

Dunn, L. (1993b). Ratio and group size in day care programs. *Child and Youth Care Forum, 22,* 193–226.

Essa, E. L. (1998). When, how, and why child caregivers respond to children's behaviors. *Early Child Development and Care, 141*, 15–29.

Essa, E. L., Favre, K., Thweat, G., & Waugh, S. (1999). Continuity of care for infants and toddlers. *Early Child Development and Care, 148*, 11–19.

Federal Register. (1980, March 19). Part V: Department of Health, Education, and Welfare, HEW Day Care Regulations (pp. 178870–17885). Washington, D.C.: U.S. Government Printing Office.

Field, T. (1980). Preschool play: Effects of teacher/child ratios and organization of classroom space. *Child Study Journal, 10*, 191–205.

Field, T. (1991). Quality infant day-care and grade school behavior and performance. *Child Development, 62*, 863–870.

Field, T. M., Vega-Lahr, N., & Jagadish, S. (1984). Separation stress of nursery school infants and toddlers graduating to new classes. *Infant Behavior and Development, 7*, 277–284.

Fiene, R. (1984). *Child development program evaluation scale and COFAS*. Washington, D.C.: Children's Services Monitoring Consortium.

Fiene, R., & Nixon, M. (1985). Instrument-based monitoring and the indicator checklist for child care. *Child Care Quarterly, 14*, 28–46.

File, N., & Kontos, S. (1993). The relationship of program quality to children's play in integrated early intervention settings. *Topics in Early Childhood Special Education, 13*, 1–18.

Gallagher, J., & Clifford, R. (2000). The missing support infrastructure in early childhood. *Early Childhood Research and Practice, 2*(1), 1–22.

Gallagher, J. J., Rooney, R., & Campbell, S. (1999). Child care licensing regulations and child care quality in four states. *Early Childhood Research Quarterly, 14*, 313–333.

Ghazvini, A. S., & Readdick, C. A. (1994). Parent-caregiver communication and quality of care in diverse child care settings. *Early Childhood Research Quarterly, 9*, 207–222.

Goelman, H., & Pence, A. R. (1987). Effects of child care, family, and individual characteristics on children's language development. In: D. A. Phillips (Ed.), *Quality in child care: What does the research tell us?* (pp.89–104). Washington, D.C.: National Association for the Education of Young Children.

Hagekul, B., & Bohlin, G. (1995). Day care quality, family, and child characteristics and socioemotional development. *Early Childhood Research Quarterly, 10*, 505–526.

Harms, T., & Clifford, R. M. (1980). *Early Childhood Environmental Rating Scale*. New York: Teachers College Press, Columbia University.

Harms, T., & Clifford, R. M. (1989). *The Family Day Care Rating Scale*. New York: Teachers College Press.

Harms, T., Cryer, D., & Clifford, R. M. (1987). *The Infant and Toddler Environmental Rating Scale*. New York: Teachers College Press, Columbia University.

Hausfather, A., Toharia, A., LaRoche, C., & Engelsmann, F. (1997). *Journal of Child Psychology and Psychiatry and Allied Sciences, 38*, 441–448.

Helburn, S. (Ed.) (1995). *Cost, quality, and child outcomes in child care centers: Technical report*. Denver: University of Colorado at Denver

Helburn, S. W., & Howes, C. (1996). Child care cost and quality. *The Future of Children: Financing Child Care, 6*(2), 62–82.

Hestenes, L. L., Kontos, S., & Bryan, Y. (1993). Children's emotional expression in child care centers varying in quality. *Early Childhood Research Quarterly, 8*, 395–307.

Hofferth, S. L., & Deich, S. G. (1994). Recent U.S. child care and family legislation in comparative perspective. *Journal of Family Issues, 15*, 424–448.

Holloway, S. D., & Reichhart-Erickson, M. (1988). The relationship of day care quality to children's free-play behavior and social problem-solving skills. *Early Childhood Research Quarterly, 3*, 39–53.

Howes, C. (1987). Social competency with peers: Contributions from child care. *Early Childhood Research Quarterly, 2*, 155–167.

Howes, C. (1988). Relations between early child care and schooling. *Developmental Psychology, 24*, 53–57.

Howes, C. (1990). Can the age of entry into child care and the quality of child care predict adjustment in kindergarten? *Developmental Psychology, 26*, 292–303.

Howes, C. (1997). Children's experiences in center-based child care as a function of teacher background and adult:child ratio. *Merrill-Palmer Quarterly, 43*, 404–425.

Howes, C., Galinsky, E., Shinn, M., Gulcur, L., Clements, M., Sibley, A., Abbott-Shim, M., & McCarthy, J. (1998). *The Florida Child Care Quality Improvement Study: 1996 report.* New York: Families and Work Institute.

Howes, C., & Hamilton, C. E. (1992a). Children's relationships with caregivers: Mothers and child care teachers. *Child Development, 63*, 859–866.

Howes, C., & Hamilton, C. E. (1992b). Children's relationships with child care teachers: Stability and concordance with parental attachments. *Child Development, 63*, 867–878.

Howes, C., & Hamilton, C. E. (1993). The changing experience of child care: Changes in teachers and in teacher-child relationships and children's social competence with peers. *Early Childhood Research Quarterly, 8*, 15–32.

Howes, C., & Hamilton, C. E. (1994). Child care for young children. In: B. Spodek (Ed.), *Handbook of Research on the Education of Young Children* (pp. 322–336). New York: Macmillan.

Howes, C., Hamilton, C. E., & Philipsen, L. C. (1998). Stability and continuity of child-caregiver and child-peer relationships. *Child Development, 69*, 418–426.

Howes, C., Matheson, C. C., & Hamilton, C. E. (1994). Maternal, teacher, and child care history correlates of children's relationships with peers. *Child Development, 65*, 264–273.

Howes, C., & Olenick, M. (1986). Family and child care influences on toddler's compliance. *Child Development, 57*, 202–216.

Howes, C., Phillips, D. A., & Whitebook, M. (1992). Thresholds of quality: Implications for the social development of children in center-based child care. *Child Development, 63*, 449–460.

Howes, C., Rodning, C., Galluzzo, D. C., & Myers, L. (1988). Attachment and child care: Relationships with mother and caregiver. *Early Childhood Research Quarterly, 3*, 403–416.

Howes, C., & Rubenstein, J. L. (1985). Determinants of toddlers' experience in day care: Age of entry and quality of setting. *Child Care Quarterly, 14*, 140–151.

Howes, C., Sakai, L. M., Shinn, M., Phillips, D., Galinsky, E., & Whitebook, M. (1995). Race, social class, and maternal working conditions as influences on children's development. *Journal of Applied Developmental Psychology, 16*, 107–124.

Howes, C., & Smith, E. (1995). Relations among child care quality, teacher behavior, children's play activities, emotional security, and cognitive activity in child care. *Early Childhood Research Quarterly, 10*, 381–404.

Howes, C., Smith, E., & Galinsky, E. (1995). *The Florida Child Care Quality Improvement Study: Interim report.* New York: Families and Work Institute.

Howes, C., & Stewart, P. (1987). Child's play with adults, toys, and peers: An examination of family and child care influences. *Developmental Psychology, 23*, 423–430.

Hyson, M. C., Hirsh-Pasek, K., & Rescorla, L. (1990). The Classroom Practices Inventory: An observational instrument based on NAEYC's guidelines for developmentally appropriate practices for 4- and 5-year-old children. *Early Childhood Research Quarterly, 5,* 475–494.

Kisker, E, Hofferth, S., Phillips, D. A., & Farquhar, E. (1991). *A profile of child care settings: Early education and care in 1990.* Washington, D.C.: U.S. Department of Education.

Kontos, S. J. (1991). Child care quality, family background, and children's development. *Early Childhood Research Quarterly, 6,* 249–262.

Kontos, S. J., & Fiene, R. (1987). Child care quality, compliance with regulations, and children's development: The Pennsylvania study. In: D. Philips (Ed.), *Quality in Child Care: What Does the Research Tell Us?* (pp. 57–80). Washington, D.C.: National Association for the Education of Young Children.

Kontos, S. J., Howes, C., Shinn, M., & Galinski, E. (1997). Children's experiences in family child care and relative care as a function of family income and ethnicity. *Merrill-Palmer Quarterly, 43,* 386–403.

Kontos, S. J., & Peters, D. L. (Eds) (1988). *Continuity and disconitnuity of experience in child care.* Norwood, NJ: Ablex.

Lamb, M. E., Hwang, C-P., Broberg, A., & Bookstein, F. L. (1988). The effects of out-of-home care on the development of social competence in Sweden: A longitudinal study. *Early Childhood Research Quarterly, 3,* 379–402.

Lamb, M. E., & Sternberg, K. J. (1990). Do we really know how day care affects children? *Journal of Applied Developmental Psychology, 11,* 351–379.

Love, J. M., Schochet, P. Z., & Meckstroth, A. L. (1996). *Are they in any real danger? What research does – and doesn't – tell us about child care quality and children's well-being.* Princeton, NJ: Mathematica Policy Research, Inc.

McCartney, K. (1984). Effect of quality of day care environment on children's language development. *Developmental Psychology, 20,* 244–260.

McCartney, K., & Rosenthal, R. (2000). Effect size, practical importance, and social policy for children. *Child Development, 71,* 173–180.

McCartney, K., Scarr, S., Phillips, D., Grajek, S., & Schwartz, J. C. (1982). Environmental differences among day-care centers and their effects on children's development. In: E. F. Zigler & E. W. Gordon (Eds), *Daycare: Scientific and Social Policy Issues* (pp. 126–151). Boston: Auburn House.

McCartney, K., Scarr, S., Rocheleau, A., Phillips, D., Abbott-Shim, M., Eisenberg, M., Keefe, N., Rosenthal, S., & Ruh, J. (1997). Teacher-child interaction and child-care auspices as predictors of social outcomes in infants, toddlers, and preschoolers. *Merrill Palmer Quarterly, 43,* 426–449.

McGurk, H., Caplan, M., Hennessy, E., & Moss, P. (1991). Controversy, theory and social context in contemporary day care research. *Journal of Child Psychology and Psychiatry, 34,* 3–23.

Melhuish, E. C. (2001). The quest for quality in early day care and preschool experience continues. *International Journal of Behavioral Development, 25.* 1–6.

Mocan, H. N., Burchinal, M., Morris, J. R., & Helburn, S. W. (1995). Models of quality in center child care. In: S. W. Helburn (Ed.), *Cost, Quality, and Child Outcomes in Child Care Centers: Technical Report* (pp. 279–295). Denver, CO: University of Colorado at Denver.

NICHD Early Child Care Research Network (1996). Characteristics of infant child care: Factors contributing to positive caregiving. *Early Childhood Research Quarterly, 11,* 269–306.

The NICHD Early Child Care Research Network (1997a). Child care in the first year of life. *Merrill-Palmer Quarterly, 43*, 340–360.
The NICHD Early Child Care Research Network (1997b). The effects of infant child care on infant-mother attachment security: Results of the NICHD Study of Early Child Care. *Child Development, 68*, 860–879.
The NICHD Early Child Care Research Network (1997c). Familial factors associated with the characteristics of nonmaternal care for infants. *Journal of Marriage and the Family, 59*, 389–408.
The NICHD Early Child Care Research Network (1998). Early child care and self-control, compliance, and problem behavior at twenty-four and thirty-six months. *Child Development, 6*, 1145–1170.
The NICHD Early Child Care Research Network (1999). Child outcomes when child care center classes meet recommended standards for quality. *American Journal of Public Health, 89*, 1072–1077.
The NICHD Early Child Care Research Network (2000). The relation of child care to cognitive and language development. *Child Development, 71*, 960–980.
Peisner-Feinberg, E. S. (1998, November). The longitudinal effects of child care quality on early school outcomes for children. Paper presented at the meeting of the National Association for the Education of Young Children, Toronto, Canada.
Peisner-Feinberg, E. S., & Burchinal, M. R. (1997). Relations between preschool children's child-care experiences and concurrent development: The Cost, Quality, and Outcomes Study. *Merrill-Palmer Quarterly, 43*, 451–477.
Peisner-Feinberg, E. S., Burchinal, M. R., Clifford, R. M., Culkin, M., Howes, C., Kagan, S. L., Yazejian, N., Byler, P., Rustici, J., & Zelazo, J. (2000). *The children of the Cost, Quality, and Outcomes Study go to school. Technical Report.* Chapel Hill: University of North Carolina at Chapel Hill, Frank Porter Graham Child Development Center.
Phillips, D. (1996). Reframing the quality issue. In: S. L. & N. E. Cohen (Eds), *Reinventing Early Care and Education* (pp. 43–64). San Francisco: Jossey-Bass.
Phillips, D., & Howes, C. (1987). Indicators of quality in child care: Review of research. In: D. Phillips (Ed.), *Quality in Child Care: What Does the Research Tell Us?* (pp. 1–20). Washington, D.C.: National Association for the Education of Young Children.
Phillips, D., Lande, J., & Goldberg, M. (1990). The state of child care regulation: A comparative analysis. *Early Childhood Research Quarterly, 5*, 151–179.
Phillips, D., McCartney, K., & Scarr, S. (1987). Child care quality and children's social development. *Developmental Psychology, 23*, 537–543.
Phillips, D. A., Voran, M., Kisker, E., Howes, C., & Whitebook, M. (1994). Child care for children in poverty: Opportunity or inequity? *Child Development, 65*, 472–492.
Phillipsen, L. C., Burchinal, M. R., Howes, C., & Cryer, D. (1997). The prediction of process quality from structural features of child care. *Early Childhood Research Quarterly, 12*, 281–303.
Phillipsen, L., Cryer, D., & Howes, C. (1995). Classroom process and classroom structure. In: S. W. Helburn (Ed.), *Cost, Quality, and Child Outcomes in Child Care Centers, Technical Report* (pp. 125–139). Denver, CO: University of Colorado at Denver.
Prescott, E., Kritchevsky, S., & Jones, E. (1972). *The day care environmental inventory.* Washington, D.C.: U.S. Department of Health, Education, and Welfare.
Reifel, S. (1997). A changing early childhood education community: Where next? *Journal of Early Childhood Teacher Education, 18*, 1–8.

Rescorla, L., Hyson, M., Hirsh-Pasek, K., & Cone, J. (1990). Academic expectations in mothers of preschool children: A psychometric study of the Educational Attitude Scale. *Early Education and Development, 1,* 167–184.

Roggman, L., Langlois, J., Hubbs-Tait, L., & Rieser-Danner, L. (1994). Infant day-care, attachment, and the "file drawer problem." *Child Development, 65,* 1429–1435.

Rubenstein, J., & Howes, C. (1979). Caregiving and infant behavior in day care and in homes. *Developmental Psychology, 15,* 1–24.

Scarr, S. (1995). The two worlds of child care. *National Forum, 75*(3), 39–41.

Scarr, S. (1997). Why most child care has little impact on most children's development. *Current Directions in Psychological Science, 6,* 143–148.

Scarr, S. (1998). American child care today. *American Psychologist, 53,* 95–108.

Scarr, S., & Eisenberg, M. (1993). Child care research: Issues, perspectives, and results. *Annual Review of Psychology, 44,* 613–644.

Scarr, S., Eisenberg, M., & Deater-Deckard, K. (1994). Measurement of quality in child care centers. *Early Childhood Research Quarterly, 9,* 131–151.

Scarr, S., Phillips, D., McCartney, K., & Abbott-Shim, M. (1993). Quality of child care as an aspect of family and child care policy in the United States. *Pediatrics, 91,* 182–188.

Schindler, P. J., Moely, B. E., & Frank, A. L. (1987). Time in day care and social participation of young children. *Developmental Psychology, 23,* 255–261.

Snow, C. W., Teleki, J. K., & Reguiero-de-Atiles, J. T. (1996). Child care center licensing standards in the United States: 1981 to 1995. *Young Children, 51,* 36–41.

Stipek, D., Feiler, R., Daniels, D., & Milburn, S. (1995). Effects of different instructional approaches on young children's achievement and motivation. *Child Development, 66,* 209–223.

Vandell, D. L., & Corasaniti, M. A. (1990). Variations in early child care: Do they predict subsequent social, emotional, and cognitive differences? *Early Childhood Research Quarterly, 5,* 555–572.

Vandell, D. L., Henderson, V. K., & Wilson, K. S. (1988). A longitudinal study of children with day-care experiences of varying quality. *Child Development, 59,* 1286–1292.

Vandell, D. L., & Powers, C. P. (1983). Day care quality and children's free play activities. *American Journal of Orthopsychiatry, 53,* 493–500.

Vandell, D. L., & Wolfe, B. (2000). Child care quality: Does it matter and does it need to be improved? Washington, D.C.: U.S. Department of Human Services.

Washington, V. (1997, October 24). Comments made at the White House Conference on Child Care. Washington, D.C.

Whitebook, M., Howes, C., & Phillips, D. A. (1990). *Who cares? Child care teachers and the quality of care in America:* (Final Report of the National Child Care Staffing Study). Oakland, CA: Child Care Employee Project.

Whitebook, M., Sakai, L., Gerber, E., & Howes, C. (2001). *Then and now: Changes in child care staffing, 1994–2000. Technical Report.*Washington, D.C.: Center for the Child Care Workforce.

Wilcox-Herzog, A., Fortner-Wood, C., & Kontos, S. (1998). Quality in child care: How much does it matter? In: C. Seefeldt & A. Galper (Eds), *Continuing Issues in Early Childhood Education* (2nd ed., pp. 87–102). Columbus, OH: Merrill.

Zaslow, M. J. (1991). Variation in child care quality and its implications for children. *Journal of Social Issues, 47,* 125–138.

PROFESSIONAL DEVELOPMENT AND THE QUALITY OF CHILD CARE: AN ASSESSMENT OF PENNSYLVANIA'S CHILD CARE TRAINING SYSTEM

Joyce Iutcovich, Richard Fiene, James Johnson, Ross Koppel and Frances Langan

ABSTRACT

The education and training of child care workers are viewed as keys to improving classroom/caregiver dynamics and the overall quality of child care. This assessment of the Pennsylvania Child Care/Early Childhood Development Training System offers an analysis of this hypothesis. The research was designed for dual purposes: to identify training needs for Pennsylvania child care providers and to assess the impact of training and work environment on the quality of care. The results highlight specific areas where there are needs for training and reveal a clear association between opportunities for professional growth and the quality of care.

INTRODUCTION

The care of children and concerns about their future are of great importance to our society. The current trend in public opinion and political action highlights our concern about children and their welfare. According to public polls "the fastest growing segment of the electorate is the one concerned about protecting children and helping parents be good parents" (McAllister, 1997, p. 36). Further, we have seen new research on the impact of a child's early experiences on how his or her brain is "wired." In an effort to bring attention to this important new research on brain development and its implications for public policy, the Families and Work Institute initiated the Early Childhood Public Engagement Campaign. A White House Conference on Early Childhood Development and a television special, *I Am Your Child*, launched this campaign in early 1997. Another White House Conference on child care was held in October 1997. Politicians have been quick to notice that children's issues strike a special chord with Americans – hence the plethora of new initiatives aimed at the young.

All of this attention on children's issues is heartening in an era of budget reallocation, welfare reform, and the move to eliminate *Big Government*. However, the extent to which all this *talk* will be translated into *action* is yet to be determined. Regardless, this public attention has brought into focus an area of critical need in our society – quality child care. With the dramatic rise in the number of mothers with small children in the labor force, the need for child care services and the maintenance of quality programs throughout the nation cannot be denied (Katz, 1994). In response to this increased demand there has been a significant rise in the number of licensed child care centers and home-based child care providers – not to mention unregulated child care settings. Welfare reform legislation has also resulted in an increase in mothers needing child care services as they move into the labor force. Some welfare-to-work mothers have been encouraged to provide home-based child care to neighbors and relatives to help meet this increased demand for child care.

Thus, as the need increases and child care facilities spring up to meet the growing demand – both regulated and unregulated – the concern over quality becomes more pressing. A study conducted by Mathematica Policy Research for the U.S. Department of Education (1990) reports that the quality of care will be jeopardized with the trend of serving more children with fewer workers. More recent studies have determined that there is far too little good child care in the United States. Only 14% of center care, 12% of family child care, and an even lower percentage of infant care can be rated as good in this country (Galinsky et al., 1994; Helburn et al., 1995).

Given this state of affairs, research on child care and factors associated with quality care are very important, particularly if they have implications for public policy. State regulations play a key role in ensuring that programs comply with minimum standards regarding structural features and staff qualifications. But minimum standards related to child/staff ratios and educational level of staff are not enough. Other dimensions found to be associated with quality care are classroom/caregiver dynamics (including caregivers' sensitivity and use of developmentally appropriate practice) and staff characteristics such as specialized education, training, and experience (Love, Schochet & Meckstroth, 1996; Barbour, Peters & Baptiste, 1995).

Education and training of child care workers are viewed as keys to improving classroom/caregiver dynamics and quality of care. But not all education and training are equally effective. The Center for Career Development in Early Care and Education at Wheelock College (newly named as the Wheelock College Institute for Leadership and Career Initiatives) has emphasized the importance of professional development programs for child care providers. The model developed by the Center focuses on linkages between education and training and development of new career opportunities for early childhood practitioners (Morgan et al., 1993). Having all training opportunities build on one another, offering incentives for practitioners to obtain training, and specifying a core body of knowledge for all early childhood care and education practitioners are particularly important elements of a model program for career development. Additionally, the Center posits that a comprehensive, coordinated system of training and education should include the following features: quality control over training content and trainers; a system for assessing training needs and offering training based on those needs; a system to make information about training easily accessible and widely distributed; a vehicle for tracking provider training; a linkage between training and compensation; and an expanded and coordinated plan for funding training – preferably through public/private partnerships.

The Study of Pennsylvania's Child Care/Early Childhood Development Training System

Pennsylvania has recognized the need to offer training opportunities for child care workers as a means to improve the quality of care. Training for various segments of the child care provider population has been available for over ten years. In 1992 a number of separate training programs were integrated into one system – *The Pennsylvania Child Care/Early Childhood Development Training System* (PA CC/ECD). The Pennsylvania Department of Public Welfare (DPW) was instrumental in the development of this training system and has supported the establishment of an affordable and flexible training system that is based on the principles of early childhood education and child development.

Pennsylvania's child care training initiative began in the early 1990s, as did other statewide training systems. States utilized the program quality portion of the Child Care and Development Block Grant (CCDBG)[1] to fund the development and implementation of such training systems (Fiene, 1995). The PA CC/ECD Training System was implemented in January 1992 after lengthy public hearings regarding the Child Care and Development Block Grant. Child care advocates expressed a definite need for a comprehensive early childhood training system throughout the state. Advocates felt that a comprehensive training system was a cost-effective way to improve the quality of early childhood programs throughout Pennsylvania.

The PA CC/ECD Training System has experienced a number of system changes since 1992 and several evaluations with the presently described study as just one of these. For example, prior to 1992, the only training available to child care providers was through a home-based, voucher training program. This program proved to be very popular with providers because it gave them ultimate flexibility in the selection of training opportunities. As the training system evolved, the home-based voucher program became part of the overall PA CC/ECD Training System by 1995. However, this program provides very little structure related to course sequencing or focus on core competencies for child caregivers.

Four school-age technical assistance and capacity building projects also existed prior to 1992, but their major focus was not on training. After 1992 this changed and their focus turned to training. In 1995 the four school-age training projects were incorporated into the overall PA CC/ECD Training System. By 1995, all training for center-based, home-based, and school-age providers were under the umbrella of the PA CC/ECD Training System.

The Early Childhood Education Linkage System (ECELS), the program responsible for health and safety training and technical assistance to Pennsylvania's child care providers, presented the American Red Cross Child Care Course throughout the state from 1992 until 1995. In 1995 this course was incorporated into the PA CC/ECD Training System. This completed the coordination of all training activities related to early childhood and child care under the umbrella of PA CC/ECD with the exception of Head Start and early intervention training.

Since 1992, over 50,000 early childhood providers have received an average of three hours of training on an annual basis. The training opportunities offered to providers include workshops, seminars, videos, learn-at-home materials, conferences, satellite teleconferences, mentoring, vouchers for college coursework, and a number of other training opportunities outside the PA CC/ECD system. The PA CC/ECD Training System is a diverse system of training modalities and funding mechanisms. Several of the PA CC/ECD Training System components have been recognized as innovative. For example, the home-based voucher program and ECELS were recognized in *Making a Career*

Professional Development and The Quality of Child Care 119

of It, a report by the Center for Career Development in Early Care and Education at Wheelock College (Morgan et al., 1993). However, a concern was expressed that training opportunities, albeit comprehensive, were not coordinated to lead an individual on a career path. Therefore, several research studies have been undertaken to determine the effectiveness of the overall system and its implementation.

A Penn State University evaluation research initiated in 1992 helped to delineate the need for additional training opportunities for staff. The accumulative amount of training taken over three years was the key variable that predicted positive developmentally appropriate changes in the classroom (Johnson, 1994). However, this study left unanswered questions about what other factors and features of training are associated with child care quality.

There were overlapping concerns, although different purposes, for two studies initiated in 1996. Wheelock College (Stoney et al., 1997) conducted one study, an assessment of the various early childhood training systems in Pennsylvania, to determine how to coordinate the existing PA CC/ECD Training System with other training systems in an effort to develop a full-fledged early childhood career development system within Pennsylvania.

The other study initiated in 1996 is the one reported herein. Recognizing the importance of tracking the impact of this training system on the quality of care, this research was designed for dual purposes: to identify the training needs for Pennsylvania child care providers and to assess the impact of training and work environment on the quality of care in child care sites. In addition, the results of this research effort are compared to earlier Pennsylvania studies that examined the quality of child care. Within these overarching research goals, this study examined the specific research questions delineated below.

Research Questions Related to Training Needs
- What are the perceived needs for training? Do various provider groups have different needs (e.g. center teachers, center directors, group providers, and family providers)?
- What are the observed needs for training as indicated through the site observations of quality of care?
- What are the most important factors affecting the selection of training? How does the director impact this?
- How do providers evaluate the training? What are their perceptions regarding appropriateness, usefulness, applicability, and effectiveness of training in achieving learning objectives? What is their level of interest in training? And how do they think it applies to their work?

- What are the barriers to training? Are the barriers different for the various provider groups?

Research Questions Related to Quality of Care
- How has the quality of care in Pennsylvania child care changed over the years?
- What factors are significant in predicting the quality of care as observed in child care classrooms?
- To what extent do staff background characteristics (e.g. current education, educational goals, age, years in field, and salary) impact the classroom's quality of care?
- How are features of a caregiver's training experience related to classroom quality of care? To what extent does the level of training impact quality? What is the impact of the training's perceived appropriateness, usefulness, applicability, and effectiveness in achieving learning objectives?
- To what extent do teachers' perceptions of organizational climate impact the quality of classroom care?
- What is characteristic of the quality of work life in child care centers in terms of organizational climate, summary of worker values, overall commitment, how the environment resembles an ideal, the importance of educational goals and objectives, and the degree of influence of teaching staff?
- To what extent is a center's organizational climate associated with director background characteristics, aggregate teacher characteristics, site turnover, accreditation status, size of site, and average hours of training per site?
- To what extent is a center's overall quality of care associated with director background characteristics, aggregate teacher characteristics, organizational climate, and other site-level features (e.g. size of center, accreditation status, turnover rate, and average hours of training per year)?

This study seeks to answer these specific research questions. The following sections present a review of the literature related to professional development systems and factors associated with the quality of child care; an overview of the conceptual framework and methodology used to guide the study; a summary of the results; and the implications of the findings for public policy.

PROFESSIONAL DEVELOPMENT AND QUALITY OF CARE

Staff development research and studies on factors associated with the quality of child care always share the same long term goal, typically hold the same theoretical orientation, and often have variables in common within their research designs. Ultimately, the goal is to bring about optimal experiences for children in child care. Descriptive and explanatory knowledge about early childhood inservice education or staff development and about program quality is needed to achieve this aim. Other related goals can be served at the same time when research adds to an understanding of quality experiences for children in child care, the value of inservice training for staff development, and the relationship between the two.

Staff Development Research

Current education literature addresses a number of issues related to the ongoing professional development of teaching staff. One very important issue concerns the application of knowledge or the ability to transfer learning into practice. What are the most effective strategies used to guarantee the transfer of knowledge into practice? Numerous reasons are provided as to why staff who participate in educational programs do or do not apply in practice what is learned through education. The perception of program participants about the value and practicality of program content, the presence or absence of follow-up strategies, and supervisory attitudes toward changes required to apply what has been learned are all critical in the transfer of learning (Caffarella, 1994).

The value and practicality of a program implies that a training curriculum should be problem-centered and site-specific. According to Jorde-Bloom and Sheerer (1992), training programs should address real issues and concerns that participants face in their work setting on a daily basis; staff development efforts should facilitate interaction between colleagues; staff developers should "take into account the distinctly different orientations, needs, and interests" of program participants; and training content should focus on bridging the gap between theoretical ideas and the practical realities of the work setting. Jorde-Bloom (1998) also emphasizes the importance of staff becoming active participants in identifying program strengths and areas in need of improvement.

Discussions about the characteristics of effective staff development programs have resulted in some key themes. Holt-Reynolds (1995) maintains the importance of being aware of the rationales underlying the use of particular teaching practices. Rather than focusing on skill training as so many pre-service and in-service teacher development programs have done, staff development must be aimed at uncovering and dealing with lay beliefs, attitudes, behaviors, and decision-making strategies that teachers bring to the classroom.

VanderVen (1994) suggests a contextual model of professional development that enhances the current linear model, which is structural. The contextual model recognizes that early childhood is age-specific and is integrated across the domains of care, education, and development; the contextual model is generic and calls for situational application of multidisciplinary knowledge. VanderVen believes that professional development programs should facilitate constructivism and articulation of theory into practice. Knowledge is gained by doing, then reflecting and dialoguing about it – a constructivist model for learning (Jones, 1993).

In light of the contextual model, outcomes-based educational programs have also been recommended as more effective than the competency-based training programs of the past. Representing a paradigm shift, outcome-based programs focus on demonstrating application of knowledge in contextual settings and quality performance of integrated tasks. Simply acquiring knowledge and demonstrating competencies in isolation is not authentic and does not address the importance of making connections between pre-service training (development) and practice (performance).

Finally, as all this relates to teachers in child care settings, career mobility and advancement is seen as a *sine qua non* of professional development programs (Morgan, 1994). According to Morgan a professional development training system should include these components:

Make training count: when substantial training of good quality is offered, it should carry college credit or be transformed to college credit that can be applied to certificate or degree programs.

Improve access to credit-bearing training for practitioners who are already employed, particularly people of color and individuals from low-income populations.

Articulate programs: accept the Child Development Associate (CDA) Credential to count toward an associate degree program; and allow the associate degree program to count in full toward a bachelor's degree program (Morgan, 1994, p. 138).

Given this background on staff development, we now raise the fundamental question: What constitutes a high quality professional development program? The National Association for the Education of Young Children (NAEYC) has provided leadership in professional development models for early childhood educators (Bredekamp & Willer, 1994). One of NAEYC's top priorities is improving professional preparation programs for the diverse staff who care for young children. A current NAEYC initiative, the National Institute for Early Childhood Professional Development, is a system designed to address the complexity of developing staff involved in the care and education of young children, improving the quality and consistency of professional pre-service and in-service programs, and linking them with improvements in practice.

Through the work of NAEYC, Willer (1994, pp. 17–19) has identified these principles from the work of Epstein (1993) and Modigliani (1993) that lay the foundation for effective professional development processes; they include:

- Professional development is an ongoing process.
- Professional development experiences are most effective when grounded in sound theoretical and philosophical base and structured as a coherent and systematic program.
- Professional development experiences are most successful when they respond to individuals' background, experiences, and the current context of their role.
- Effective professional development opportunities are structured to promote clear linkages between theory and practice.
- Providers of effective professional development experiences have an appropriate knowledge and experience base.
- Effective professional development experiences use an active hands-on approach and stress an interactive approach that encourages students to learn from one another.
- Effective professional development experiences contribute to positive self esteem by acknowledging the skills and resources brought to the training process as opposed to creating feelings of self-doubt or inadequacy by immediately calling into question an individual's current practices.
- Effective professional development experiences provide opportunities for application and reflection, and allow for individuals to be observed and receive feedback upon what has been learned.
- Students and professionals should be involved in the planning and design of their professional development program.

This attention to early childhood professional development comes at a critical time. Research on the background and skills of child caregivers paints a bleak picture. There is significant concern that child caregivers lack the skills, knowledge, and education to appropriately address the developmental needs of children. "Six out of seven child care centers provide care that is mediocre to poor. One in eight might actually be jeopardizing children's safety and development" (Children's Defense Fund, 1998).

The 1993 National Child Care Staffing Study cited low wages as one factor that accounts for poor quality care. Low wages make recruitment and retention of qualified personnel difficult. Another reason for the low quality of child care is inadequate staff training. "Staff education and training are among the most critical elements in improving children's experiences in child care" (Children's Defense Fund, 1998, p. 39). Regardless, many states do not require pre-service training for teachers in licensed or regulated child care centers. Further, a majority of states require only 12 or fewer hours of annual training (Children's Defense Fund, 1994). Research has shown that a threshold for training to show some impact is around 18 hours (Howes, Smith & Galinsky, 1995).

Quality of Child Care Research

Reviews of the research on the factors related to child care quality (Phillips, 1987; Love et al., 1996; Chung & Stoney, 1997) group the studies into several categories. Some studies address global assessments of child care quality while others focus on the structural dimensions of quality or the dynamic measures of classroom quality. For our purposes, studies focusing on global assessments and structural dimensions of quality care are of particular importance.

Research from the late 80s (Phillips, 1987) identifies the following as key indicators of quality child care:

- The program is licensed.
- The child's interaction with the caregiver is frequent, verbal, and educational, rather than custodial and controlling.
- Children are not left to spend their time in aimless play.
- There is an adequate adult-child ratio and reasonable group size.
- The caregiver has a balanced training in child development, some degree of professional experience in child care, and has been in the program for some period of time.

More recent studies (Helburn, 1995; Phillips, Howes & Whitebook, 1992) confirm the importance of these indicators and identify other factors that are important. In addition, the following features of high-quality child care for preschool children include:

- **Space:** the indoor environment is clean, in good repair, and well-ventilated; classroom space is divided into richly equipped activity areas; fenced outdoor play space is equipped with swing, climbing equipment, tricycles, and a sandbox.
- **Children's activities:** most of the time children work individually or in small groups; children select many of their own activities and learn through experiences relevant to their own lives; caregivers facilitate children's involvement, accept individual differences, and adjust expectations to children's developing capacities.
- **Parent-caregiver interaction:** parents are encouraged to observe and participate in the program; caregivers talk frequently with parents about children's behavior and development.

For infants, the following signs of high-quality child care are in addition to the key indicators identified by Phillips (1987):

- Play materials are appropriate for infants and toddlers and stored on low shelves within easy reach.
- Daily schedule includes time for active play, quiet play, naps, snacks, and meals; it is flexible rather than rigid, to meet the needs of individual children; and the atmosphere is warm supportive, and children are never left unsupervised.
- Caregivers respond promptly to infants' and toddlers' distress; hold, talk, sing, and read to them; interact with children in a contingent manner that respects the individual child's interests and tolerance for stimulation.
- Parents are welcome anytime; caregivers talk frequently with parents about children's behavior and development.

In light of this overview, the underlying theme is the consistency in which the above factors, as indicators of quality, appear in the research findings.

Conceptual Framework

As previously indicated, staff development research and studies on the quality of child care share the same long term goal and typically the same theoretical orientation. The present study, with its twofold purpose of investigating the PA CC/ECD Training System with respect to user perceptions and the relationship between training and program quality, intersects with the current research literature. Accordingly, its long range purpose, its conceptual underpinnings, and its choice of variables and measures are consistent with previous work in these two areas.

A socio-ecological or systems theory perspective provides a framework for this study. This perspective emphasizes reciprocal transactions between individuals and their environments. Individuals' constructions (beliefs and attitudes) of their social environments, rather than some notion of objective reality, are central to personal adaptation and behavior (Bronfenbrenner, 1979; Lewin, 1935). Child care and training workshops are dynamic, psychological entities as well as physical ones. Providers' social role behaviors and interpersonal relations relevant to the care of children are associated with the totality of factors that constitute a particular child care site (i.e. overall staff and program characteristics). Likewise, providers' role behaviors and relations within child care (staff-staff, staff-child, staff-parents) that contribute to program quality are assumed to influence and be influenced by the PA CC/ECD Training System.

The selection of variables and measures involved in this study, the rationales for the choices, how the variables are conceptually organized, and how they are consistent with previous research are described in the remaining part of this section. These variables are organized into categories as depicted in Figs A and B relevant to the two major purposes of the present study.

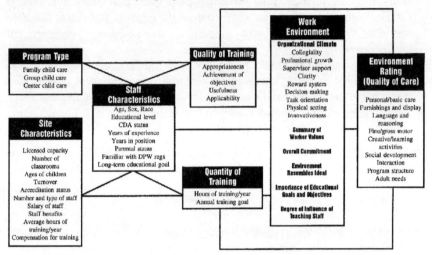

Fig. A. Quality of Care Conceptual Model.

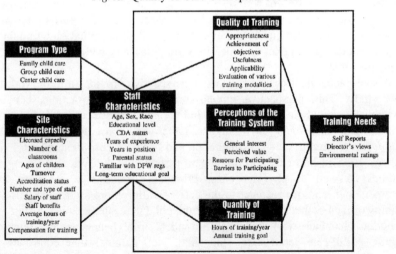

Fig. B. Training System Conceptual Model.

Figure A illustrates how the variables are conceptually organized and associated with levels of child care quality. Quality of child care is operationalized by scores from the Harms and Clifford Environment Rating Scales, while the various dimensions of a child care work environment are measured with Paula Jorde-Bloom's Early Childhood Work Environment Survey. Characteristics of the work environment are viewed as a primary set of intervening variables.

As measures of program quality, three separate environment rating scales were used in this study: the Early Childhood Environment Rating Scale (ECERS), the Infant/Toddler Environment Rating Scale (ITERS), and the Family Day Care Rating Scale (FDCRS). Although each scale has comparable areas that are assessed, the individual items composing each scale do vary depending on the type of child care site or classroom observed. Across each of these scales, the major areas that are assessed relate to furnishings and display; personal/basic care; language and reasoning; fine/gross motor activities; creative/learning activities; social development; interactions; program structure; and adult needs.

As indicators of the various dimensions of an early childhood work environment, the Early Childhood Work Environment Survey (Jorde-Bloom, 1988, 1998) includes a number of distinct conceptual areas. These areas include organizational climate, summary of worker values, overall commitment of staff to center, summary of how current work environment resembles the ideal, importance of educational goals and objectives, and degree of influence of the teaching staff. Organizational climate consists of the collective perceptions of staff regarding the dimensions of collegiality; emphasis on professional growth; degree of supervisor support; clarity of policies and procedures; fairness and equity of the reward system; degree of autonomy in decision making; goal consensus among staff; emphasis on task orientation; extent to which the physical setting facilitates programming; and degree of innovativeness.

Figure A shows an overview of the variables and how they are conceptually organized with respect to the investigation of factors related to the quality of child care. Although the left- to right-hand side ordering of the variable categories in Figure A suggests directionality of effects, it is important to keep in mind that this study is basically descriptive-correlational in nature. The non-experimental, cross-sectional nature of the research design precludes testing directional hypotheses. Program quality could be the cause or the effect of the other variable categories. Nevertheless, the original rationale for selecting this research design centered on the plausible assumption that higher levels of training of personnel in a program would go hand-in-hand with the quality of care. In addition to organizational climate, certain staff and program characteristics were also expected to show a positive and statistically significant relationship with the quality of care.

The variable categories of program type and program variables shown on the far left-hand side of Figure A are best viewed as moderating variables. These variables suggest data-based comparisons but are not seen as predictors or mediators of quality (with the exception of accreditation status).

Figure B illustrates the relationships among variable categories pertinent to the identification of training needs and user perceptions of the training system. Of major interest, again, are the comparisons involving program type (family child care, group child care, and centers) and type of staff (directors versus teachers). The model included these specific areas: site characteristics, staff characteristics, quality of training, perceptions of the training system, and quantity of training. Training needs and interests were also identified. Questionnaires administered to child care staff were used to identify perceived training needs and interest areas via teacher and director self reports as well as directors' views of staff interests and needs. Needs (as opposed to interests) were also gleaned from information obtained from the environmental rating scales.

METHODOLOGY

As previously indicated, this is a cross-sectional study that collected data from child care sites throughout Pennsylvania. Child care sites were sampled and trained fieldworkers conducted site visits to observe the quality of care in classrooms and to administer questionnaires to child care staff. Specific details about the sampling process, fieldworker training, and data collection instruments are described in the following sections.

Sampling

A stratified systematic sampling process was utilized to identify child care sites for this study. At the time in which we drew the sample, there was a population of approximately 4,144 family child care sites, 590 group homes, and 3,067 child care centers (the *registered* family child care sites and *licensed* group homes and centers). Within each of these separate lists, we then created sampling frames stratified by geographic region. From these stratified lists, we randomly selected a number of child care sites of each type within the various geographic regions of the state. Our sampling design called for a disproportionate number of sites in each category: 30 family child care homes, 30 group child care homes, and 60 child care centers.

We decided on these numbers for several reasons. First, limited resources and time would not allow us to conduct more than the 120 site visits. Second, to have a sufficient number of group homes to analyze, the number of group child care sites in the sample had to be disproportionate to what they represented in the population. Given the disproportionate nature of the sample, weights were used in any analysis that involved more than one type of child care site.

To encourage voluntary participation in this study, we implemented a number of procedures. First, we initially sent a letter to selected sites to explain the purpose of the study and the importance of the findings for *improving the child care training system in Pennsylvania.* In this letter we explained the advantages of provider participation in the study: receipt of a voucher to purchase children's books/toys from Gryphon House ($100 for centers and $50 for family and group homes); an opportunity to have an early childhood professional visit their site and provide some feedback regarding the environment rating scales; and receipt of a certificate acknowledging participation in the study. A follow-up call to the sites was made to further explain the study and encourage their participation. Once a confirmation was received from the site, a fieldworker was assigned to the site to establish a date for a site visit.

A number of the sites initially drawn for the sample were not included in the final total (some were no longer in business, some refused, some could not be visited due to scheduling difficulties). In each case, another randomly drawn site was used as a replacement. Our analysis of the data confirms that the resulting sample was not biased as a result of this replacement; the indicators of quality vary in the expected manner and other site level characteristics reflect known data. The final sample size consisted of 29 family child care homes, 30 group homes, and 60 centers.[2]

Fieldworker Identification and Training

The importance of having trained observers in a study of this nature cannot be underestimated. For this reason, we took care to identify fieldworkers who were familiar with the Harms and Clifford Environment Rating Scales or with the validation procedures used by the National Association for the Education of Young Children when conducting accreditation site visits. Once fieldworkers were identified, they were sent the training materials (video and manual) for the

Harms and Clifford Environment Rating Scales. Subsequently, a training session was held to review these materials and other procedures to be used in setting up and conducting the site visits. A fieldworker manual was prepared and distributed to everyone; monitoring of their work and progress was conducted from the research office; and inter-rater reliability was determined for a small percentage of each fieldworker's observations for the environment rating scales. The high inter-rater reliability scores indicate consistent use of the scales.[3] Furthermore, the overall quality of the data gathered by the fieldworkers attests to their ability.

Data Collection Instruments

The operationalization and measurement of two key areas in this study were previously discussed. Quality of child care was measured through the three Harms and Clifford Environment Rating Scales: ECERS, ITERS, and FDCRS. The scale value for each of the items assessed on these instruments ranges between 1 and 7, where 1 = inadequate, 3 = minimal, 5 = good, and 7 = excellent.[4]

The dimensions of the child care work environment were measured with Paula Jorde-Bloom's Early Childhood Work Environment Survey (ECWES). There are six separate conceptual areas assessed through this instrument, as identified earlier. For each of the *organizational climate* dimensions, a score of 0 to 10 is calculated by averaging the staff responses to 10 items for each dimension. The *summary of worker values* is indicated by the percentage of staff (0 to 100%) that identify an organizational climate dimension as one of the three most important aspects of their work. *Overall commitment* has a range of values between 0 and 10 where 0 = not committed and 10 = highly committed. Staff perceptions of *how their current work environment compares with their ideal* ranges between 1 = not like my ideal and 5 = like my ideal. The *importance of educational goals and objectives* is indicated by a priority ranking, ranging from 1 = low priority to 6 = high priority. Finally, the *degree of influence of teaching staff regarding organizational decisions* is assessed on a scale of 0 to 10 where 0 = very little influence and 10 = considerable influence.[5]

In addition to these standardized instruments, we developed a series of questions to gather background and training information from both directors and teachers within the child care sites. Although the questions were comparable for directors and teachers and across the type of sites, there were some items that applied only to one or the other. Given this, separate instruments were developed. One instrument was for family providers; one for directors of small sites (group homes and some small centers); one for directors of centers; one for teachers of small sites; and one for teachers from centers. In the end, we analyzed the data in terms of the type of site (family, group, or center) as well as type of respondent (director or teacher).

The comprehensive background information gathered with these questionnaires included:

- **Director and/or Teacher Background**: age, sex, race, education, years in early childhood field, years with current employer, employment status, salary, long-term educational goal, CDA status, and parental status.
- **Training Background and Assessment**: number of training hours in past three years, annual training goal, evaluation of training system (appropriateness, achievement of goals/objectives, usefulness, applicability), helpfulness of additional training, specialized training, assessment of specific training modalities, decisions about staff training, presence of staff development plans, compensation for training, factors affecting the selection of training, barriers to training, interest in training, and need for additional training in selected topic areas.
- **Site Characteristics**: age of children in facility, type of facility, licensed capacity, number of classrooms, change in licensed capacity in past year, number of paid staff, number of new staff in current year; presence of assistant director, and accreditation status.

FINDINGS

The results of this study address a number of specific research questions within the context of identifying training needs and assessing the factors associated with the quality of care. In presenting the results of the data analysis, we first provide an overview of the background data for each of the provider groups, followed by the findings for the specific research questions.

Background Data on Provider Groups and Child Care Facilities

The socio-demographic characteristics of the provider groups, their training background, and various site characteristics are summarized in Tables 1–3 to give a better understanding of the child care providers and facilities included in this study.

As Table 1 shows, the socio-demographic characteristics of this sample are typical of what we find in national statistics. As expected, the vast majority of providers are female. Their average age is between 34.8 and 45.8 with directors slightly older than teachers. A majority of providers are parents (between 59 to

Table 1. Background Characteristics of Provider Groups.*

Characteristic	Center Directors (N=60)	Center Teachers (N=561)	Group Directors (N=30)	Group Teachers (N=70)	Family Providers (N=44)
SEX (% female)	98.3	98.0	100	95.7	93.2
AGE (mean)	41.6	34.9	45.8	34.8	38.8
RACE/ETHNICITY					
White	88.1	82.6	76.7	80.0	66.7
Black	8.5	14.4	20.0	17.1	31.0
Other	3.4	3.0	3.3	2.9	2.4
PARENTAL STATUS					
(% yes)	72.9	59.1	93.3	70.0	90.9
EDUCATION:					
High school	3.3	32.4	33.3	55.7	54.8
Some college	1.7	22.4	40.0	27.1	33.3
Associate degree	13.3	10.8	13.3	2.9	4.8
Bachelors degree	33.3	23.7	6.7	10.0	4.8
Some graduate	30.0	6.4	3.3	4.3	2.4
Masters degree	13.3	3.1	0.0	0.0	0.0
Post masters	3.3	1.1	3.3	0.0	0.0
Doctorate	1.7	0.0	0.0	0.0	0.0
YEARS IN EARLY CHILDHOOD FIELD (mean)	13.7	6.8	13.1	6.5	7.2
YEARS WITH PRESENT EMPLOYER (mean)	8.6	3.7	9.5	4.2	6.5
EMPLOYMENT STATUS:					
Full-time (35+ hrs)	93.3	62.6	93.3	40.6	80.5
Part-time	6.7	37.4	6.7	59.4	19.5
SALARY					
(approx. average)	$19,900/yr	$6.40/hr	$17,250/yr	$5.89/hr	$12,500/yr
BENEFITS					
Pension		18.5		2.1	
Vision		15.5		2.1	
Dental		32.6		2.1	
Health	N.A.**	48.7	N.A.	2.1	N.A.
Life insurance		23.5		0.0	
Paid maternity		3.2		0.0	
Disability		16.3		2.1	
Education reimbursement		25.9		17.0	

* Percentages are reported except where otherwise noted.
** N.A. = Question not asked of this provider group.

Table 2. Training Background of Provider Groups.*

Characteristic	Center Directors (N=60)	Center Teachers (N=561)	Group Directors (N=30)	Group Teachers (N=70)	Family Providers (N=44)
LONG TERM EDUCATIONAL GOAL					
GED/High school	0.0	4.3	6.9	7.4	4.8
Non-credit adult education	1.8	5.8	6.9	13.2	9.5
Early childhood certification	0.0	12.2	13.8	16.2	9.5
Associate degree	1.8	6.8	13.8	4.4	16.7
College degree	5.4	15.9	13.8	8.8	14.3
Graduate degree	57.1	17.6	10.3	10.3	4.8
No long term goals	33.9	37.5	34.5	39.7	40.5
SEEKING CDA CERTIFICATE					
Yes	1.8	16.9	20.8	16.9	25.6
No	85.5	75.8	79.2	74.6	71.8
Already have	12.7	7.3	0.0	8.5	2.6
TRAINING IN PAST 3 YRS (mean hours)	43.1	18.5	40.3	20.5	20.2
ANNUAL TRAINING GOAL					
6 hours	40.7	67.3	31.0	63.9	53.7
12 hours	27.1	19.0	31.0	21.3	14.6
12+ hours	32.2	13.7	37.9	14.8	31.7
PERSONAL CAREER DEVELOPMENT PLAN (% yes)	N.A.**	55.1	N.A.	55.0	71.1

* Percentages are reported except where otherwise noted.
** N.A. = Question not asked of this provider group.

93%) with center teachers least likely to hold this status. Center directors hold the highest levels of education while group teachers and family providers have the lowest levels. The directors for both centers and group facilities have been in the field of early childhood education longer than the other provider groups (on an average of thirteen years for directors in comparison to approximately seven years for child care teachers and family providers). Center teachers have the least amount of time with their current employer when compared with their total number of years in the field. The vast majority (over 93%) of directors for

Table 3. Facility Characteristics.

Characteristic	Centers (N=60)	Group Homes (N=30)	Family Homes (N=29)
Licensed Capacity (mean)	76.23	13.8	6.6
Number of Classrooms (mean)	4.93	2.1	N.A.**
Number of Children Enrolled (mean)	68.73	15.9	7.2
Age of Children (% of facilities with):			
Birth to 12 months	55.0	66.7	52.3
13–24 months	71.7	80.0	68.2
25–36 months	83.3	90.0	72.7
3–5 years	96.7	96.7	88.6
6–8 years	63.3	60.0	43.2
9+ years	48.3	33.3	22.7
Special needs (% yes)	61.7	16.7	11.4
Number of Paid Staff (mean)	10.93	3.6	N.A.
Assistant Director (% yes)	37.3	35.0	N.A.
Turnover Rate	0.22	0.31	N.A.
Accreditation Status (% yes)	26.3	10.0	22.5

** N.A. = Question not asked of this provider group.

both centers and group facilities are full-time, while a majority of group teachers (59.4%) are part-time. Regarding compensation, group teachers are also the lowest paid (approximately $5.89/hour), while center directors, on the average, earn the highest salaries – just under $20,000 per year. Benefits are also not prevalent in the field, although center staff are more likely to have some benefits than are home-based providers. Health benefits are the most common, yet less than half (48.7%) of the center teachers report having this benefit.

Table 2 summarizes the responses to questions that are indicators of the extent to which providers are motivated to pursue additional as well as higher levels of education and training. Over one-third of each provider group indicate that they have no long-term educational goals. However, center directors are more likely to express a desire for higher education, with 57.1% indicating that a graduate degree is a long-term educational goal. As far as other child care training, a substantial percentage of providers do *not* have a Child Development Associate (CDA) certificate, but center directors (12.7%) are more likely to have the CDA than are other provider groups. Furthermore, directors of both centers and group facilities have, on the average, twice the number of training hours than do teacher and family provider groups. Over the past three years, directors averaged over 40 hours of training, while teachers and family providers averaged around 20 hours (just slightly higher than what is required to meet the state regulations of 6 hours per year). The emphasis on only meeting state requirements is further evidenced by the responses from providers when asked to indicate their annual training goal. A majority in each provider group, except group directors, indicates that completing 6 hours is their goal. Both directors of centers and group facilities, as well as family providers show greater interest in education/training beyond the minimum required. The final indicator of a provider's educational interest and motivation is revealed when asked, "Do you have a plan for your individual career development in early childhood care and education?" More than half of teachers and family providers indicate they have a personal career development plan. This appears to be a higher percentage than expected, given the responses to the other questions related to educational interest and motivation. However, this question did not ask if the plan was written and/or formalized; as such, the responses to this question may include individuals who at a minimum have *thought* about their plans for further training and education.

Characteristics of the sample sites are shown in Table 3. On average, centers have a licensed capacity for 76 children, just under five classrooms, and an enrollment of 69 children. While centers have fewer enrolled children than they are licensed for, both group and family homes have more (probably due to school-age children or children who might not be enrolled for full-time child care). As for the age of children served, children age two through five are most likely to be enrolled in child care. Special needs children are most likely served by centers, not group or family homes. Staffing patterns are also consistent with common knowledge – centers average just under eleven paid staff, while group homes average just fewer than four. Approximately one-third of both centers and group homes have an assistant director. The turnover rate, indicated by the ratio of new staff to total number employed, is slightly higher for group child care (0.31) than it is for centers (0.22). Centers are most likely to be accredited (26.3%) while group homes are least likely (10%).

Training Needs and Perceptions of Current Training System

Perceived Training Needs

The survey instrument distributed to child care staff asked both directors and teachers to identify the need for training in specified training topics.[6] They were also asked to base their assessment on the *need for training* for child care providers, not just the importance of the topic alone. Table 4 summarizes the responses of these provider groups: center directors, center teachers, group providers,[7] and family providers.

In examining Table 4, if we rank order the topics in terms of perceived priority, we see that the general topic area of **supervision, motivation, and discipline/guidance of children** is considered an area of very serious need for training. This topic is ranked at the top for all provider groups except family providers who rank it as the second most needed area of training. Family providers identify **fostering social development (e.g. dealing with conflict)** as the top priority for training. These two topics are closely related in that they both deal with the issue of behavior management – a serious concern for providers that is repeatedly expressed by them. A concern over behavior management is further supported by the data. All provider groups rank both topics as either first or second priority for training.

When all topics are listed in rank order (from topics that are a very serious need to topics that are not a priority), there is a high degree of consistency across all provider groups – for center directors and teachers as well as home-based providers. The four areas consistently ranked as priority training topics are:

- supervision, motivation, and discipline/guidance of children
- social development (dealing with conflict)
- child development
- developmentally appropriate practice

In addition, family providers identify **nutrition** and **infant/child development** as important areas of training. Regardless of relative importance and rank order position, providers view none of the training topics specified on the research instrument as *unimportant*. The average scale value for these topics ranged between 1.28 and 2.53–thus, there is no topic area that is viewed as *not a priority* for training.

Training Needs as Observed via the Environment Rating Scales

In addition to the identification of training needs through the self-reports of child care staff, we are able to provide a more objective measure via the Harms

Table 4. Perceived Need for Training in Selected Topic Areas.*

Training Topic	Center Directors (N=60)	Center Teachers (N=546)	Group Providers (N=100)	Family Providers (N=44)
Child care business, management	2.19 (16)	2.36 (17)	2.43 (19)	2.03 (17)
Child care program development	1.77 (8)	1.87 (9)	2.04 (12)	1.67 (9)
Child development	1.63 (4)	1.57 (3)	1.72 (2)	1.59 (5)
Child/staff health	2.05 (14)	1.77 (5)	1.89 (7)	1.80 (12)
Development appropriate practice	1.43 (3)	1.65 (4)	1.81 (4)	1.66 (8)
Emergent literacy, children's literature or literacy-based socio-dramatic play	1.84 (9)	2.04 (15)	2.05 (13)	1.92 (14)
Emergent numeracy, science for young children	1.74 (6)	1.92 (11)	2.12 (16)	2.00 (16)
Fostering social development (e.g. dealing with conflict)	1.39 (2)	1.56 (2)	1.76 (3)	1.44 (1)
Inclusive/special needs education issues	1.74 (7)	1.78 (6)	2.05 (14)	1.69 (10)
Infant/Toddler child development/programming	1.88 (10)	1.78 (7)	1.89 (6)	1.51 (3)
Multicultural, gender sensitivity in programming for young children	1.93 (11)	2.02 (14)	2.01 (10)	1.95 (15)
Music, dance, movement for young children	1.98 (13)	1.93 (12)	2.02 (11)	1.89 (13)
Nutrition	2.27 (17)	1.99 (13)	1.88 (5)	1.57 (4)
Personal care routines (naptime, toileting, grooming)	2.46 (19)	2.12 (16)	2.11 (15)	1.60 (6)
Play	1.97 (12)	1.91 (10)	1.96 (8)	1.71 (11)
Supervision, motivation discipline/guidance of children	1.28 (1)	1.51 (1)	1.55 (1)	1.46 (2)
Working with parents/community services	1.73 (5)	1.85 (8)	2.00 (9)	1.64 (7)
Statewide conference on multiple topics	2.48 (20)	2.53 (20)	2.48 (20)	2.22 (20)
Regional conference on multiple topics	2.41 (18)	2.53 (19)	2.36 (18)	2.18 (19)
Mentoring, multiple topics	2.18 (15)	2.45 (18)	2.30 (17)	2.08 (18)

* Perceived need is indicated by the mean score for the provider group on a scale of 1 = a very serious need, 2 = important but not critical, 3 = more would be helpful, and 4 = not a priority; in addition, a rank order of training needs for each provider group is indicated in parentheses.

and Clifford Environment Rating Scales. By identifying areas where child care sites are weak (e.g. where average scores are less than 5), we can specify needed training topics. Table 5 summarizes the average scores for the individual items included in each of the environment rating scales (FDCRS, ITERS, and ECERS).

In analyzing the set of individual items on the three different Harms and Clifford Environment Rating Scales, we see that there are a number of areas that receive a very low rating – below a scale value of 4.00. Items rated this low indicate areas where special attention should be placed in the design and delivery of training. Across all three scales – FDCRS, ITERS, and ECERS – these items are consistently rated low: **cultural awareness, personal grooming, dramatic (pretend) play,** and **sand and water play**. Furthermore, these areas are rated low in two out of the three environment rating scales: **displays for children** (FDCRS and ITERS), **space alone** (FDCRS and ECERS), **helping infants/toddlers understand language** (FDCRS and ITERS), **art** (ITERS and ECERS), and **blocks** (FDCRS and ITERS).

Overall, the ECERS reveals fewer areas of serious concern (only 16% of the items on this scale have a score below 4.00), while the ITERS reveals the most (46% of the ITERS' items have a score below 4.00). This is consistent with national data on the environment rating scales (Phillips, 1987; Scarr, 1994). Indeed, if we compare the overall average score for each scale, (FCDRS = 4.47; ITERS = 4.26; ECERS = 4.63), the ITERS has the lowest average score. This indicates a need for particular focus on infant/toddler training, a finding that is consistent with anecdotal evidence and comments.

On the other end of the continuum, there are a number of items on each of these scales that score above 5.00, indicating an assessment in the *good* range. Keeping in mind that there are not comparable items across all three scales,[8] we consistently see these areas rated highly: **nap/rest time, discipline/supervision, provision for parents, informal use of language with infants/toddlers,** and **health practice and/or policy**. Consistent with our analysis of the items rated poorly, the ECERS fares the best. It has the highest percentage of items (38%) receiving a score above 5.00 (ITERS only has 26% of the items scoring above 5.00, while FDCRS has 23%). There are several points of interest in our examination of these ratings. First, it is noteworthy that the health area received such a positive evaluation. No doubt, concerns about health and safety are of primary importance to parents as well as officials who regulate child care. Second, the high rating for discipline/supervision is paradoxical given the consistent identification of this area by caregivers as one in which they most need training. This illustrates that caregivers are performing better in this area than they perceive; it also reveals that discipline/supervision

Table 5. Average Score on Individual Environment Rating Scale Items.

Scale Item	FDCRS (N = 67)*	ITERS (N = 36)	ECERS (N = 57)*
Furnishings and Display	**4.10**	**4.09**	**4.57**
Furnishings for routine care	4.87	4.53	5.75
Use of furnishings for learning activities	–	4.44	4.14
Furnishings for relaxation and comfort	4.71	3.69	4.13
Room arrangement	–	3.86	4.74
Child-related display	3.01	3.92	4.08
Indoor space arrangement	4.11	–	–
Active physical play	4.55	–	–
Space to be alone			
a. Infants/toddlers	3.43	–	–
b. 2 years and older	3.76	–	–
Personal/Basic Care	**4.84**	**4.66**	**4.59**
Arriving/departing	6.15	5.56	4.71
Meals/snacks	4.72	3.93	4.40
Nap/rest	5.07	5.10	5.64
Diapering/toileting	4.13	3.62	5.05
Personal grooming	3.78	3.71	3.23
Health practice	5.17	4.21	–
Health policy	–	5.63	–
Safety practice	4.86	5.40	–
Safety policy	–	5.46	–
Language and Reasoning	**4.54**	**4.37**	**4.82**
Informal use of language			
a. Infants/toddlers	5.01	5.00	–
b. 2 years and older	4.90	–	4.89
Helping children understand language			
a. infants/toddlers (books & pictures)	3.47	3.74	–
b. 2 years and older	4.29	–	5.02
Helping children use language	4.45	–	4.99
Helping children reason	4.35	–	4.36
Fine/Gross Motor	**N.A.**	**N.A.**	**5.11**
Fine motor			5.41
Supervision (FM)			5.10
GM space			5.02
GM equipment			4.66
GM time			5.21
Supervision (GM)			5.44

Table 5. Continued.

Scale Item	FDCRS (N=67)*	ITERS (N=36)	ECERS (N=57)*
Creative/Learning Activities	**4.12**	**3.39**	**4.46**
Eye-hand coordination	4.48	4.67	–
Active physical play	–	3.53	–
Art	4.08	3.81	3.81
Music and movement	4.76	4.19	5.20
Sand and water play	2.60	3.07	3.75
Dramatic (pretend) play	3.74	3.07	3.62
Blocks	3.88	3.21	4.44
Use of T.V.	4.19	–	–
Schedule of daily activities	4.59	–	4.93
Supervision of play indoors and outdoors	4.79	–	5.51
Cultural awareness	–	1.75	–
Social Development	**4.72**	**N.A.**	**4.20**
Tone	5.73		5.34
Discipline	5.59		–
Cultural Awareness	2.85		2.96
Space (alone)	–		3.60
Free play	–		4.53
Group time	–		4.33
Exceptional provisions	–		4.69
Interaction	**N.A.**	**4.98**	**N.A.**
Peer interaction		4.93	
Adult-child interaction		4.99	
Discipline		5.01	
Program Structure	**N.A.**	**4.53**	**N.A.**
Schedule of daily activities		3.75	
Supervision of daily activities		4.71	
Staff cooperation		4.98	
Provisions for exceptional children		5.30	
Adult Needs	**5.17**	**4.28**	**4.80**
Adult personal needs	–	3.31	4.11
Opportunities for professional growth	4.79	3.57	4.50
Adult meeting area	–	4.94	5.10
Provisions for parents	–	5.35	5.51
Relationships with parents	5.37	–	–
Balancing personal and caregiving responsibilities	5.28	–	–

* This is the weighted N since there were observations made in more than one type of child care (i.e. family, group, or center).
** N.A. = Question not applicable for this environment rating scale. Spaces where there are no applicable scores are indicated by "–".

is perhaps one of the most challenging areas in child care and something for which caregivers think they need constant help and support.

Selection of Training

Providers were asked to indicate the importance of a number of factors in their selection of training.[9] In Table 6 we see, again, there is a high degree of consistency across all provider groups. Providers indicate that their selection is based primarily on their **interest in a topic** and if a topic **helps in understanding children**. Furthermore, center staff (directors and teachers) identify **opportunities for professional development** as important. All provider groups, except center directors, rank **training that offers practical solutions** within the top five on their list of factors that are important in the

Table 6. Factors Affecting the Selection of Training.*

Selection Factors	Center Directors (N = 60)	Center Teachers (N = 546)	Group Providers (N = 100)	Family Providers (N = 44)
Location/convenience	1.19 (5)	1.27 (5)	1.29 (3)	1.08 (3)
Session length	1.51 (10)	1.70 (12)	1.71 (11)	1.54 (10)
Meet state requirements	1.24 (7)	1.40 (8)	1.33 (5)	1.34 (6)
Quality of previous training	1.41 (8)	1.39 (7)	1.49 (8)	1.49 (8)
Cost of training	1.53 (11)	1.66 (11)	1.50 (9)	1.58 (11)
Scheduled times of training	1.13 (4)	1.31 (6)	1.36 (6)	1.03 (1)
Interest in topic/contents	1.10 (1)	1.15 (2)	1.15 (1)	1.21 (4)
Networking opportunities	1.75 (12)	1.81 (13)	1.82 (13)	1.66 (12)
Training organization	1.76 (13)	1.65 (10)	1.72 (12)	1.69 (13)
The trainer	1.48 (9)	1.51 (9)	1.66 (10)	1.54 (9)
Offers practical solutions	1.19 (6)	1.25 (4)	1.32 (4)	1.24 (5)
Helps understand children	1.12 (3)	1.09 (1)	1.16 (2)	1.08 (2)
Professional development	1.10 (2)	1.22 (3)	1.37 (7)	1.35 (7)
Sent by director	N.A.**	1.91 (14)	1.87 (14)***	N.A.**

* Importance of factors in the selection of training is indicated by the mean score for the provider group on a scale of 1 = a very important, 2 = somewhat important, and 3 = not important. In addition, the rank order of the factors in terms of importance is indicated in parentheses.
** N.A. = Not asked of this provider group.
*** This represents the response from the group teachers only.

selection of training. Center directors mention **scheduled times for training** as important. Similarly, home-based providers mention **scheduled times for training** or **location/convenience** as important factors affecting their selection of training. These priority rankings are congruent with the role responsibilities of center directors and teachers and home-based providers. Directors are responsible for the scheduling of staff at their child care facility, while home-based providers must participate in training that is offered during nonbusiness hours – hence the importance of when training is scheduled. On the other hand, teachers deal with the day-to-day child care activities for which they want practical guidance.

However, all of the factors that might affect selection of training are considered at least *somewhat important* by child care providers. (Note that none of the factors have a mean score above 2.0.) But, in terms of priority, the factors having the least priority across all provider groups are **networking opportunities, training organization, session length, cost of training,** and **trainer**. The relative unimportance of the cost of training is to be expected. The Pennsylvania child care training system provides training opportunities at no cost, or for a minimal registration fee, therefore cost is not a critical issue. As for the trainer and training organization, it may be that providers are satisfied with current training organizations and trainers (as expressed elsewhere in these data and also in the participant evaluation forms completed for each training session). These data indicate the trainer or training organization may not be as important as other factors in the selection of training.

What is of interest is the relative unimportance of **networking opportunities**. Anecdotally, we often hear that the opportunity to meet and talk with other child care providers is highly valued. On closer inspection, we see that family providers (the provider group that is most isolated from peers), are more likely to consider networking opportunities as important than are the other provider groups. Half of the family providers indicate that networking is a very important factor in their selection of training, while only around one-third of the other provider groups indicate this.

Center teachers also were asked to indicate the importance of **being sent by the director** in selection of training. In comparison to other factors, **being sent by the director** is relatively unimportant – it is ranked at the bottom. Regardless, approximately one-third of the teachers in centers indicate that **being sent by the director** is a *very important* factor. Ideally, directors of child care centers should be working with staff to establish professional development plans that meet the individual needs of workers. However, this question, as asked, does not identify the reason why a director sends staff to a particular training – i.e. whether the selected training corresponds with professional development needs of staff or whether training is offered at a convenient time and place.

When directors were asked about how decisions are made regarding staff training, just under half of center directors (46.6%) indicated that they "guide the selection but the staff make the final decision." Whereas, in the group child care situation, 60% of group directors indicated this.

Having a personal plan for career development is related to this decision-making process and the selection of child care training. Whether or not staff have such plans was assessed by asking directors "What percentage of your child care staff have personal plans for career development in early child care and education?" Center directors, on average, indicate that over half (51.9%) of staff have personal plans. In group child care, directors report that only 24.2% of staff have personal career development plans. A much higher percentage (71.1%) of family providers have a plan for personal development as a child care provider. This question does not ask for specific details, therefore the interpretation of what constitutes a *plan* probably varies considerably.

Evaluation of Training

The appropriateness, usefulness, applicability, and effectiveness of training in achieving learning objectives, as perceived by providers, were used as one means to evaluate the training system. Providers were asked for an *overall* assessment of training in which they participated, knowing that many have participated in a number of training opportunities over the past few years (see Fig. 1).

A majority of all provider groups consider the training to be either very appropriate or somewhat appropriate. In comparison, group providers are more likely than the others to consider the training appropriate (94.6%), while family providers are least likely (89.2%).

Providers also positively assess the usefulness of training. More than four-fifths of each provider group consider the training somewhat or very helpful. Comparatively, home-based providers are most likely to consider the training useful (group = 89.1% and family = 89.2%), while center directors are least likely (87.5%).

The applicability of training (or the knowledge and skills learned) to the work environment should be an important feature of any training system if it is to have an impact. It is impressive that a substantial majority of providers (over 90%) indicate that they could apply *all, a lot,* or *some* of what they learned in the training to current work.

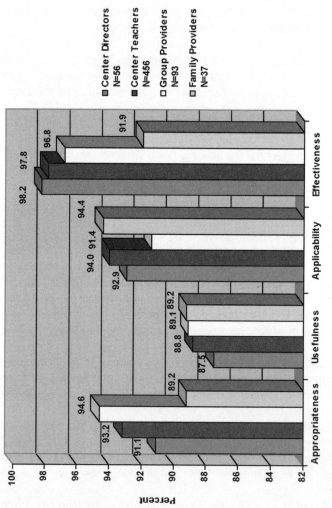

Fig. 1. Perceived Appropriateness, Usefulness, Applicability, and Effectiveness of Training.*

* For each factor, these data represent the percentage of providers who indicate that:
  - The training was "very appropriate" or "somewhat appropriate" (*Appropriateness*)
  - The training was "very helpful" or "somewhat helpful" in their work (*Usefulness*)
  - They can apply "all," "a lot," or "some" of what the learned in training (*Applicability*)
  - The training goals were "achieved" or "somewhat achieved" (*Effectiveness*)

To assess the perceived effectiveness of training, providers were asked to indicate the extent to which the training goal(s) were achieved, that is, the extent to which they learned the material.[10] As with applicability of training, almost all providers (over 90%) respond that they learned at least some of the material. A slightly smaller percentage of family providers indicate this (91.9% vs. over 96% for the other provider groups).

Overall, the training system is viewed positively by provider groups, as evidenced by response to questions about appropriateness, usefulness, applicability, and effectiveness in achieving learning objectives. The providers consider the training appropriate for their level of knowledge and skill, find it helpful in their current work, indicate they are able to apply what they have learned, and feel training goals have been achieved.

This positive assessment also corresponds with their response when asked about level of interest in training and if more training would be (see Fig. 2).[11] As with the other evaluative factors, the level of interest is high among teachers, with over 80% of center and group teachers indicating they are either *interested* or *very interested* in taking training. Furthermore, directors are on target in assessing levels of interest of their staff. As further evidence of the positive evaluation of the training by providers, a substantial percentage (86–100%) indicate that attending more workshops or training will help them in their work.

A final evaluative measure used in assessing the current training system asked about the perceived helpfulness of the various training methods used in the Pennsylvania Child Care/Early Childhood Development Training System. Only directors and family providers were asked about this.[12] Table 7 summarizes the responses for center directors, group directors, and family providers. **On-site training** ranks as the most helpful method by the directors of centers and group homes, while family providers rank it as second most helpful. Center directors and family providers also positively assess **workshops**. While **satellite** and **video** methods of training may be cost effective and efficient in reaching providers in more rural areas, both these methods of training are viewed as less helpful than other methods. Interestingly, both family providers and group directors express a more positive view of these two methods than do center directors.

Barriers to Training

Several factors may limit child care providers from attending training. Providers were asked to indicate the importance of a number of factors that might prevent them from attending training or workshops (see Fig. 3).

146   JOYCE IUTCOVICH ET AL.

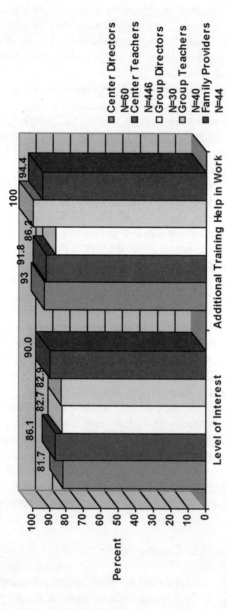

Fig. 2. Level of Interest and Perceived Helpfulness of Additional Training.*

* For *level of interest*, numbers represent:
 • The percentage of *family providers, group teachers,* and *center teachers* who indicate they are "interested" or "very interested" in taking workshops or training.
 • The percentage of *center directors* and *group directors* who indicate their staff are "interested" or "very interested" in taking workshops or training.

For *helpfulness of additional training* in one's work numbers represent:
 • The percentage of providers who indicate that attending additional workshops of training will "somewhat" or "very much" help them in their work.

Table 7. Perceived Helpfulness of Training Methods.*

Training Method	Center Directors (N = 60)	Group Directors (N = 30)	Family Providers (N = 44)
Workshop	1.26 (2)	1.52 (2)	1.16 (1)
Satellite	2.33 (6)	2.07 (6)	1.84 (6)
Video	1.92 (5)	1.56 (5)	1.60 (4)
On-site Training	1.21 (1)	1.39 (1)	1.42 (2)
Conference	1.57 (4)	1.48 (3)	1.55 (3)
Mentoring	1.38 (3)	1.47 (4)	1.78 (5)

* Perceived helpfulness is indicated by the mean score for those who have experienced a method of training, on a scale of 1 = very helpful, 2 = somewhat helpful, and 3 = not helpful. In addition, a rank order of the methods is indicated in parentheses.

Lack of child care for their own children while attending training is considered important as a barrier only by family providers. This is another expected finding since family providers are most likely to have to attend training outside of work hours, necessitating the need to find care for their own children while attending training.

Contrary to what we might expect given the current lack of status and minimal reward system for child care providers, having **no long term gains or rewards** for training is not considered a very important barrier by provider groups. However, center directors, in comparison to other provider groups, were more likely to perceive this as an important barrier.

Having **no one to watch the children during the child care hours** is seen as the most significant barrier to training by all provider groups. Center directors and family child care providers, however, are more likely to indicate this as a very important factor than are teachers and group providers. This is to be expected, since directors and family providers are responsible for finding substitutes in their child care settings.

In identifying other barriers to training, we can also examine the reward system attached to training. Providers were asked, "Do you receive any compensation for attending relevant training?" Figure 4 shows the types of compensation received by the center and group child care providers. Few providers receive any type of compensation, i.e. being paid while in training, receiving compensatory time, or being reimbursed for expenses. Center directors appear to fare better than other provider groups – 55.9% indicate that they are paid while in training. This can be interpreted that they are more likely to attend relevant training during the work hours.

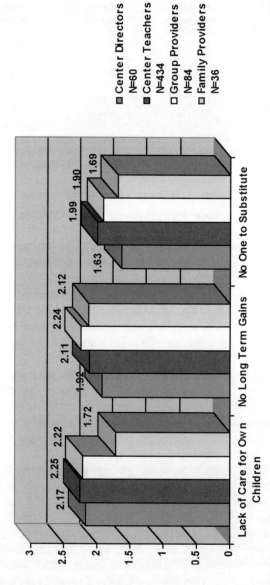

Fig. 3. Barriers to Training.*

* Importance of barrier is indicated by the mean score of the provider group on a scale of 1-very important, 2 = somewhat important, and 3 = not important.

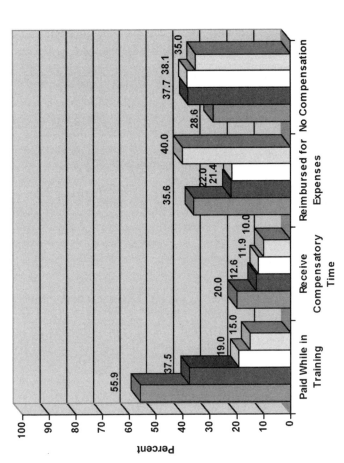

Fig. 4. Compensation for Attending Relevant Training.*

* Percent indicating "yes" for each of these types of compensation that they receive for relevant training.

The Work Environment of Child Care Facilities

The quality of work life is not only an indicator of one type of quality within a child care environment, but it is an important factor that can influence the overall quality of care for young children, as well. As stated previously, we have used Jorde-Bloom's Early Childhood Work Environment Survey (ECWES) to assess a number of dimensions of the work environment within child care centers. The ECWES[13] includes measurements of:

- Ten dimensions of *organizational climate* (collegiality, professional growth, supervisor support, clarity, reward system, decision making, goal consensus, task orientation, physical setting, and innovativeness);
- The importance that staff assign to each dimension (*summary of worker values*);
- The staff's *overall commitment* to the center;
- How the current *work environment resembles the staff's ideal*;
- The importance of various *educational goals and objectives*;
- The *degree of influence of the teaching staff* regarding various organizational dimensions.

Organizational Climate

The ten dimensions of organizational climate are shown in Fig. 5. In analyzing the scale values, which can range between 0 and 10, we see that the dimension of **professional growth** ranks at the bottom (3.94), followed by **reward system** (5.88) and **clarity** (5.91). This indicates that overall, staff in centers do not perceive many opportunities for professional growth, they do not feel that pay and fringe benefits are fair and equitably distributed, and they feel that communication about policies and procedures is unclear. These results are similar to national data where professional growth opportunities and reward systems are evaluated poorly by most child care staff (Jorde-Bloom, 1996).

It is important to determine what factors, if any, are associated with these ten dimensions of organizational climate. Table 8 provides a summary from an analysis of relationships between each of the organizational climate dimensions and a series of factors. A number of **director characteristics** are examined first. In addition, **characteristics of teachers** (aggregated per site) and overall **site characteristics** are analyzed.

Overall, the average age of teachers is significantly related to all dimensions of organizational climate. Centers with older workers have a more positive work environment. Correspondingly, two other factors that are closely related

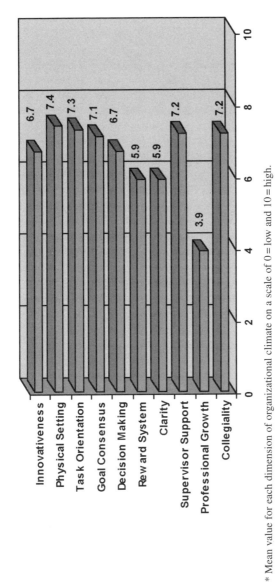

Fig. 5. Organizational Climate at Child Care Centers.*

* Mean value for each dimension of organizational climate on a scale of 0 = low and 10 = high.

Table 8. Factors Associated with Organizational Climate.

Factor	Collegiality	Professional Growth	Supervisor Support	Clarity	Reward System	Decision Making	Goal Consensus	Task Orientation	Physical Setting	Innovativeness
Director Background										
Age	-0.005	0.27*	-0.09	0.02	0.10	-0.14	0.05	0.03	0.11	0.05
Educational level	0.01	0.17	0.15	0.08	0.14	0.03	0.02	0.17	0.14	0.14
Years in field	0.08	0.44***	0.20	0.18	0.30*	0.24	0.19	0.26*	0.30*	0.26*
Years in present job	0.26*	0.40**	0.23	0.17	0.40**	0.19	0.12	0.26*	0.30*	0.14
Full-time/part-time	-0.23	-0.08	-0.06	-0.18	-0.32*	0.03	-0.13	-0.11	-0.03	-0.16
Salary	-0.14	0.34**	-0.07	0.13	0.15	0.04	-0.12	0.03	0.03	-0.00
Long term educational goal	0.06	0.20	0.07	-0.02	0.01	-0.12	-0.03	0.00	-0.06	0.03
CDA Status	-0.13	0.02	-0.05	0.15	-0.12	0.02	0.04	-0.15	-0.03	0.13
Training hours/year	-0.01	0.18	-0.05	-0.08	0.08	-0.01	-0.02	0.12	0.07	0.06
Annual training goal	-0.20	0.18	-0.14	-0.05	-0.06	-0.14	-0.07	-0.11	-0.17	0.02
Aggregate Teacher Characteristics										
Average age of teachers/site	0.39**	0.45***	0.29**	0.28*	0.43***	0.26*	0.34**	0.38**	0.32**	0.29*
Average educational level	-0.03	0.25	0.04	0.16	-0.07	0.01	-0.03	0.09	-0.08	0.00
Average years in field	0.24	0.36**	0.18	0.30*	0.27*	0.12	0.17	0.18	0.04	0.10
Average years in present job	0.36**	0.38**	0.28**	0.26*	0.28*	0.25	0.26*	0.29*	0.17	0.20
% of teachers working full-time	0.04	0.10	-0.13	-0.04	0.00	-0.07	0.15	-0.11	-0.05	0.09
Average teacher salary	0.15	0.53***	0.13	0.31*	0.27*	0.15	0.23	0.29*	0.18	0.24
Average long-term educational goal	-0.10	-0.06	-0.07	-0.08	-0.13	-0.13	-0.02	-0.14	0.05	-0.09
% teachers with annual goal 12+ hrs	0.01	0.43***	-0.01	-0.02	-0.01	-0.06	0.06	0.05	0.08	0.14
Average training hours/year	-0.09	0.11	0.23	-0.22	0.09	-0.18	-0.13	-0.10	-0.01	-0.20
Average teacher interest in training	-0.00	0.00	-0.14	-0.06	-0.03	-0.07	-0.09	-0.11	-0.08	-0.06
Site Characteristics										
Licensed capacity	-0.08	0.18	0.07	-0.11	0.08	0.02	-0.11	0.01	0.22	0.11
Number of classrooms	-0.09	0.33**	-0.05	0.05	0.02	0.11	0.00	0.06	0.11	0.15
Turnover	-0.31*	-0.33**	-0.40**	-0.26	-0.20	-0.30*	-0.39**	-0.24	0.28*	-0.26
Accreditation status	-0.18	-0.33**	-0.16	-0.24	-0.36**	-0.13	-0.28*	-0.25	-0.29*	-0.20
Average training hours/year	-0.07	0.17	-0.16	-0.18	0.14	-0.13	-0.12	-0.06	0.03	-0.12

* $p \leq 0.05$; ** $p \leq 0.01$; *** $p \leq 0.001$.

Professional Development and The Quality of Child Care   153

to each other (i.e. average number of years in current position and site turnover) are also significantly related to a number of the dimensions of organizational climate. In addition, a number of these organizational climate dimensions are more positive in centers that have older, more experienced directors. Hence, an older and more stable workforce is closely associated with a positive organizational climate. The causal link between these factors cannot be determined from this analysis, but it is plausible that there is a reciprocal effect – a positive organizational climate results in a more stable workforce and vice-versa.

Summary of Worker Values
The previous analysis of organizational climate gives us a picture of how child care centers fare on each of these dimensions. We see that opportunities for professional growth are particularly poor while at the other end of the continuum, the physical setting is viewed very positively by workers. These perceptions, however, are tempered by the degree to which child care workers value these aspects of their work environment.

Figure 6 gives us an indication as to the overall value placed on each of the 10 dimensions of organizational climate. Center staff identified the three most important aspects of their work from the list of organizational climate dimensions. The most highly valued aspect is **collegiality and co-worker relations** – over 60% of child care center staff identify this as one of the three most important aspects of their work. **Reward system** – fairness in pay and benefits-is second most important (48.2%) and **supervisor support** is third (40.9%) most important. The dimension of **opportunities for professional growth** comes in fourth with 30.5% of the caregivers identifying it as important. Therefore, even though staff do not perceive many opportunities for professional growth, this aspect is not as highly valued as other areas. Those areas least valued are **goal consensus** (13.5%), **clarity** (15.2%), and **innovativeness** (20.5%).

How these organizational climate dimensions are rated compared to the value placed on them gives us an indication where to focus improvement efforts. **Improvement in the reward system** will probably accrue the most lasting results since it is very poorly rated, yet highly valued. **Improving opportunities for professional growth** is also an area in which to focus attention since it is the most poorly rated area and ranked fourth in importance. On the other hand, **collegiality** is very important to workers, but given its positive assessment as a dimension of organizational climate, there is no need to improve it.

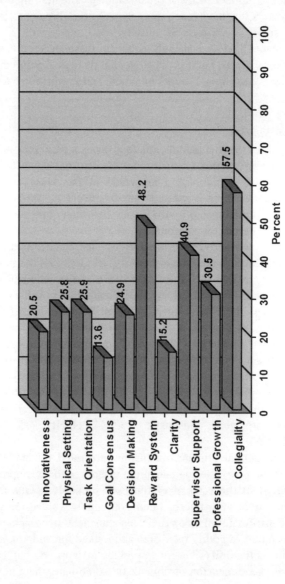

Fig. 6. Summary of Worker Values.*

* Percent of center staff (averaged across sites) that identify each dimension of organizational climate as important.

Professional Development and The Quality of Child Care 155

Summary of How Current Work Environment Resembles Ideal
As a way of understanding the discrepancy between ideal and real work conditions, child care workers were asked, "If you could design the ideal job, how close would your present position resemble this ideal position with respect to the following?" Responses range between *not at all like my ideal* to *is my ideal*. Based on this assessment, Figure 7 illustrates that the greatest discrepancy is in the **reward system**. There is a wide gap between what child care workers are paid versus what they think they should be paid. Given their current low salaries, this is an accurate appraisal on their part. The autonomy of staff to **make decisions** or express opinions on important issues is another area where child care staff feel that work environments least resemble the ideal.

The smallest gap between the ideal environment and the real one experienced by child care workers is in the areas of **collegiality** and **supervisor support**. As far as **opportunities for professional growth**, the discrepancy between *ideal* and *real* falls mid-range on the continuum.

Importance of Educational Goals and Objectives
Early childhood programs can have a number of educational goals and objectives – but the priority given to each can vary across programs. Figure 8 shows how these educational goals and objectives are ranked in Pennsylvania child care centers. Consistent with developmentally appropriate practice in the early childhood field, the greatest emphasis is on **helping children to develop positive self concepts and self esteem** while the least emphasis is placed on helping children develop concepts needed for reading and math.

Degree of Influence of Teaching Staff Regarding Organizational Decisions
Perceptions of workers regarding the degree of influence of teaching staff with respect to various organizational decisions provides a fuller understanding of the decision making dimension of organizational climate. Staff were asked how much influence they have (very little to considerable influence) in ordering materials and supplies, interviewing and hiring staff, determining program objectives, training new aides or teachers, and planning daily activities. Figure 9 depicts the difference between what directors perceive is the degree of influence versus what teachers perceive is the degree of influence. Not unexpectedly, teachers do not perceive that they have as much influence as what directors say they do. This discrepancy also points to an area where improvement efforts can be focused.

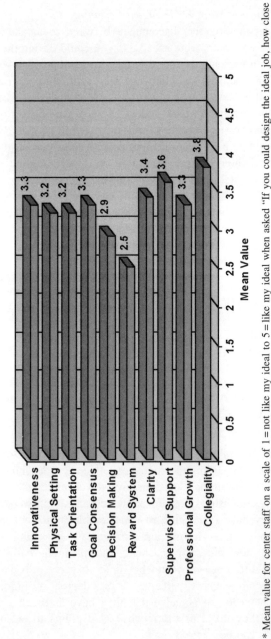

Fig. 7. Summary of How Current Work Environment Resembles Ideal.*

* Mean value for center staff on a scale of 1 = not like my ideal to 5 = like my ideal when asked "If you could design the ideal job, how close would your present position resemble your ideal work environment with respect to the following?"

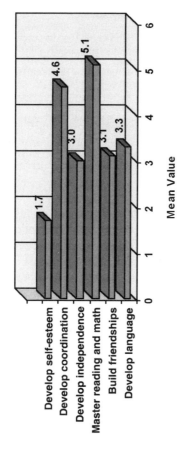

Fig. 8. Importance of Organizational Goals.*

* Mean value of importance on a scale of 1 = most important and 6 = least important.

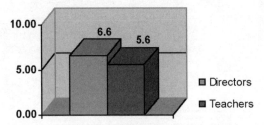

* Mean value on a scale of 0 = very little influence to 10 = considerable influence when asked, "How much influence does teaching staff currently have in each of the following areas below:

a) ordering materials/supplies
b) interviewing/hiring new staff
c) determining program objectives
d) training new aides/teachers
e) planning daily schedule of activities."

Fig. 9. Degree of Influence of the Teaching Staff Regarding Various Organizational Decisions.*

Overall Commitment

All of the characteristics that have been discussed provide an understanding of specific areas where attention can be paid in intervention efforts to improve child care work environments. The commitment scale provides a summary of overall commitment of child care staff to their centers. Individuals who feel deeply committed tend to put extra effort into their work and take pride in their centers. In such environments, turnover is generally lower. Commitment among Pennsylvania child care staff is relatively high. Figure 10 indicates that

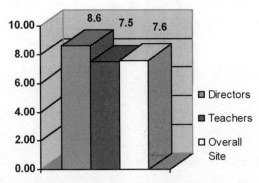

* Mean value on a scale of 1 = not committed to 10 = highly committed.

Fig. 10. Overall Commitment.*

directors have a slightly higher level of commitment than teachers – 8.6 for directors compared to 7.5 for teaching staff. Together this gives us an overall value of 7.6 for child care centers.

Quality of Care

Quality of Care as Assessed through the Environment Rating Scales
A previous section discussed results from observations of child care sites and the scale values for each individual item on the FDCRS, ITERS, and ECERS. That discussion focused on these data as a means to assess training needs. Our focus now shifts to the question about overall quality of care and its changes over the years, based on earlier studies conducted in Pennsylvania in 1984 (Kontos & Fiene, 1987) and in 1989 (Fiene & Melnick, 1991).

Overall, average ECERS scores have improved through the years, although the changes are not statistically significant, increasing from 3.78 in 1984 to 4.27 in 1989 to 4.63 in 1996. An analysis of FDCRS scores shows a marked (statistically significant, $p<0.05$) improvement from 1989, increasing from 3.80 in 1989 to 4.47 in 1996. Several observations can be made in comparing 1996 with 1989 and 1984 data sets. While program quality scores on the ECERS and FDCRS have improved over the 12-year time frame, the bad news is that quality scores, on average, are still at the mediocre level. The ITERS score in 1996 is even worse (4.23) and is a major concern. There are no comparable data for the ITERS from the 1989 or 1984 research studies. Overall, Pennsylvania child care has improved, but it is still not in the good or excellent range. National and international data from research studies are very similar with ranges from 3.70 for family child care homes with little training to 5.22 for child care centers that are accredited (Phillips, 1987).

What are some reasons for the improvements? Two major interventions occurred during this 12-year time period. Both occurred at approximately the same time so it is difficult to determine the contribution of each to the overall improvement in quality. In 1992 new child care regulations were promulgated and the new comprehensive PA CC/ECD Training System was implemented. New regulations were an improvement over existing regulations, but the regulations deal primarily with basic health and safety issues. Although this will contribute to overall quality, it will not be a major contributor (Fiene & Melnick, 1991). What has been and continues to be a major contributor is the training system that has been implemented. When data are compared from the ECERS and FDCRS, family child care homes improved significantly more than child care centers. The home-based training system has been in place for twice as long as the center-based system. This is a very encouraging result.

Factors Associated with the Quality of Child Care

The previous analysis presents an overall picture of the quality of child care in Pennsylvania and the progress made in improving quality. Analysis presented in this section examines the current data to determine what factors are significantly related to the quality of care. Part of this analysis will be based on a data set that has matched the environment rating scales with the child care providers that were observed. This data set establishes the most direct link between an indicator of quality and the set of factors that might be associated with it (e.g. background characteristics of the caregiver, level of training of the caregiver, and caregiver's assessment of organizational climate). Other parts of this analysis will be based on a site level data set where aggregate values for most of the variables have been created to represent the site, overall. Where necessary, data have been weighted to adjust for the different probabilities of sample selection (i.e. the FDCRS included both family and group homes while the ECERS included both centers and group homes, necessitating that these analyses be based on weighted data). This analysis, which will include both bivariate correlation and multivariate regression analyses, will be presented separately for each of the three environment rating scales.

Table 9 provides a summary of results of a series of bivariate correlations between the measure of quality (i.e. either the FDCRS, ITERS, or ECERS average score) and a set of factors hypothesized to be related to quality (e.g. caregiver background characteristics, training experience, and assessment of organizational climate; for family providers, in lieu of organizational climate, an indicator of their *connectedness* to a child care network is used).

Bivariate analysis of the FDCRS finds four factors that are significantly correlated with quality of family child care. Family caregivers that are younger and have higher long-term educational goals are more likely to provide a higher quality of care. The other factors are measures of a family provider's assessment of the current training system. Providers who evaluate the current system of training as **inappropriate to their skill level** and **not useful for their work** as a family caregiver are more likely to provide a higher quality of care. This is not as unexpected as it sounds. It is likely that those providers who are already providing quality care do not find as much benefit from the current training system that focuses most of its attention on entry level skills. In an attempt to further analyze this unusual finding, we examined the relationship between hours of training and evaluation of training by providers. We see that providers who have more hours of training are also more likely to rate the current system positively in terms of **goal achievement** ($B = -0.24$, $p \leq 0.07$), **appropriate skill level** ($B = -0.28$, $p \leq 0.03$), and **usefulness** ($B = -0.45$, $p \leq 0.000$).[14]

Table 9. Factors Associated with Quality of Care.

Factor	FDCRS (N = 67)†	ITERS (N = 36)	ECERS (N = 57)†
Caregiver Background			
Age	–0.30**	0.13	0.05
Educational level	0.003	0.14	0.11
Years in field	–0.20	0.25	0.04
Salary	0.19	0.43**	0.36**
Long term educational goal	0.27*	0.07	–0.09
Training Characteristics			
Annual educational goal	0.12	0.19	0.14
Training hours per year	–0.03	0.02	–0.09
CDA status	0.05	0.005	0.03
Training helpful in work	0.08	–0.16	–0.03
Evaluation of training system			
a. appropriateness	0.29**	–0.01	0.03
b. goal achievement	0.11	0.07	–0.10
c. usefulness	0.28**	0.08	–0.05
d. applicability	–0.02	–0.16	–0.08
Organizational Climate	N.A.††		
Overall commitment		0.01	0.14
Collegiality		–0.31	–0.005
Professional growth		0.15	0.41**
Supervisor support		–0.29	0.22
Clarity		0.06	0.32*
Reward system		–0.27	0.29*
Decision making		–0.24	0.17
Goal consensus		–0.18	0.32*
Task orientation		0.05	0.30*
Physical setting		0.20	0.21
Innovativeness		–0.07	0.23
Connectedness		N.A.	N.A.
(Family Child Care Only)	–0.10		

† This is the weighted N since there were observations made in more than one type of child care (i.e. family, group, or center).
†† N.A. = Not applicable
* $p \leq 0.05$;   ** $p \leq 0.01$.

Our bivariate analysis of the ITERS finds only **salary level** of the caregiver to be significantly related to quality of infant/toddler care. Caregivers with higher salaries provide higher quality infant/toddler care. The bivariate analysis of the ECERS reveals a number of factors that are significantly correlated with quality of child care: salary and organizational climate factors of **professional growth, clarity, reward system, goal consensus, and task orientation**. Thus, the caregivers that provide high quality early childhood care are more likely to:

- have higher salaries;
- indicate that their center has opportunities for professional growth;
- feel that communication at their center is good and that work schedules, job descriptions, and rules are clear and well-defined;
- indicate that the pay and fringe benefits are fair and equitably distributed in their center;
- indicate that staff at their center agree on school philosophy, are united in their approach, and are committed to program goals and objectives;
- believe that they work hard but still have time to relax, that program procedures are efficient, and that meetings are productive.

In an analysis of the site level data set, we created an overall *quality of care* variable as an indicator of child care quality.[15] In a multivariate analysis of these data, we then determined what site level factors significantly contribute to the variance in *quality of care* at the site level. Initially, we did not include any of the work environment variables (Bloom items) since that would result in excluding all home-based providers from the analysis. Our analysis reveals that **size** and **turnover** are significant factors and explain 19% of the variance in quality of care. Results from the regression analysis are: **class number** ($B = 0.11$, $p \leq 0.0053$) and **turnover** ($B = -1.08$, $p \leq 0.0216$). Thus, sites with more classrooms and lower turnover have a higher quality of care.

When we add the Bloom items on organizational climate (thereby eliminating all home-based providers from the analysis), we find that only **opportunity for professional growth** independently contributes to the variation of quality of care at child care sites (the majority of which are centers). The results from the regression analysis for **professional growth** are $B = 0.45$, $p \leq 0.0000$. Forty percent of the variation is explained by this factor. As we hypothesize, child care facilities that have more opportunities for professional growth have a higher overall quality of care.

CONCLUSIONS AND POLICY IMPLICATIONS

What can we conclude as a result of this research and what are the implications for public policy? There are numerous issues addressed and volumes of data analyzed. First, we can examine the overall conclusions with regard to the current training system – how it is evaluated and what the training needs are.

Overall, the training system is viewed positively by the provider groups, as evidenced by their response to questions about appropriateness, usefulness, applicability, and effectiveness in achieving learning objectives. Furthermore, providers express a high level of interest in the training and, for the most part, feel that additional training will help them in their work. When directors of centers and home-based providers are asked about the particular methods of training that they perceive to be most helpful to them and their staff, training methods that provide direct contact with a trainer (e.g. on-site training and workshops) are viewed as most helpful. Methods where the contact is indirect (e.g. video and satellite training) are viewed as least helpful. This is understandable, although the policy implications of this are not to abandon some of the more indirect, yet very cost effective methods of training like the learn-at-home videos and satellite training. A balance of methods is important in a system as massive as this one, where the ability to access training varies tremendously across providers and where resources are limited.

Regarding training needs, there is a high degree of consistency across provider groups in terms of areas they perceive as most critical. They identify supervision/discipline of children, social development (dealing with conflict), child development, and developmentally appropriate practice as areas with the highest priority. Although, providers do not identify any topic area as *not a priority* for training.

On the other hand, if we use the environment rating scales as an indicator where there are weaknesses in child care settings (hence, an area in need of training), we see a slightly different picture. Given the overall low score for the infant/toddler area (ITERS), any training in this area can be viewed as a priority. In addition, these items are consistently ranked low on all three environment rating scales: cultural awareness, personal grooming, dramatic (pretend) play, and sand and water play; furthermore, these areas are rated low in two out of the three environment rating scales: displays for children (FDCRS and ITERS), space alone (FDCRS and ECERS), helping infants/toddlers understand language (FDCRS and ITERS), art (ITERS and ECERS), and blocks (FDCRS and ITERS). Indeed, a number of these items from the environment rating scales fall under the broader categories of social development, child development, and developmentally appropriate practice. The information from the environment rating scales offers more specific areas of need.

One particularly interesting finding is the contradictory information related to the area of supervision/discipline of children. While providers identify this as a high priority area for training, the environment rating scales indicate an assessment in the *good* range for the discipline item. Again, this shows that providers are performing better in this area than they think and it reveals the extent to which this is viewed as one of the most challenging areas in child care.

Turning now to the issue of quality of care and the factors that are associated with it, we find that our results that examine the relationship between the level of training at a site and quality of care are not as predicted. We do not find that the number of hours of training is a significant predictor of quality. What we do see, however, is that the most significant change in the quality of care since 1989 has occurred in family child care sites. Although we are unable to definitively conclude that the training system has been instrumental in improving quality of care in family homes, we do note that the home-based training system has been in existence for the longest period of time – over 10 years. Furthermore, the intervention effort (i.e. hours of training per year that staff at a site average), is still considerably low – 8.5 hours on average, with 98% of sample sites averaging fewer than 18 hours per year. A threshold for training to show some impact is around 18 hours according to other research (Howes, Smith & Galinsky, 1995). Given this, it is not unexpected that we do not find a significant relationship between number of hours of training and quality of care – there simply is not enough intervention (i.e. training hours) to determine impact.[16] What does this mean for public policy? While limitations of the research design and measurements do not allow for definitive conclusions, there are some tentative policy implications that can be drawn from this study (Kagan & Wechsler, 1998). One concerns the number of hours of training that are mandated in the state regulations for child care. It indicates the need to increase the number of hours of training for child care providers if a significant impact of training is to be detected.

This is further supported when we see the strong association between the organizational climate dimension of opportunities for professional growth and overall quality of care at the site level. Centers where staff report more opportunities for professional growth have a higher quality of care and this factor, alone, explains a considerable portion of the variation in quality (40%). This finding substantiates the importance of fostering professional growth opportunities for child care providers. But it also implies the importance of making sure that these opportunities are linked to a model of career development and progression – not just a few hours of training that providers haphazardly take because they have to or because they are offered at a time that fits their schedule.

Overall, these data have given us some solid evidence to guide the development of the training system in Pennsylvania. We have highlighted some very specific areas where there is a need for training and we have shown the clear association between opportunities for professional growth and the quality of care. Although there are some anomalies in the data and some unexpected findings, as a whole, these data are supportive of the efforts to implement a training system that fosters career development in the prediction that these efforts will improve the quality of care for children in Pennsylvania.

NOTES

1. This included a set aside of 6.25% of $731,915,000 in 1991 federal funds for program quality initiatives.
2. A decision was made not to extend the data collection process for one additional family site after we had difficulty in scheduling the final site visit.
3. There were 26 paired observations analyzed to determine inter-rater reliability. For the ECERS, the rank order correlation was 0.90; for the ITERS it was 0.95; for FDCRS, it could not be calculated since there was only one paired observation; however, a visual inspection of the FDCRS data shows a high degree of consistency across observers.
4. The instruments include specific descriptions of what to look for in assigning a value of 1,3,5 and 7 for each of the items assessed. A mid-point rating of 2,4 or 6 is given when all the lower and part of the higher description applies. The internal consistency scores (Cronbach's Alpha) for each of the environment rating scales is 0.83 for ECERS, 0.83 for ITERS, and between 0.70 and 0.93 for the individual subscales of the FDCRS.
5. The total scale alpha coefficient for internal consistency for the ten dimensions of organizational climate is 0.95. The specific details on the reliability and validity for other components of ECWES can be found in P. Jorde-Bloom (1996), *Improving the Quality of Work Life in the Early Childhood Setting: Resource Guide and Technical Manual for the Early Childhood Work Environment Survey*. Wheeling, Illinois: The Early Childhood Professional Development Project.
6. They were asked to indicate if there is a need for training, based on a scale of 1 to 4 where 1 = a very serious need and 4 = not a training priority.
7. For the purpose of this analysis on training needs, both the directors and teachers within a group home have been combined into one category, representing group child care providers. This decision was made because the child care setting is usually small and a distinction cannot always be made between a "director" and "teacher" within the group site.
8. For example, the ECERS doesn't assess health and safety areas and the discipline item is spread across a number of supervision items.
9. Each factor was assessed by providers as 1 = very important, 2 = somewhat important, or 3 = not important.
10. It is important to keep in mind that when providers indicate that they learned the material, this is based on their subjective assessment, and the extent to which they actually did learn the material is not objectively measured through this question.
11. Teachers in both group and center settings were asked, "In general, how interested are you in taking workshops or courses on teaching and/or caring for

children?" Directors were asked to indicate their perception of interest on the part of their staff.

12. Family providers were asked, "Based on your experience, what method(s) of training are most helpful for you?" Center directors, were asked, "Based on your experience, what method(s) of training are most helpful for your staff?"

13. The analysis in this section only includes child care centers since we used Bloom's instrument only in facilities that had more than three staff, as recommended. There were a total of 60 centers included in the data set, however, due to missing data from some centers, only 55 are included in the analysis presented herein.

14. These coefficients are negative since a lower value on each of the evaluative factors indicates a more positive assessment.

15. If there were two classrooms observed, the new overall *quality* measure was an average of the two scores (regardless of the type of classroom observed). If only one classroom was observed, then that score became of the site's overall *quality* score.

16. There may also be measurement problems related to the *quantity of training* variable. Issues related to recall on the part of the provider and definitions of what constitutes training may vary – both of which can affect the reliability and validity of the reported hours of training.

ACKNOWLEDGEMENTS

Keystone University Research Corporation would like to thank the many individuals who contributed to the completion of this study. First, the Department of Public Welfare, Office of Children, Youth, and Families, Bureau of Child Day Care is commended for their commitment to child care and efforts to improve quality. Without their funding of this project, it could not have been completed. The support of Richard Fiene, Ph.D., the DPW Project Officer, has been particularly instrumental in moving this research forward and guaranteeing that the policy implications of the findings are not loss. In addition, others within DPW – Kathryn Holod, Director, Bureau of Child Day Care and Jo Ann R. Lawer, Esq., Deputy Secretary for Children, Youth, and Families – are thanked for their continued support of this effort.

Thanks are also extended to the family, group, and center child care facilities that participated in this study as sample sites. Without their willingness to participate in this study and to be open and honest with the fieldworkers, this study would not have been possible. Furthermore, Lisa Heintzelman, Carla Thompson-John, Patricia Robinson-Burns, Ronnie Braun, Debora Reiff, Robin Eckert, Judy Peterson, Barbara Marchese, Majida Mehana, Kristi Hannan, Winifred Feise, Marilyn Albert, Jeanette Twombly, and Genevieve Mann are thanked for their quality efforts as fieldworkers for this study.

Finally, the individuals who contributed countless hours to assist the principal investigators in implementing this project – Pauline Kraus and Scott Johnson of Keystone University Research Corporation – are thanked for their perseverance over the long haul. The tasks of organizing the data collection process, monitoring the work of fieldworkers, organizing the data sets, and assisting in its analysis were complicated and required considerable attention to detail.

The overarching goal of this study is to contribute to our knowledge about how to improve the quality of care for children in Pennsylvania. We hope that the knowledge gained as a result of this study will take us one step further in that direction.

REFERENCES

Barbour, N. E., Peters, D. L., & Baptiste, N. (1995). The child care associate credential program. In: S. Reifel (Ed.), *Advances in Early Education and Day Care (Vol. 7): Social Contexts of Early Development and Education.* Greenwich, CT: JAI Press.

Bredekamp, S., & Willer, B. (1994). Introduction. In: J. Johnson & J. McCracken (Eds), *The Early Childhood Career Lattice: Perspectives on Professional Development.* Washington, D.C.: National Association for the Education of Young Children.

Bronfenbrenner, U. (1979). *The ecology of human development.* Cambridge, MA: Harvard University Press.

Caffarella, R. (1994). *Planning programs for adult learners: A practical guide for educators, trainers, and staff developers.* San Francisco, CA: Jossey-Bass.

Children's Defense Fund (1998). *The state of America's children.* Washington, D.C.: Author.

Children's Defense Fund (1994). *The state of America's children.* Washington, D.C.: Author.

Chung, A., & Stoney, L. (1997). *Using research to improve child care for low-income families.* Washington, D.C.: Child Care Bureau, Administration for Children, Youth and Families, Department of Health and Human Services.

Fiene, R. (1995). Utilizing a statewide training system to improve child day care quality. *Child Welfare, 74*(6), 1189–1201.

Fiene, R., & Melnick, S. (1991). Quality assessment of early childhood program: A multi dimensional approach. Paper presented at the annual meeting of the American Educational Research Association, Chicago, Illinois.

Galinsky, E., Howes, C., Kontos, S., & Shinn, M. (1994). *The study of children in family child care and relative care: Highlights of findings.* New York, NY: Families and Work Institute.

Helburn, S. (Ed.) (1995). *Cost, quality, and child outcomes in child care centers.* Denver, CO: Center for Research in Economics and Social Policy, Department of Economics, University of Colorado.

Holt-Reynolds, D. (1995). Preservice teachers and coursework: When is getting it right wrong? In: M. O'Hair & S. Odell (Eds), *Educating Teachers for Leadership and Change.* California: Corum Press.

Howes, C., Smith, E., & Galinsky, E. (1995). The Florida child care quality improvement study: Interim report. New York, NY: Families and Work Institue.

Johnson, J. (1994). *Child care training and developmentally appropriate beliefs and practices of child care employees in Pennsylvania.* Harrisburg, PA: Center for Schools and Communities.

Jones, E. (Ed.) (1993). *Growing teachers: Partnerships in staff development.* Washington, D.C.: National Association for the Education of Young Children.

Jorde-Bloom, P. (1988). Assess the climate of your center: Use the early childhood work environment survey. *Day Care and Early Education*, (Summer), 9–11.

Jorde-Bloom, P. (1998). Using climate assessment to improve the quality of work life in early childhood programs. In: S. Reifel (Ed.), *Advances in Early Education and Day Care* (Vol. 10). Greenwich, CT: JAI Press.

Jorde-Bloom, P., & Sheerer, M. (1992). The effect of leadership training on child care program quality. *Early Childhood Research Quarterly*, 7, 579–594.

Kagan, S. L., & Wechsler, S. (1998). Changing realities – changing research. In: S. Reifel (Ed.), *Advances in Early Education and Day Care* (Vol. 10). Greenwich, CT: JAI Press.

Katz, L. (1994). Perspectives on the quality of early childhood programs. *Phi Delta Kappan*.

Kontos, S., & Fiene, R. (1987). Child care quality, compliance with regulations, and children's development: The Pennsylvania study. In: D. Phillips (Ed.), *Quality in Child Care: What Does Research Tell Us?* Washington, D.C.: National Association for the Education of Young Children.

Lewin, K. (1935). *A dynamic theory of personality: Selected paper.* New York: McGraw-Hill.

Love, J., Schochet, P., & Meckstroth, A. (1996). *Are they in any real danger? What research does – and doesn't – tell us about child care quality and children's well being.* Princeton, NJ: Mathematica Policy Research, Inc.

McAllister, J. (1997). The Children's Crusade. *Time*, (August 25), 36.

Modigliani, K. (1993). Readings in family child care professional development: Project-to-project compiled. Boston, MA: Wheelock College Family Child Care Project.

Morgan, G., Azer, S., Costley, J., Elliott, K., Genser, A., Goodman, I., & McGimsey, B. (1994). Future pursuits: Building early care and education careers. In: J. Johnson & J. McCracken (Eds), *The Early Childhood Career Lattice: Perspectives on Professional Development*. Washington, D.C.: National Association for the Education of Young Children.

Morgan, G., Costley, J. B., Genser, A., Goodman, I. F., Lombardi, J., & McGimsey, B. (1993). *Making a career of it: The states of the states report on career development in early care and education.* Boston: The Center for Career Development in Early Care and Education, Wheelock College.

Phillips, D. A. (1987). *Quality in child care: What does research tell us?* Washington, D.C.: National Association for the Education of Young Children.

Phillips, D. A., Howes, C., & Whitebook, M. (1992). The social policy context of child care: Effect on quality. *American Journal of Community Psychology*, 20, 25–51.

Stoney, L., Elliott, K., Chung, A., Genser, A., & Raggozzine, D. (1997). *Common Threads: Weaving a Training and Career Development System for 21st Century Pennsylvania.* Boston, Massachusetts: The Center for Career Development in Early Care and Education, Wheelock College.

U.S. Department of Education (1990). *A profile of child care settings: Early Education and care in 1990.* A final report by Mathematica Policy Research. Washington, D.C.: author.

VanderVen, K. (1994). Professional development: A contextual model. In: J. Johnson & J. McCracken (Eds), *The Early Childhood Career Lattice: Perspectives on Professional Development*. Washington, D.C.: National Association for the Education of Young Children.

Willer, B. (1994). A conceptual framework for early childhood professional development. In: J. Johnson & J. McCracken (Eds), *The Early Childhood Career Lattice: Perspectives on Professional Development*. Washington, D.C.: National Association for the Education of Young Children.

PROFESSIONAL CARING AS MOTHERING

Noelene McBride and Susan Grieshaber

ABSTRACT

This chapter provides a case study of one woman who works as a professional child carer in a capital city of Australia. It details the ways in which she cares professionally for young children and shows how connections have been made between this caring and the ways in which the carer was mothered, or cared for as a child. Cultural feminist perspectives are adopted to theorise the lived experiences of mothering and caring that are depicted. The case study provides insights into connections that existed between the carer's experiences of growing up in relationships with her mother and the caring philosophies that continued to mediate her professional caring practices with young children.

INTRODUCTION

An overview of literature related to professional caring and mothering is the starting point for this chapter. It then discusses the theoretical perspectives on which the research was based and provides some methodological details about the study. Following this, data from one of the six child care workers (Molly) in the study are presented to show the connections between the ways in which Molly cared professionally for the young children for whom she was responsible and the ways in which she was mothered. The paper concludes with notions about Molly's motivations to care and makes some links to training for those who care professionally.

Early Education and Care, and Reconceptualizing Play, Volume 11, pages 169–201.
Copyright © 2001 by Elsevier Science Ltd.
All rights of reproduction in any form reserved.
ISBN: 0-7623-0810-9

This study explored the early memories of caring experiences of women who cared professionally. It made connections between these early caring experiences and the current attitudes and actions of professional carers. Although caring is integral to work with young children, how adults learn to care has not been a focus of research. The impact of environment on children's early development has been well researched (Bronfenbrenner, 1979, 1986; Malaguzzi, 1993; Honig, 1995, 1997) and the power of adults within the environment is also well documented (Belsky, 1981, 1986; Henry, 1996; Honig & Brophy, 1996; Vygotsky, 1978, 1962). Much discussion has related to child care standards, qualifications and services to families and children, but little attention has been focused on those who provide care for children attending child care centers. Currently, research about child care around the world focuses on adults' or teachers' responsibilities to be responsive to children (Abbott & Moylett, 1997; Gammage, 1997, 1998; Honig, 1997; Hutchins, 1995, Moylett, 1997, 1999) without acknowledging the reciprocal nature of responding or the complexities of interactions. Leavitt (1994) and Leavitt & Power (1997) have observed that professional carers' interactions lack warmth and responsiveness. This work presents a strong case to know more about professional carers' motivations to care that relate to prior experiences of being cared for and their training in care.

Women's Life Experiences: Mothering and Caring

In Australia, women comprise the majority of professional carers who work mainly with and are responsible for children from birth to two or three years of age. This is similar to the situation in the United States of America where 98% of the adults who work with young children are women (Goldstein, 1994). Yet, Martin (1994) claimed when she wrote about the need to reclaim women's voices in education that women's stories have only begun to be told recently. Recent writing about life histories of women in education by Collins, Miller and Razey (1997), Gaskell (1992), Hite (1990), and Martin (1994) opens possibilities for understanding better how personal experiences influence the formation of gendered identities, beliefs and values that infiltrate teaching and caring. Such writing offers insights into the way that personal experiences of women may infiltrate their professional practice with children.

Martin (1994) wrote that women in education were involved in what she called "using female traits of caring" (p. 115) and that they took feminine qualities of "care, concern and connection" (p. 116) into teaching. Hence, images of women who chose to be prepared to teach were largely role models or exemplars of women who were caring, concerned and connected, and who would reproduce nurturance and love for children. Grumet (1988a) argued that the identification of teaching with nurturing and mothering behaviour was an enduring legacy of very mixed blessings, as it may add to the oppression of women. That is, women teachers' status may be lower than mens' status because they are associated with mothering. Such a proposition also has implications for other women involved in caring professions. Little is known about how being a woman and having the experiences of women influences professional decisions and actions.

Research from the teaching profession has demonstrated that teachers' life experiences generated personal knowledge that filtered into professional behaviour and influenced their decisions and actions (Clandinin & Connelly, 1996; Halliwell, 1992; Spodek, 1988). Beach (1992) showed that nurturance was gender-specific in early childhood educators by comparing three generations of teachers. She studied how each generation of women's experiences of nurturance varied according to their distinctive life and career histories, and institutional change in the nature of teaching. Beach (1992) presented insights into the hierarchy of values that teachers act on within their practice about the cultural expectations of teachers. Responses indicated that teachers displayed a maternal quality (specifically, nurturance) in their professional work and that they aspired to act in a nurturing manner. Little work of a similar nature has been undertaken with professional carers.

While curriculum ignores what Gilligan (1982) referred to as feminine experiences of teachers, professional care that revolves around caring, nurturing and relationships with children continues to be marginalised. Life experiences of women lead them to know about nurture and care in many ways that spill over into their teaching (Martin, 1994; Grumet, 1988a). Grumet (1988a) argued that women in teaching care for children and have been depicted as nurturant. Goldstein (1994, 1998a) described teachers as fulfilling duties often synonymous with mothering and involving love. She used the word love as a label to encapsulate Martin's (1994) collective of "care, concern and connection" (Martin, 1994, p. 116). Martin (1994) argued that these same qualities of care, concern and connection that are deemed desirable qualities of nurturance have been problematic for the status of women and teaching, situating women teachers as "soft" and with non-rigorous characteristics.

Another reason for examining the life experiences of female professional carers is raised by Biklen (1993) who asserted that all women's occupations appeal more to the heart than the mind. This is important because women in child care are operating in *loco parentis*, in roles that may resemble mothering more than teaching. How much of their caring behaviour is a function of what is perceived to be "maternal instinct", may be crucial to discourse about women carers' ways of understanding and acting on experiences. Thus, caring and nurturing experiences contribute to professional care in ways that may prove critical in shaping practice but are as yet unidentified. As most carers are women, their ideologies of mothering are shaped by their experiences of mothering, mother-daughter relationships and cultural influences (Chodorow, 1978, 1989, 1995a, b, 1997 and 1999; Wearing, 1984; Woollett & Phoenix, 1991a, b). To what extent professional carers' caring actions can be attributed to experiences of being mothered and being in a mother-daughter relationship is an unknown variable of caring. However, Grumet (1988b) argued that our relations with other people's children are inextricably tied to influences that we attribute to our own affiliations. She stated that there might be contradictions between women's own experiences of childhood and mothering, and the children and curricula "that we teach" (p. 28). What these contradictions mean for carers needs to be explored.

Caring Work

Abbott and Moylett (1997) and Moylett (1999) have investigated how adult carers in the United Kingdom viewed their work of care. Moylett (1999) wrote that many carers did not see their role as being close to any child. Carers' perceptions of their roles raise questions about what the carers meant by being "close to" children and why they felt this way. Moylett (1999) expressed a concern with these findings about carers, stating that there was a need for loving to be part of professionalism.

While Moylett (1997) recognised that "values and beliefs influence relationships and interactions ... more significantly than any other factor" (p. 20), she did not suggest ways for carers to reconcile the messages from their experiences with their practice of care. Questions about origins of values and beliefs that underpin the actions of those who work in child care, or life experiences that infiltrate professional practice remain unanswered. There is a need to know more about values and beliefs that play an important part in the choices and actions that carers take and that shape them as professional carers. Moylett's (1999) writing indicates gaps in our knowledge about carers' attitudes and actitivies of care. For example, she stated that children need loving relationships with parents and "carers" (p. 7), and that it is through these relationships with adults that they learn to respect each other. An examination of the personal experiences or attitudes of carers to relationships or loving that is part of the way in which carers engage with young children is needed.

How carers know about giving care and love may be linked to experiences of being cared for and loved. If this is so, the need to explore their personal experiences has been overlooked. For example, Honig (1997) wrote that adults can help infants and toddlers become loving, caring persons from an early age by showing love and concern to the child and other persons. She argued that early childhood professionals required qualities of warmth, patience, and kindness but did not refer to carers' personal experience as a possible filter that might convey meaning to such terms according to different experiences and contexts. Honig and Brophy (1996) argued that a child who receives gentleness from adults could be gentle with others. Thus they appear to presume that adults know how to be gentle without acknowledging the power of personal experiences of receiving love or giving love. Hence, the power of experiences and the meanings that carers attribute to care and love is an issue for investigation.

What Messages of Mothering Might Mean

Few links are made between women's experiences of being mothered and the influence of mothering experiences on women's attitudes to caring for others, especially those who work as professional carers with young children. Extensive literature exists about being a mother, mothering experiences, and mother-daughter relationships. In this literature, links are made between experiences of being mothered and attitudes to being a mother.

Understanding carers' distinctive personal experiences of being mothered emerges through literature that examines characteristics of mothers as well as influences of being mothered and mother-daughter relationships. Critiques that focus solely on mothering roles as oppressive are not explored as the purpose of this investigation is to locate the personal experiences of being mothered that permeated professional caring. Therefore, depictions of mothers included refer to mothers as idealised (Phoenix & Woollett, 1991a, b), as "good" (Hite, 1990; Wearing, 1984) and as women engaging in self-sacrifice (Wearing, 1984). Research that provides perspectives on women as daughters (Chodorow, 1978, 1989, 1995a, b, 1997, 1999; Hite, 1990) offered insights into experiences in relationships that may influence daughters' ways of caring for others.

Wearing (1984) has disputed the notion of an ideal mother, arguing that it reflected the mixture of sentiment surrounding motherhood. Woollett and Phoenix (1991a, b) argued that because motherhood has been idealised it was viewed as an essential stage in women's adult development, providing them with a central identity as women and as adults. This approach has little to say about how motherhood intersects with women's other identities, about structural factors, and the circumstances in which women bring up children. Therefore, idealised views may fail to isolate factors such as mothers' attitudes and behaviours or to recognise different approaches to motherhood, women's identity as mothers and their understanding of their positions and functions as women. Moreover, as Phoenix and Woollett (1991b) suggested, ideal is synonymous with normal and good in terms of being a mother, and allows for only a limited range of maternal behaviours that accentuate women's role in caring for children. Phoenix and Woollett (1991b) argued that ideal or good mothers were rarely considered as having an existence of their own. Mothers were deemed responsible for the environment that they provide for children and the ways in which this may influence children's development. That is, "good" mothering involved a concern with children's "needs", often to the exclusion of mothers' views of themselves and their own needs for the family context in which they bring up children (Woollett & Phoenix, 1991b, p. 28).

Based on research with 150 Australian mothers, Wearing (1984) produced a list of characteristics of "good" mothers. A "good" mother was defined as "one who is basically concerned with the needs of her children both physical and emotional, and who will make every endeavour to satisfy these needs" (p. 51). Wearing (1984) stated that being the good mother meant a woman was involved in hard but rewarding work; put her children first, and was unselfish; was always available to her children; spent time with her children; guided, supported, encouraged and corrected her children; loved and cared for children physically; and was responsible for her children's cleanliness (p. 72). While Wearing (1984) provided generalised attributes that are invaluable in considering the roles of mother, she does not provide an understanding of individual experiences of women. However, characteristics and attributes of mothers that are documented by Wearing (1984) also presented ways of exploring women's experiences of being mothered. For instance, women may have experienced mothers as "unselfish, self-sacrificing, loving" (p. 32) people who make efforts to spend time in the interests of children. Women may also have experienced mothers who had engaged in long hours of work each day for which there is "no monetary gain" and where "gentleness and advice [are] freely and generously given" (p. 9). Therefore, women's experiences with mothers could be useful in understanding professional carers' current ways of caring.

Wearing (1984) argued that because mothers are imaged as unselfish, women continue to be ensnared with "core beliefs that motherhood and womanhood are intermeshed" (p. 72). What Wearing (1984) described as being a "good" mother resembled what Hays (1996) called intensive mothering. Intensive mothering was a term used by Hays (1996) to describe mothers being there to satisfy the needs of the child and with few, if any, needs of their own. She argued that in intensive mothering, mothers were expected to:

give the baby all the warmth, comfort and cuddling that he [sic]seems to need; to meet his wishes in the matter of satisfying and appropriate food; to adjust our habit training to his individual rhythm; and to see that he has an opportunity to exercise each new accomplishment as it emerges (p. 46).

Hays (1996) argued that intensive mothering involved mothers spending a tremendous amount of time, energy and money raising their children and that not to do so, is deemed selfish and insensitive. She also argued that intensive mothering was negative as it created a feeling of obligation for women to be "good" mothers. According to Hays (1996), good mothers engaged in appropriate child rearing based on love. In her critique, Hays (1996) said sensitive mothering valorised mothers as "keepers of morality" and as "sensitive and emotional women" (p. 41). She expressed concern with intensive mothering as it viewed constant nurturing as the child's right, meaning that mothers should focus on what is best for the child. Such an attitude to care meant that mothers had few rights themselves because they must focus on "what she [they] can do for the child rather than what is convenient" (pp. 114–116) for them. Hays (1996) explained that people are involved in experiences that become part of their way of knowing and thinking and that, as a consequence, they are no longer able to separate from these experiences. Therefore experiences of intensive mothering may lead women to act in a manner that replicates their experiences of intensive mothering. This was likely, as Hays (1996) suggested, for much of a social model is so "fully internalised that it seems a matter of intuition and common sense" (p. 73). What Hays (1996) describes as internalisation of experiences in women resembles Ruddick's (1989) maternal thinking.

Ruddick's (1989) understanding of maternal thinking draws on ideas of engaging in self-sacrifice, of mothering and idealised mothering. Ruddick (1989) suggested that maternal thinking was developed socially and teaches messages about mothering. For example, a mother does everything to protect others or "not to endanger [other] people" (p. 23). She described maternal thinking as concentrating on preserving the life of children and fostering their growth. Ruddick (1989) defined mother as "a person who takes on responsibility for children's lives and for whom providing child care is a significant part of her ... working life" (p. 40). Therefore, such notions have meaning for professional carers and their work. Aspects of maternal thinking that are evident in caring work need investigation to understand current caring actions and concern for others and their relation to aspects of idealised mothering and good mothers. For, as Ruddick (1989) said, it may be that for carers, the "work of mothering is a central symbol of care" (p. 46).

While Wearing (1984) did not draw attention to the differences in mothering, Hays (1996) discussed the complex and different backgrounds that exist for mothers, acknowledging the individual differences that make it difficult to universalise women's experiences. She drew attention to differences based on the "complex map of class position, race, ethnic heritage, religious background, political beliefs, sexual preferences, physical abilities and disabilities, workplace environment, formal education . . . experiences of being raised" (p. 76).

Exploration of whether individual professional carers see their work as like mothering may begin to address Phoenix and Woollett's (1991a, b) critique of motherhood. They (1991a) argued that mothering was "assumed" (p. 2), that motherhood was women's ultimate fulfilment and that such assumptions tended to universalise women. An exploration of professional carers' experiences of being mothered moves past assumptions of knowledge about mothering to locate distinctive experiences of being mothered that influence caring work.

Mother-Daughter Relationships as a Factor in Care

Mother-daughter relationships form part of professional carers' experiences that require attention to understand the influence of mothers on daughters. Mother-daughter relationships that are explored through the works of Wearing (1984), Chodorow (1978, 1989, 1995a, b, 1997, 1999) and Hite (1990) are shown to be complex and very different from mother-son relationships. For instance, Wearing (1984) claimed that mothers usually identified "more with daughters and experience them as being less separate than sons" (p. 109). Chodorow (1978, 1989, 1995a, b, 1997, 1999), a feminist psychoanalyst, argued that daughters learn about mothering by being mothered.

Hite's (1990) focus on studying women's experiences as daughters to find out whether the feminist movement had changed women's life circumstances adds another dimension to mother-daughter relationships, from the position of daughters. Daughters reported positive and negative feelings for mothers that suggested that emotion was part of what daughters had learnt "subtly and convincingly, by example" (p. 30). Hite (1990) reported that most daughters described their mothers as givers, and 30% saw this as positive in their mothers. Positives were expressed in quotes such as "what I like most about my mother is her giving, loving nature. She's never cross. She's seldom unfair" (p. 20). However, for many daughters, mothers' ways were "whimpy" (p. 30), but still "internalised and unquestioned" (p. 30). That is, the ways of being a mother were so much a part of daughters that daughters did not question caring ways that resembled their mothers' ways. Many women had a "terrific fear of being like their mothers" (p. 13), while many "muse[d] over how much they are like"

(p. 14) their mothers. Some daughters were pleased to be like their mothers, while other daughters were not happy to be like their mothers but recognised the similarities. Hite's (1990) study showed mother-daughter relationships had the potential to influence daughters' behaviour as adults and how aspects of mother-daughter relationships were enduring. Seemingly, experiences endured because of the nature of continuous experiences that were ongoing between mothers and daughters. Such findings were similar to Chodorow's (1989) claims that mother and daughter maintained elements of their primary relationship and thus they felt alike in fundamental ways.

Links exist between Hite's (1990) findings and Chodorow's (1989) theories about mother-daughter relationships that offer ways of thinking about the influence of mothers on daughters' relationships with and motivations to care for others. Chodorow (1989) theorised that women's sense of relation, connection and embeddedness in a social life associated with mothers provided daughters with a type of security. She described a connection to mother's identity as a normal outcome of being involved in a positive relationship with mother. Chodorow (1978, 1995a, b) claimed that a girl connects with her mother's identity, learns to be a woman through socialization and is connected to the world of mothering by virtue of mother-daughter relationships. Seventy three percent of the daughters in Hite's (1990) study felt a very "deep love and tie" (p. 15) to their mothers. The tie that Hite (1990) reported between mother and daughter is similar to Chodorow's (1978, 1989) description of connection where the daughter identifies with the mother.

Chodorow (1978, 1989) described a daughter's lack of connection to her mother's identity as separation. Chodorow's (1978, 1989) understanding of separation is similar to Hite's (1990) report that some daughters felt of a sense of "ambiguity about their mothers" (p. 14) and had distressed and confused feelings about their mothers. These daughters mentioned disappointment and anger about their mothers' "subservience, passivity" (p. 15). Although Hite (1990) did not interpret the behaviour of daughters, her descriptions of daughters' anger are similar to Chodorow's (1978, 1989) description of daughters who felt hostile to their mothers and who differentiated from their mothers, causing Chodorow to argue that daughters were separated from their mother's identities. When a girl differentiates from her mother rather than connects with her mother's identity, she is described as separating from her mother's identity (Chodorow, 1978, 1989). However, according to Chodorow (1989, 1995a), healthy self-definition requires a balance of connection and separation. Neither Hite (1990) nor Chodorow (1978, 1989) explored whether a "tie" or connection, or ambiguities in mother-daughter relationships affected attitudes and interactions of daughters who worked as professional carers with children.

Chodorow's (1978) theorizing about the reproduction of mothering begins with early involvement of mother and daughter and a girl's subsequent identification with her mother. Ultimately, this close involvement has conscious and unconscious effects. The normal outcome of identification results in mother-daughter connection (Chodorow, 1978). The way this connection occurs is explored by Chodorow to foreshadow how a daughter's relationship with her mother is foundational for care.

Early involvement of girls with their mothers is described as a differentiating experience in female development that arises from the fact that "women, universally, are largely responsible for early childcare and for (at least) later female socialization" (Chodorow, 1995a, p. 199). For a woman, mother-daughter relationships are of central importance as a basis for feminine identification. Feminine identification is based on "gradual learning of a way of being familiar in everyday life, and exemplified by the person (or kind of people/women) with whom she was most involved" (1995a, p. 204). That is, daughters reproduce mothering through being mothered (Chodorow, 1978). Reproduction of mothering relies on the daughter identifying with her mother and the belief that the early mother-child relationship, built over time, is joyful (Chodorow, 1978). It also relies on mothers being perceived as good, caring, responsive and without needs of their own. To Chodorow (1978, 1989), this meant that the child aimed to be loved and satisfied without being under any obligation to give anything in return. Early involvement is the mechanism through which continuous identification with the mother takes place. Mothers contribute to this experience because they treat the same sexed child as an extension of themselves and are unable to distinguish between their own needs and those of the child. Thus, "features of the mother-daughter relationship are internalized as basic elements of feminine ego structure" (1995b, p. 199).

Recently Chodorow (1997) argued that where female identity formation took place in a context of ongoing relationship with mothers, girls developed a stronger basis for experiencing another's needs. The close involvement between mothers and daughters enables the development of a capacity within daughters to empathise with another's needs to the extent that daughters can share others' feelings "as one's own" (p. 167). Gilligan (1997) challenged Chodorow's (1997) notions about being able to feel another's feelings, stating that it might be a case of "thinking that one is so experiencing another's needs and feelings" (p. 200). Gilligan's (1997) argument is provocative and is based on her approach that focuses on *differences* among women. Chodorow (1997) seeks ways to explain the *patterns* of what happens with women as mothers and daughters and her propositions enable a basis for exploring the strength of daughters' experiences of mothering that could lead to understanding of how another person felt.

Double identification is another aspect of identification by the daughter with the mother. Chodorow (1995b) argued that a girl's "experience of mothering... involves a double identification [where she] identifies with her own mother and [later] through identification with her child, she (re) experiences herself as a cared-for child" (p. 201). One explanation for double identification might be what Chodorow (1995a) described as conscious and unconscious effects of a daughter's early involvement with her mother. Conscious and unconscious effects of early involvement and of being cared for by a female permeate aspects of individual women's beliefs, values and perceptions that derive from mother-daughter relationships. For example, girls "retain ... concern with early childhood issues in relation to their mother, and a sense of self involved with these issues" (1989, p. 260).

When Chodorow (1999) asserted that there was an ongoing attempt in a girl's adult life to reactivate the fundamentally positive experience of being a daughter in a mother-daughter relationship, she identified the strength of early positive experiences on later motivations to interact with others. Chodorow's (1997) explanation of identity and connection presents ways to understand how a woman is encultured into maternal attitudes and activities of care. A mother's relationship with a daughter therefore becomes instructive for the daughter as a mother: the daughter has been prepared to mother. Chodorow's (1978, 1989, 1995a, 1999) work rests on the assumption that when a girl's own mother is a good, competent and identification-worthy person, a girl does not need to construct an alternative or better maternal person. In such cases the mother is powerful.

Some aspects of a daughter's enculturation into maternal ways concerned Chodorow (1978, 1989, 1995a, b). She argued that aspects of mother-daughter relationships socialised girls in ways that burden them with responsibility of care. For instance, a mother's inability to distinguish her own needs from her child may lead to women feeling that they must care for others as though they have no needs of their own. She concluded that a burden exists for women and mothers when care is felt as a conscious and unconscious sense of responsibility and that a burden created an "inescapable connection to and responsibility for her children" (p. 210). Her argument that a burden of care exists is similar to Gilligan's (1982, 1988) notion that women may be burdened by a duty to care. Chodorow (1978, 1989) postulated that the burden of care resulting from mothering experiences is potentially oppressive for girls and women.

Care According to Gilligan and Noddings

The feminine ethic of care that Gilligan (1982, 1988, 1995, 1997) identified provides a way to explore women's actions of care that abound in different social experiences. In addition, the feminine ethic of care offers explanations for women's moral reasoning about care and the influences on their instinctive and intuitive behaviour. Gilligan (1982, 1988, 1995, 1997) showed that women's ways of thinking about care and concern were different from men, establishing that women's care and concern is valuable in its own right. Her work is significant as it highlights women's motivations or obligations to care.

Gilligan (1982) emphasised relationality that is based on a responsive, caring, and empathetic connection to others as a hallmark of being female. She (1982) also focused on female connectedness to others in relationship and the importance for women of recognising self-other relationships. Her concern is with the extent to which women can voice their own experiences and see things in their own terms as well as the child's. Because Gilligan's (1982, 1995) feminine ethic of care places women and responsibilities to care as central, she takes Chodorow's (1978) celebration of traditional female virtues a step further. For instance, Gilligan (1982, 1988) included nurturing traits associated with mothers as strengths.

Recognition that women "know in a world that extends through an elaborate network of relationships" (Gilligan 1982, p. 147) evolved from the women in her study. Women's responses are depicted by Gilligan (1982) as complicating the morality of any decision and removing the possibility of a clear or simple solution. Women's morality, rather than being opposed to integrity or tied to an ideal of agreement, is aligned with an integrity and a responsibility to make choices and decisions after "working through everything [they] think is involved and important in the situation" (p. 147). Responsibilities to care are also related to traits that traditionally defined the "goodness" (Gilligan, 1997, p. 208) of women and a sensitivity to the needs of others (Gilligan, 1997). Thus, when Gilligan (1997) argued that "women not only define themselves in a context of human relationship but also judge themselves in terms of their ability to care" (p. 207), she offered a way to understand women's perceptions of responsibilities to care for others.

Gilligan (1997) related women's conceptions of morality to the activity of care, responsibility and relationships. Thus, women's moral reasoning and moral responsibilities to care can be understood as contextual and inclusive of women's connections and different experiences. Gilligan also (1982, 1988) argued that women's maintenance of social ties and experiences of being mothered or mothering held value in their own right and that they should be viewed as such.

According to Gilligan, (1982, 1988), women who act through a sense of responsibility to care for others are guilty of excluding themselves from the provision of care. Thus, women can become enmeshed in a feminine ethic of care or in "an ethic of special obligations and interpersonal relationships built on selflessness or self sacrifice" (Gilligan, 1995, p. 2). Potentially, women are restricted by women's ways of caring or maternal ways of care. This is so because women measure "their strength in the activity of attachment ('giving to', 'helping out', 'being kind', 'not hurting')" (p. 2). They seem unable to act selfishly. However, Gilligan (1997) suggested that a sensitivity to the needs of others leads women to assume responsibility for taking care that de-emphasises women's attention to their own needs for care.

Noddings' (1984) work on an ethic of care focuses on care as concrete, as an attitude and activity and involving feelings and engrossment, and her theories of care (1984, 1992, 1995, 1998) are instructive for exploring links between care received early in life and care given to others later in life. Noddings (1984) extended Gilligan's (1982) understanding of care by focusing on caring activities and attitudes associated with women, such as tending and responding, that involve more than an obligation to care for others. Noddings (1984) also identified a reciprocal relationship between the cared for and carer. A reciprocal relationship involved carers' feelings about caring and being cared for as influential in responsibilities to care. Like Gilligan (1982), Noddings (1984) considered that carers also had needs. However, Noddings (1984) moved from acknowledgment of needs to exploring carers' needs as an influence on their activities and attitudes of care.

In Noddings' (1984, 1992, 1995, 1998) understanding, an ethic of care focuses broad attention on the need for community, care and social values and is often associated with being female. While Noddings (1998) referred to a female ethic of care she offered a disclaimer that the use of female or feminine is intended only to point to a difference in experience, "not a biological difference" (Noddings, 1998, p. 155). She asserted that her feminine perspective meant "rooted in receptivity, relatedness and responsiveness" (p. 2). Attitudes to protection, attending to needs and reasons for provision of material, emotional support and nurturance are issues that are foundational to Noddings' (1984, 1992, 1995, 1998) theories of an ethic of care.

Noddings (1995) stated that her ethic of care formulated a call for each person to foster caring relationships in which the caregiver is a recipient as well as a giver of care. The strength of the care ethic is its capacity to move us beyond an "often mechanical and vindictive enforcement of rules towards a thinking that includes all parties affected by our actions within its purview" (Noddings, 1998, p. 40). Noddings (1995) questioned unidirectional care as it situated carers as self-sacrificing and she argued that, in caring, it is the relationship that is primary. That is, being in relationships involves being caring and nurturing in ways that are receptive and responsive to others' needs rather than concentrating on an adult care giver's own need for care.

By way of summary, Table 1 identifies the aspects of the theories of Chodorow, Gilligan and Noddings that are drawn on in this chapter.

Investigations of professional carers' experiences of mothering are needed to provide information about personal experiences of care such as warmth and comfort. Carers engage in dimensions of caring, nurturing and intimacy that have previously been the domain of mothers (Honig, 1997). The way that carers understand nurture may be evident in their distinctive care. Exploration of mothering experiences of carers may enable understanding of how they experienced the world of care. It may also explain both their caring conduct and the way that they see themselves as carers. That is, investigating the influence of carers' caring experiences may disclose the source of views about and actions of professional caring. It may also offer explanations about the way continuous childhood experiences of daughters with their mothers, aunts and grandmothers, have the capacity to endure.

FEMINIST HERMENEUTICS

Feminist methodologies were used in this study to capture the core of meanings, identify contradictions and use the language of women to explain their experiences (Ribbens & Edwards, 1998). Feminist methodologies are complex as they strive for methods that fit the emerging issues that represent the public and private spheres of lives, with particular emphasis on the lives of women (Mauthner & Doucet, 1998; Ribbens, 1998; Miller, 1998; Ribbens & Edwards, 1998). In addition, feminist methodologies acknowledge and value personal and professional knowledge and experiences (Martin, 1994). In the context of this study, feminist methodologies held the potential to awaken insights (van Manen, 1990) about the enduring nature of women's lived experiences in their work as carers and to understand individual formations of care that are constituted by people's historical reality.

The aim was to arrive at an understanding of the past and present and the meaning people ascribed to experiences and realities, rather than describing the reality itself. In feminist hermeneutics, the researcher, together with the participants, construct a reality. However, there is a need to be mindful of whose reality is being constructed or even if it is possible to construct a reality.

Table 1. Aspects of Chodorow, Gilligan and Noddings' Theories Used to Explain Mother-Daughter Relationships, Nurture and Care.

Theorist and Theory	Elements used in analysis
Chodorow's theories present ways to understand how women are encultured into maternal attitudes and activities of care. She offers ways of understanding generational reproduction as a factor of socialization. She depicts women as emotionally complex and recognizes the importance of women's experiences.	Reproduction of mothering occurs through early involvement and feminine identification with mother, particularly if experiences are positive. Development of mother-daughter relationships occurs through close involvement that continues to have conscious and unconscious effects on girls' identities and as mothers. A normal outcome of mother-daughter relationship is connection that may result in restricting women's roles. In double identification a girl identifies with her mother and through identification she experiences herself as a cared-for child. When a girl differentiates from her mother she separates her identity from that of her mother. Healthy self-definition requires a balance of connection and separation.
Gilligan's theories about the feminine ethic of care values the different experiences of women and their moral reasoning in caring. In addition she argues that women link morality to responsibility and relationship. Ultimately, Gilligan stresses the importance of including self in acts of care.	Rationality is based on responsive caring and empathetic connection – a hallmark of being female. Women's responsibility or obligation to care means being sensitive to the needs of others. In showing care and concern, women need to recognize self-other relationships or be excluded from the caring process.
Nodding's theories about care see care as concrete and as an attitude and activity that includes feelings and involvement. She also alludes to links between care received early in life and care given to others later in life.	Care as an activity and attitude involves concrete situations, personal practical situations and daily choices. Care is reciprocal and complex. It occurs between the carer and the one being cared for and involves tending and responding. Care involves more than an obligation to care for others. Care is motivated by feelings. Carers need to recognize their reality as different from the reality of those being cared for.

Perhaps, as Glucksman (1994) said, "there is no one reality" (p. 159). Feminist hermeneutics directed the focus on subjectivity and the non-exploitive investigation of lived experiences through life histories with the women in this study. The aim was to move with participants into experiences within their life histories, seeking to create understandings, new expressions and perspectives of women working as professional carers.

A hermeneutic approach brings into perspective our consciousness and our memories of phenomena that are experienced "pre-reflectively" (van Manen, 1990, p. 9). Pre-reflectivity results in meanings being designated and reconstructed in the light of knowledge of the consequences of earlier action, rather than meanings beings uncovered. van Manen (1990) stated that this occurred, "through meditations, conversations, day dreams, inspirations and other interpretative acts [in which] we assign meaning to the phenomena of lived life" (p. 37). To Glucksmann (1994), feminist hermeneutics concentrate on the meaning that the action, or experience, or the memory of the action or the experience holds for the participant.

Within the feminist hermeneutic circle, there is no such thing as a detached, neutral, or objective place to stand where we know something. Thus, conversations in this research project aimed, as van Manen (1990) argued, to "turn the interviewees into participants or collaborators of the research project" (p. 63). The circle was nourished by what Casper, Cuffaro, Schultz, Silin, and Wickens (1998) referred to as conversations that are subjective, not objective, where there was no "attempt to create a seamless narrative" (p. 73). The structure of conversations resembled much more the dialogic relation of what Socrates called the situation of "talking together like friends" (van Manen, 1990, p. 100), where friends aim to bring out strength, not weakness and where the knowledge produced was important in its own right. The circle of analysis was ongoing and focused on what the women said and did, in an effort to locate connections between personal experiences and professional actions. Additionally, analysis was assisted by continuously making meaning of experiences and connections. Throughout the research, analysis was a continuous cyclic situation that identified emerging concerns and interests.

Data consisted largely of experiences of life and practice that were explored via audio taped and transcribed conversations with full acknowledgement that such procedures may be part of van Manen's (1990) "transformation of these experiences" (p. 54). He argued that all recollections of experiences, reflections on experiences, descriptions of experiences, taped interviews about experiences or transcribed conversations about experiences are already transformations of those experiences" (p. 54). van Manen's (1990) assertions that whether caught in oral or written discourse, interpretations and memories are never identical to

Professional Caring as Mothering 185

lived experience, were accepted. Lived experiences were viewed as embedded in our daily lives and an important way to understand what women carers did in the course of caring for young children, as well as why they did it.

In this study, the "incremental collection of data and the unfolding nature of the research" (Layder, 1993, p. 45) worked together to create an understanding of women in their work. Former family contexts were crucial for participants because they constituted and gave meaning to situations and offered rich insights into current professional caring. Increasingly, evidence of what Layder (1993) described as the "nature of behaviour" (p. 72) occurred within the study context where, as participants engaged in the research process, more of themselves emerged.

CASE STUDY

As this study explored the early memories of caring experiences of women who cared professionally, it made analytic connections between these early caring experiences and the current attitudes and actions of professional carers. A combination of explanatory and exploratory case study approaches (Yin, 1994) provided a description that, in turn, could be used to explain the causal links in a real life situation of caring that was too complex for survey or experimental strategies. In addition, case study is suitable because the investigator had little control over the events, given that how and why questions were being posed. This case study consists of what Yin (1994) referred to as "a 'whole' study, in which convergent evidence is sought regarding the facts and conclusions for the case" (p. 49).

Molly is one of six cases drawn from the larger study. Data for the case study of Molly consisted largely of experiences of being mothered and child care practices that were explored via a series of audio taped and transcribed conversations. Each carer worked in long day settings as a group leader and was involved in discussions on at least four consecutive occasions. Each session was between one and two hours duration and all participants engaged in a minimum of 12 hours of audio taped conversations. The sequential conversations were conducted in an interactive manner where, within conversations, an unstructured and open-ended format was used, with a focus on being open, receptive and understanding. Careful listening was essential. During each session diary notes were made and excerpts were selected from the transcribed audiotapes for use in future conversations. Prior to beginning the research, up to five days was spent in each carer's workplace engaging in informal discussions with each participant in an effort to build trusting relationships.

Participants were initially asked: *Tell me about your earliest memories*. This type of question, referred to as a descriptive question (Minichiello, Aroni, Timewell & Alexander, 1995), was chosen to assist in establishing rapport and to indicate that participants were important. The answers were probed so that participants' responses could be expanded and clarified. Conversations proceeded in an undirected fashion at first. Later, specific questions drawn from diary notes and transcripts were used to direct conversations. This approach is similar to the idea of funnelling (Minichiello et al., 1995) where the focus of questions becomes narrower. Through tracking the series of conversations it was possible to understand carers' sense of their own experiences and make meaning of these experiences. Personal narratives described past and present events in a similar way to those described by Mishler (1990), who said that narratives are "the most internally consistent interpretation of presently understood past, experienced present and anticipated future [events]" (p. 68). Carers talked, engaged in conversations and told their own stories, both personal and professional, in response to stimulus questions and probing that ultimately offered insights into the "human face" of carers and caring.

Ultimately, the case studies were viewed through a cultural feminist lens and not generalised. Yin (1994) wrote that a case study is "generalisable to a theoretical position" and not "to universes" (p. 10). The case studies are a form of inquiry that do not depend solely on ethnographic or participant-observer data to investigate a phenomenon of mothering within its real-life context. This case study of Molly explores the boundaries between phenomena and context that were not clearly evident from observation. That is, the connections between early experiences of caring and later professional caring became evident only after re-visiting the data many times.

CONNECTIONS BETWEEN MOLLY'S CARING AND THE MOTHERING SHE EXPERIENCED

Molly was a professional carer. She was in her mid 30s, married with three children aged six, eight and ten years. She was the second child in a family of four with an older sister and two younger brothers. She undertook her Diploma of Child Care, a two-year qualification, while working part-time and then full-time, and had been working as a professional carer for over 5 years. The connections between aspects of Molly's professional practice and features of her personal experiences that centred on being mothered were gradually identified over time but invariably left unexamined during conversations. In part, this occurred because most connections were only apparent on revisiting conversations and reading and rereading transcripts. Increasingly, it was possible to recognise "connections" between Molly's current caring practices and her memories of early experiences of being cared for that involved mothering or being mothered. These connections were links between early personal experiences and current professional practice that were made through detailed and repetitious analysis of transcripts and are different from Chodorow's (1978, 1989, 1995a, b) idea of connection used to theorise why daughters reproduced their mother's ways of mothering.

In what follows, the nature of Molly's relationship with her mother is explored to understand how connection, as postulated by Chodorow (1978, 1989, 1995a, b) might lead to the reproduction of her personal experiences in her professional practice. Subsequently, links between personal experiences of being mothered and Molly's mother being at home with her as a child, as influential in professional caring, are explored. The theoretical perspectives of an ethic of care (Gilligan, 1982, 1988, 1995, 1997; Noddings, 1984, 1990, 1992, 1995) are also used to analyse Molly's attitudes to, activities of and responsibilities to care.

Mother-Daughter Connections and Reproduction of Personal Experiences in Professional Practice

Molly spoke of her mother as a source of emotional support and a role model. She said she had learnt to love and to adapt to hardship from watching her mother. Molly described herself as tough because her life had not always been easy. Her mother had tried to keep the family together as her father engaged in weekend alcoholism. She remembered the weekends "with the hurt, the anger and frustration". This hurt, anger and frustration had stayed with Molly who said that she had "a great gift of being able to block a lot of things out". There were occasions when Molly admitted that sometimes she couldn't keep being positive but most of the time she was a positive person, or, at the least, tried to find the positive aspects of situations.

Molly talked about how her mother was always there and the consistent person in her life:

> What mum did from Monday to Thursday was very different from how [she] approached things on a Friday to Monday. Dad was a weekend alcoholic and you had to make the most of it. Mum had to get on with life. She was the consistent one in my family . . . my mum was always really *consistent* in my life. She was *always there*.

Molly knew about overcoming the difficulties and deficits of life from her own remembered experiences, recalling that real support had come from her mother who provided "warmth and consistency [and] an arm around her when she needed it". This made her aware of children who did not "come from a very loving home".

Molly understood her past experiences as enabling her to "become in tune with those kids [from broken homes] a lot more". She made similar statements on more than one occasion about how she could "feel that hurt", "feel their anger" and "feel the loneliness". Such feelings of anger pervaded her interactions to the extent that she said she "worked along with the kids and their feelings". For example, she stated, "when a child is angry at a situation I can understand where it came from", because "I would be angry in those situations". Molly understood that she "hadn't learnt to click off to those feelings" [the hurt, anger and frustration of the weekend] and therefore could "relate to the children as if I was part of them, so when they're going through something I feel what they feel". Thinking that she could feel what the children feel is similar to Chodorow's (1997) understanding that daughters can share others' feelings "as one's own" (p. 167), but Gilligan's (1997) understanding is that it might be a case of "thinking that one is so experiencing another's feelings and needs" (p. 200).

Because Molly felt she had overcome so much in her life, she was compelled to give love to the children with whom she worked and saw herself engaging in activities that were "like a mother ... [giving] unconditional love". Her actions with children are summed up in the following verbatim excerpt:

> I think kids are just innocent. They're beautiful. They're so honest. They're so fresh ... who else could come up to you and say: "Molly, you're back", and mean it, and then turn around and love you, you know. Who else can come up to you after you've just said, "Look, I don't like the way you were doing that. Please come over and sit on the chair". Whereas an adult would hold that grudge for years, you know. They'll [children will] turn around and say: "Okay I'm over it. Do you love me?" [And I say] "Yes, come on; let's get on with life". And you just think ... you have the future in your hands here.

Molly did not express a need for separation from her mother that Chodorow (1978, 1989) described as a normal outcome of connection between mother and daughter. Molly's connection with her mother may be attributable to Molly associating her personal safety and well being with her mother who acted as a shield and "protected the children from an alcoholic father". Molly talked about her mother as an essential part of her life and how she has learnt to do similar things as her mother did for her, such as being there when the children in her care needed her:

> Your mum is a person who is part of you ... who knows you more than anyone else can ... that you can talk to. I want to do that [with kids]. When do you ever grow out of wanting a cuddle, just to be close; to be able to talk?

Professional Caring as Mothering 189

Molly's mother could be relied on to be "there when needed". Now, Molly stated: "I've learnt to do that". Molly's depiction of her mother as effective in protecting her in her childhood is what led to Molly's confidence in being protective toward the children in her care, just as she understood what her mother did with her. This is also an example of what Chodorow (1995b) called double identification; where Molly has identified with her own mother, then later identified with the children in her care, re-experiencing herself as a cared-for child, which in turn caused her to be there for the children when they need her. As Molly's mother was her emotional support, she too felt that she was able to give emotional support to the children with whom she worked. Her talk and actions in wanting to protect children from "broken homes" showed that she had learnt from her mother to protect children in her care in various ways.

Connections as Empowering and Limiting

In Chodorow's (1978, 1989) terms, Molly experienced herself as continuous with her mother and a consequence of such a connection is a desire to be like her mother. Molly talked about her mother as self-sacrificing and without needs of her own, which was similar to the way in which Hays (1996) discussed intensive mothering and Ruddick (1989) talked about mothers engaging in self-sacrfice. Molly's relationship with her mother was based on trust because she knew her mother would protect her. Now, when she strives to be that trusted protector for others, there is a sense that she may be unable to distinguish her own needs from the needs of the children. In her critique of intensive mothering, Hays (1996) argued that mothers were expected to satisfy the needs of the child and were perceived to have few needs of their own. Molly too, provides ample evidence of what Hays (1996) understands as self-sacrifice for the children. However, Molly also talked about meeting the needs of particular children in a way that suggested she might still have similar needs of her own. If this is the case, Molly is meeting what she perceives to be the needs of the children but is also meeting some of her own needs at the same time by caring this way. Thus it may appear that Molly has few needs of her own (as per Hays' understanding) but she may in fact be meeting her own needs through caring as she does.

One outcome of connection could be emotional dependence of the mother on the child (Chodorow, 1978, 1989). Molly saw herself as central to the lives of the children with whom she worked. She said that she might be "the only person in the children's lives that they get a cuddle from every day". Such actions derived from her mother who offered emotional support of cuddling when Molly needed it. Now, Molly felt confident and justified in giving cuddles in case she was the only person who cuddled the children every day. Her own experience with her mother reinforced that she was important to children's emotional wellbeing because "if these guys have got an arm around them, maybe in the back of their minds it will still come through in those later years. Maybe they can remember an arm coming around them". Clear is Molly's memory is her mother's arm supporting her when she needed it.

A consequence of Molly's connection with the way she was mothered is that she feels secure about her role with others who need care. However, when Molly cares for another person, her focus is reproducing the type of caring she has experienced, that is, the type of care provided by her mother. Molly's identity as carer in the lives of children is drawn largely from continuously aligning herself, possibly unknowingly, with her own mother-child relationship and memories of her mother who she saw as loving and available to her. Thus patterns of care provided by her mother are trusted.

Chodorow's (1978, 1989) theory of separation as empowering for women, and connection as a normal outcome but possibly limiting in relation to self-actualisation, provides another way of viewing Molly's actions. For example, Molly is connected to her mother's identity and ways of caring. This connection can be both limiting and empowering. Molly is empowered to engage in professional care in a similar way to how she was mothered. However, experiencing herself as continuous with her mother who is above reproach restricts Molly's ability to question her own caring ways. Because Molly saw no need to question her mother, she also saw no need to question herself and confidently adopted her mother's patterns of care as her own. Molly's emotional connection to her mother may have been compounded by the physical affection that she remembered as highly significant. As a professional carer, she responded to children's overtures for cuddles and hugs without hesitation. Molly also initiated cuddling children. For instance, she said, "I really love the children and invite them to come and have a cuddle".

Molly's relationship with her mother may have created a need for her to give and receive cuddles and hugs. In discussing an incident with a child called Mal, Molly described him as "very teary, really cuddly". She had previously described Mal as a child who "was not a cuddly boy", and now perceived that he "wanted cuddles at group time". Her response was to be " there for him" and to give to him as she remembered receiving cuddles and affection. This connection with her mother suggests that she felt she had the ability to feel with others and is similar to the point made earlier about Molly relating to the children "as if I was part of them, so when they're going through something I feel what they feel". This presumption may be problematic for the child or children if her perceptions prove misguided.

Professional Caring as Mothering

Connections can also be restrictive. According to Chodorow (1995a, b), connection leads to non-separation of identity. A consequence of this may be a limited understanding of other ways of caring as well as a possible emotional dependence on children. For Molly this seems to be so. Her judgment of another person's care was based on her own experiences of care provided by a nurturing mother. As a professional carer, failure to nurture in ways that resembled Molly or her mother's care led Molly to criticise other female carers. For example, when a carer did not cuddle a crying child who was distressed when her mother left her at the centre, Molly confidently passed a judgment on the carer as non-nurturing. She said: "How that carer could say that she's nurturing a child [when she didn't cuddle the crying child] . . . it was a woman who did these things [did not cuddle the distressed child]". Here Molly attributed nurturing to women and expected that a woman, as carer, should cuddle and comfort a child in distress, because this was what Molly had experienced and what she considered the correct way to care. There are consistencies here with views that women should devote themselves entirely to self sacrificing mothering of the type critiqued by Hays (1996), and Chodorow's (1978) understanding that mothers are perceived as good, caring and responsive. Through her connection with her own mother, Molly is certain that the responsibility of a woman carer should be to nurture and comfort children. Here Molly judged another carer against her own experiences and perceptions of nurture, and children's needs for physical comfort. She did not see a need to explore any alternative ways of caring or question her own judgment. Seemingly, her continuous identification with her mother excluded other caring or responding possibilities and perhaps even created a dependence on children for her own emotional support. Despite expressing love for children regardless of what they had done, Molly did not engage in any self-reflection or feel a need to examine the reasons for her caring activities and attitudes.

Chodorow (1995b) claimed that women who did not separate from their mothers could develop insufficiently individuated senses of self and become dependent on their own children. She expressed concern that women who were thus connected would be unable to form their own identity because of what she referred to as non-separation. The expectations placed on mothers in idealised versions of mothering (Phoenix & Woollett, 1991a, b), good mothers (Hite, 1990; Wearing, 1984), intensive mothering (Hays, 1996) and mothering as self sacrifice (Wearing, 1984) include unselfish, self-sacrificing loving that potentially risks mothers becoming dependent on their own children. Non-separation existed for Molly and seemed to hinder personal awareness of her impact on interactions with children in her care. In the context of child care, non-separation might account for the lack of challenge by Molly to her mother's ways of caring. Additionally, Molly's sense of security brought about by the reproduced caring patterns may act as an inhibitor to questioning the applicability of her actions and interactions with children in her care.

Mother-Daughter Relationships as Training for Care

Molly's mother-daughter relationships acted as a preparation for attending to and caring for others. Chodorow's (1978, 1989, 1995a, b, 1997, 1999) theoretical insights about mother-daughter relationships and reproduction of care bring understanding of how mother-daughter relationships can be the origin of caring ways that are enacted when carers' care for others in professional caring situations. Extrapolating from relationships as social-relational experiences from earliest infancy, it is possible to suggest that the nature and quality of the social and emotional relationships that Molly experienced with her mother were appropriated, internalized and organised over time. Chodorow (1995a) theorised that family socialization formed the basis of role training for unconscious features of personality. Such a theory presents a way of thinking about ongoing interpersonal relationships in which caring behaviours are given meaning.

The experiences that Molly remembered such as her mother cuddling her when she needed it; empathising with children who were angry in similar situations to what Molly said she would have been; not being able to forget the feelings she associated with the weekends; and believing that she could feel what the children felt because she related to them as if she was a part of them, support Chodorow's (1995a) argument that feminine identification [or rejection] is based on "gradual learning of a way of being familiar in everyday life, and exemplified by the person (or kind of people – women) with whom she was most involved" (p. 204). Molly's actions in her professional caring seemed continuous with her early childhood identifications and attachments, just as Chodorow (1989, 1995a) had theorised. Moreover, Molly's conversations and statements support Chodorow's (1995b) exposition that a woman's close connections to her mother led to the reproduction of such care in a world in which women are like their mothers.

This analysis of a carer's mother-daughter connections shows that what Molly's mother did and the context of her caring, as well as her mother being there when she needed her, were important factors in Molly's own caring philosophies. For Molly, mothering was "caught rather than taught [as each girl grows up]" (Chodorow, 1989, p. 274). Furthermore, with Molly there is evidence that a daughter's identity can become subsumed or partially subsumed by the mother's identity (Chodorow, 1989). Molly saw her mother as a pivotal part of her life, as someone who was always there and whom Molly stated was a person who was part of her. Similarly, Molly wanted to be there for kids, physically and emotionally. And she said that she had learnt how to do that from her mother.

This mother-daughter relationship, where Molly remembered her mother positively, led to her mother's ways of caring being adopted by Molly in her own caring contexts. It also led to little or no resistance to, or questioning of, caring that was based on remembered experiences. However, this reproduction of care did not appear to create a burden of care for Molly. For example, Molly seemed not to be burdened by her "inescapable connection to and responsibility to [others'] children" (Chodorow, 1995a, p. 210).

Overall, Chodorow's (1978, 1989, 1995a, b, 1997, 1999) theories about the reproduction of mothering and connections to ways of mothering formed through mother-daughter relationships are pertinent to understanding ways that this carer cared for others. Molly's remembered experiences of her mother's caring influenced her caring practices strongly, resulting in her own caring being permeated by her mother-daughter relationship experiences and what Molly had grown up remembering. Connections with her mother that were remembered as positive empowered Molly as a carer. Confidence in caring seemed a product of the continuous identification with her mother and her mother's way of caring as this enabled ways to cater for the needs of others. However, mother-daughter connections also acted as a barrier to Molly's capacity to consider contexts of care that differed from her own. Clearly, the context that mattered for this carer was her own.

The work of Chodorow (1978, 1989, 1995a, b, 1997, 1999) provided a way of understanding reproduction of care as a feature of Molly's mother-daughter relationship. For Molly, the closer the mother-daughter connection, that is, the more Molly connected emotionally with her mother, the more likely it was that Molly's personal experiences with her mother would be reproduced in her own professional practices. In other words, the closer Molly felt to her mother emotionally, the more likely it was that selected elements of her mother's attitudes and activities of care would be reproduced in her own professional caring practices.

IMPOSED REALITY OR EMOTIONAL TUNING IN?

It is difficult to say whether Molly's assumptions and assessments of the children's situations were based on a receptivity to children's needs or on an extrapolation from her own personal needs and an assumed reality for children. Molly did seem to be imposing her reality (Noddings, 1984) on others, as according to Molly, she could "put herself easily into the shoes of the child". Her empathetic responsiveness centred on her experiences and the emotions involved in them, which she transferred to the children without question. Molly's intuition was also a feature of her caring responses and derived from her need to respond to perceived demands of the children. She stated that she could "feel the ones that just weren't in synch" and this understanding, combined with her perceptions that she could put herself easily into the shoes of the child, and that she was able to relate to the children as if she was part of them, suggests that Molly is imposing her experiences or her perceived reality onto the children, rather than attempting "to grasp or receive a reality" (Noddings, 1984, p. 22) from the children's experiences. Her identification of children's emotional needs as well as her responses to these needs seemed to mirror Molly's own experiences.

Molly's relationship and socialising experience provide ways of understanding care as situational, concerned with concrete caring that resulted in Molly's attitudes and activities of care (Noddings', 1984, 1992, 1995, 1998). For instance, Molly feelings are important, as are duty and obligation as motivations to care. Meanings that Molly attributed to experiences of being mothered or her mother being there provided the connections between aspects of her professional care and personal experiences of being mothered that were fundamental to her ways of caring for children. Motivation to tune in is driven by an obligation to care (Gilligan, 1988). For instance, Molly recalled distressing circumstances that motivated her to "tune in" to children. Tuning in was a response generated by a perceived vulnerability in children because Molly felt she could put herself in their shoes. All of these tuned-in responses were linked to personal experiences where her mother had been there and where Molly remembered pleasure, protection, tranquillity, frustration, hurt and anger. Context seemed to be another factor that influenced motivations to tune-in to children. For instance, if the situation of professional caring was seen to be similar to a personal experience with her mother, Molly appeared to transfer the solutions that she had experienced to professional practice. As she sought to give children the same sense of safety and security that her mother's protective actions gave her, she displayed caring and nurturing that she associated with being mothered. This is not surprising. Chodorow (1978, 1995a, b, 1997), Gilligan (1982, 1988, 1995, 1997) and Noddings (1984, 1990, 1992, 1995) described relationships as the primary focus of care and nurture, especially family life and other close personal relationships.

Perhaps because Molly's certainty about the children's feelings emerged from her personal reality that was charged with feelings of "anger and frustration", any need to question her responses and actions was effectively nullified. Molly's approach with children is similar to descriptions of an ethic of care as the emotional-cum-moral state of engrossment in another person's reality (Noddings, 1984, 1992, 1995). Molly depicted care as an activity that involved her personal attitudes to care and thus perceived children's needs as her own reality and dealt with them based on her own experiences of being mothered.

Molly's professional practice is overwhelmingly about adult responsibilities and responses to care that involve known caring sentiments. It seems that Molly's aim in caring is to involve the cared-for in receiving and accepting her care. For example, she described an incident in the sandpit: "He was laughing and chuckling and we got on the horse and then he put his head on my shoulder. The minute I just brushed my hand against his, he just came alive. It was just incredible to see". This incident depicts Molly in a "relationship[s] of genuine caring" (Noddings, 1984, p. 20) that involves her receiving affection and feeling emotion through the child. There is a sense that she needs to feel that she is offering the child emotional support. Noddings (1984) described such care as ensuring that the "cared for glows, grows stronger" (p. 20). Molly seemed to judge her caring on her ability to comfort children and detect their emotional needs. She showed an approach to care that is interdependent, contextual and responsive. Perceptions of a child needing to be comforted, assisted or distracted by a carer seemed important in triggering a duty to care to respond to children. Gilligan's (1982, 1988) model of development is helpful in understanding caring grounded in obligation to care for children selflessly, as seems the case with Molly. Molly appears to be within Gilligan's (1982) conventional level where goodness is equated with caring for others and not with caring directly for oneself, as that is considered selfish. She did not include herself. She perceived that she gave care and not that she received it. Gilligan (1982, 1988) argued that a person in this stage was psychologically inadequate and unstable and that it was desirable for women to include the self in the equation of care. Because of Molly's lack of reflection on her carer role, there is no consideration of the origins or contextual nature of care that unselfishly gives love and affection.

CONCLUSION

Molly had built her own ideas of what constituted "good" caring from her early experiences of being mothered, though these were probably supplemented by continuing contacts with her mother. Connections to her mother's identity as discussed by Chodorow (1978, 1989) resulted in her mother's, and thus Molly's ideas about caring for children being central to her professional practice. The enduring and unquestioned nature of Molly's personal experience is significant. For Molly, notions of caring were linked intimately with particular moments of being mothered and how she remembered feeling. From the past came memories of enjoyable and sometimes unhappy experiences of care. These gave a distinctive emphasis to her personal caring philosophy. Experiences of being mothered lived on in further experiences and influenced every day professional behaviour in Molly's distinctive ways of caring.

The degree to which Molly was committed to meeting children's demands for love, nurturance, and protective care, or perceived that children were in need of care for survival, was defined by personal distinctive maternal experiences and by practices acceptable to Molly because of her experiences. Additionally, the way that Molly perceived and responded to children's demands was permeated by her own past experiences and personal perceptions of vulnerability. How she recalled being cared for by her own mother underpinned professional caring judgments, values, and strategies of protection.

For Molly, double identification (Chodorow, 1995b) was significant. She identified initially with her own mother, then with the children in her care, (re) experiencing herself as a cared for child. This impacted substantially on her professional caring. A return to experiencing herself as a cared for child, re-visiting childhood passions and needs may be related in a complex way to the fact that professional carers are trained to assume a child's-eye view in the interests of acting effectively with and on behalf of children (Honig, 1997). Another explanation is that these experiences have endured but their meaning becomes understood only as carers talk about their experiences of being mothered and the ways in which they care professionally for others.

Tucked away beneath the decisions that Molly spoke of making, was a constellation of feelings and ideas that could be said to be continuing into the next generation, a distinctive way of acting as carer (Belenky, Clinchy, Goldberger & Tarule, 1986). The power of a caring philosophy was such that Molly felt secure in her actions, secure because through her own experiences she was convinced that she knew how to "be there" and how to protect children from emotional hurt. Through visiting and re-visiting the transcripts it was possible to show that Molly described how patterns of care that resembled personal experiences of being mothered, mediated her professional caring practices.

Molly may not be the mother of the children in her care but her caring philosophy is imbued with messages of being mothered, where her mother unselfishly put the needs of children before her own to protect the children from an alcoholic father. Molly remembered a mother who was there physically and who displayed little regard for herself. The children came first. Now, Molly felt vindicated in her caring approach by positioning herself as central to the care and protection that children received. Her manner of interacting and her concern for each child's emotional well being flowed-on from her own experiences with her mother. It was not simply a matter of this carer wanting or not wanting to be like her mother. Chodorow (1989, 1999) has described the power of relationship between mother and daughter and theorised that when this relationship led to connection, girls would reproduce their mothers' ways. Molly reproduced many of her mother's fundamental attitudes and activities of care.

In sum, Molly's personal experiences of being mothered endured to become a distinct caring philosophy. The way in which Molly cared about the well being of children, took an interest in, was concerned for, or had regard for children is evidence of reproduction of mothering. Reproduction resulted from a connected mother-daughter relationship that seemed to preclude a capacity for reflecting on professional care. Molly's identification with her mother's care was such that Molly experienced herself as continuous with her mother. Her unquestioning approaches to care meant that she operated confidently from an experiential base of caring that was mediated from unexamined experiences of being mothered.

Implications for Training

Molly's remembered personal experiences of mothering and care underpinned her caring philosophy, her activities and attitudes to care. Although there can be no definitive recommendations for training from Molly, it is possible to suggest that other professional carers have experiences of care that pervade their activity of and attitude to care. Certainly, for Molly, caring for children involved more than taught knowledge about human development. It was largely caught knowledge, that is, her identity as a female and her feelings about being cared for and caring. Consequently, courses aimed at professional carer training may need to find ways of acknowledging previous experiences, especially of mothering that are positive and negative. A focus on the carer in professional carer training courses would offer scope for carers to examine how and why they are motivated to care about the well being of children, take an interest in, are concerned for, or have regard for children. In addition, a focus on the experiences that carers bring with them would provide ways of gaining insights into what aspects of caring practice reproduce personal experiences of mothering.

New Understandings of Care

The cultural Feminist theories of Chodorow, Gilligan and Noddings have been adapted and modified to understand Molly's motivations to care as well as her feelings about being a carer. While their existing theories have enabled a focus on the how and why of Molly's care rather than attempting to suggest "normal" ways of caring that exist within human or life span development, all three perspectives were needed to offer the broadest understanding of Molly. For example, while Chodorow's focus was on mother-daughter relationship and reproduction, her work has been extended through this study of Molly to show that reproduction of mothering and identities of what it is to care, also occur with a daughter when she cares for and forms relationships with the children of others. Gilligan's notion that women need to recognize self-other relationships or be excluded from the caring process is inadequate to understand Molly without Chodorow's perspective of reproduction and Nodding's view about the cared-for and the carer interactions that replicate early experiences. Because Molly's identification with her mother and her mother's way of caring endure, they seem to preclude her from examining herself in relationship and to situate herself in the equation of care. All theorists provide scope for professional carers to explore their activities of and attitudes to care but one without the others may lead to a sense of guilt or inadequacy rather than offering understanding and opportunities to rethink.

REFERENCES

Abbott, L., & Moylett, H. (Eds) (1997). *Working with the under-3s*. London: Running Head.

Beach, B. B. (1992). Teaching nurturing and gender across three generations of early childhood teachers. *Early Childhood Research Quarterly, 7*(3), 463–481.

Belsky, J. (1981). Early human experience: A family perspective. *Developmental Psychology, 17*, 3–28.

Belsky, J. (1986). Infant day care: A cause for concern? *Zero to Three, 6*, 1–7.

Biklen, S. (1993). Feminism, methodology and point of view in the study of women who teach. In: L. Yates (Ed.), *Feminism and Education* (pp. 10–22). Bundoora, Vic: La Trobe University.

Bronfenbrenner, U. (1979). *The ecology of human development*. Cambridge, MA: Harvard University.

Bronfenbrenner, U. (1986). Ecology of the family as a context for human development: Research perspectives. *Developmental Psychology, 22*, 723–742.

Casper, V., Cuffaro, H., Schultz, S., Silin, J., & Wickens, E. (1998) Toward a most thorough understanding of the world: Sexual orientation and early childhood education. In: N. Yelland (Ed.), *Gender in Early Childhood* (pp. 72–97). New York: Routledge.

Chodorow, N. (1978). *The reproduction of mothering: Psychoanalysis and the sociology of gender*. Berkeley, CA: University of California Press.

Chodorow, N. (1989). *Feminism and psychoanalytic theory.* London: Basil Blackwell.
Chodorow, N. (1995a). Becoming a feminist foremother. *Women & Therapy, 17*(1/2), 114–154.
Chodorow, N. (1995b). Family structure and feminine personality. In: N. Tuana & R. Tong (Eds), *Feminism and Philosophy: Essential Readings in Theory, Reinterpretation, and Application* (pp. 199–216). Boulder: Westview Press.
Chodorow, N. (1997). The psychodynamics of the family. In: L. Nicholson (Ed.), *The Second Wave: A Reader in Feminist Theory* (pp. 181–197). New York: Routledge.
Chodorow, N. (1999). Gender personality and the reproduction of mothering. In: A. Elliot (Ed.), *The Blackwell Reader in Contemporary Social Theory* (pp. 259–262). Oxford, U.K.: Blackwell Publishers.
Collins, M., Miller, E., & Razey, M. (1997). *Conflict in the mother/daughter relationship* [Online]. URL: http:// www.bates.edu
Gammage, P. (1997). *Early childhood care and education.* Paris: OECD.
Gammage, P. (1998). *Child care and the growth of love: Preparing for an unknown future.* Paper presented at the 1998 Australian Association for Research in Education Conference, Macquarie University. January 5–10.
Gaskell, J. S. (1992). *Gender matters from school to work.* Milton Keynes, Bucks: Open University Press.
Gilligan, C. (1982). *In a different voice: Psychological theory and women's development.* Cambridge, MA: Harvard University Press.
Gilligan, C. (1995). Hearing the difference: Theorising connection. *Hypatia, 10*(2), 120–127.
Gilligan, C. (1997). Woman's place in man's life cycle. In: L. Nicholson (Ed), *The Second Wave: A Reader in Feminist Theory* (pp. 198–215). London: Routledge.
Gilligan, C. (Ed.) (1988). *Mapping the moral domain: A contribution of women's thinking to psychological theory and education.* Cambridge, MA: Harvard University Press.
Glucksmann, M. (1994). The work of knowledge and the knowledge of women's work. In: M. Maynard & J. Purvis (Eds), *Researching Women's Lives From a Feminist Perspective* (pp. 149–165). London: Taylor and Francis.
Goldstein, L. (1994). *What's love got to do with it? Feminist theory and early childhood education* Paper presented at annual meeting of the American Educational Research Association, New Orleans. (ERIC Document Repoduction Service No. ED 375 956).
Goldstein, L. (1998). *Negotiating identities: Am I a teacher or am I a mother?* Paper presented at the 8th Reconceptualising the early childhood curriculum conference. Honolulu, Hawaii. 6–19 January.
Grumet, M. R. (1988a). *Bitter milk: Women and teaching.* Amherst, MA: University of Massachusetts.
Grumet, M. R. (1988b). Women and teaching: Homeless at home. In: W. F. Pinar (Ed.), *Contemporary Curriculum Discourses* (pp. 513–539). Scottsdale, Arizona: Gorsuch Scarisbrick.
Halliwell, G. (1992). *Dilemmas and images: Gaining acceptance for child responsive classroom practices.* Unpublished doctoral dissertation, University of Queensland, Brisbane.
Hays, S. (1996). *The cultural contradictions of motherhood.* New Haven: Yale University Press.
Henry, M. (1996). *Young children, parents and professionals: Enhancing the links in early childhood.* New York: Routledge.
Hite, S. (1990). I hope I'm not like my mother. In: J. Price Knowles & E. Cole (Eds), *Woman-defined motherhood* (pp. 13–30). New York: Haworth.
Honig, A. S. (1995). Choosing child care for young children. In: M. Bornstein (Ed.), *Handbook of Parenting* (Vol. 4, pp. 411–435). Hillsdale, NJ: Lawrence Erlbaum.

Honig, A. S. (1997). *Quality Infant Caregiving Workshop.* Syracuse University, New York. June 16–20.

Honig, A. S., & Brophy, H. (1996). *Talking with your baby: Family as the first school.* Syracuse, NY: Syracuse University.

Hutchins, T. (1995). *Babies need more than minding: Planning programs for babies and toddlers in group settings.* Canberra: Australian Early Childhood Association.

Layder, D. (1993). *New strategies in social research.* Oxford: Blackwell.

Leavitt, R. L. (1994). *Power and emotion in infant-toddler day care.* Albany, NY: State University of New York.

Leavitt, R. L., & Power, M. B. (1997) Civilizing bodies: Children in day care. In: J. Tobin (Ed.), *Making a Place for Pleasure in Early Childhood Education* (pp. 39–75). New Haven: Yale University Press.

Malaguzzi, L. (1993). *Your image of the child: Where teaching begins.* Paper presented at Reggio Emilia, Italy.

Martin, J. (1994). *Changing the educational landscape: Philosophy, women, and curriculum.* New York: Routledge.

Mauthner, N., & Doucet, A. (1998). Reflections of a voice-centred relational method: Analysing maternal and domestic voices. In: J. Ribbens & R. Edwards (Eds), *Feminist Dilemmas in Qualitative Research: Public Knowledge and Private Lives* (pp. 147–170). London: Sage.

Miller, T. (1998). Shifting layers of professional, lay and personal narratives: Longitudinal childbirth research. In: J. Ribbens & R. Edwards (Eds), *Feminist Dilemmas in Qualitative Research: Public Knowledge and Private Lives* (pp. 58–71). London: Sage.

Minichiello, V., Aroni, R., Timewell, E., & Alexander, L. (1995). *Indepth interviewing: Principles, techniques, analysis* (2nd ed.). Melbourne: Longman Cheshire.

Mishler, E. G. (1990). Validation in inquiry-guided research: The role of exemplars in narrative studies. *Recent Trends in Qualitative Research: Harvard Educational Review,* 69(4).

Moylett, H. (1997). It's not nursery but it's not just being at home – A parent and childminder working together. In: L. Abbott & H. Moylett (Eds), *Working With the Under 3's* (pp. 11–34). Milton Keynes, Bucks: Open University Press.

Moylett, H. (1999, April). *Working with people under three: Being professional – 'messing up a lot of positive things'.* Paper presented at the 3rd Warwick International Early Childhood Conference, University of Warwick, U.K.

Noddings, N. (1984). *Caring: A feminine approach to ethics & moral education.* Berkley, CA: University of California Press.

Noddings, N. (1992). *The challenge to care in schools: An alternative approach to education.* New York: Teachers College Press.

Noddings, N. (1995). *Philosophy of education.* Boulder, CO: Westview.

Noddings, N. (1998). Feminist fears in ethics. In: M. Gatens (Ed.), *Feminist Ethics* (pp. 149–157). Aldershot: Dartmouth Publishing Company.

Phoenix, A., & Woollett, A. (1991a). Introduction. In: A. Phoenix, A. Woollett & E. Lloyd (Eds), *Motherhood: Meanings, Practices and Ideologies* (pp. 1–11). London: Sage.

Phoenix, A., & Woollett, A. (1991b). Motherhood: Scoail construction, politics and psychology. In: A. Phoenix, A. Woollett & E. Lloyd (Eds), *Motherhood: Meanings, Practices and Ideologies* (pp. 13–27). London: Sage.

Ribbens, J., & Edwards, R. (Eds) (1998). *Feminist dilemmas in qualitative research: Public knowledge and private lives.* London: Sage.

Ribbens, J. (1998). Hearing my feeling voice? An autobiographical discussion of motherhood. In: J. Ribbens & R. Edwards (Eds), *Feminist Dilemmas in Qualitative Research: Public Knowledge and Private Lives* (pp. 24–38). London: Sage.

Ruddick, S. (1989). *Maternal thinking: Towards a politics of peace.* Boston: Beacon Press.

Spodek, B. (1988). Conceptualising today's kindergarten curriculum. *The Elementary School Journal, 38,* 13–22.

van Manen, M. (1990). *Researching lived experience.* Ontario, Canada: Althouse.

Vygotsky, L. S. (1962). *Language and thought.* Cambridge MA: Massachusetts Institute of Technology Press.

Vygotsky, L. S. (1978). *Mind in society.* Cambridge, MA: Harvard University Press.

Wearing, B. (1984). *The ideology of motherhood: A study of Sydney suburban mothers.* Sydney: Allen and Unwin.

Woollett, A., & Phoenix, A. (1991a). Afterword: Issues related to motherhood. In: A. Phoenix, A. Woollett & E. Lloyd (Eds), *Motherhood: Meanings, Practices and Ideologies* (pp. 216–229). London: Sage.

Woollett, A., & Phoenix, A. (1991b). Psychological views of mothering. In: A. Phoenix, A. Woollett & E. Lloyd (Eds), *Motherhood: Meaning, Practices and Ideologies* (pp. 28–46). London: Sage.

Yin, R. (1994). *Case study research: Design and methods* (2nd ed.). London: Sage.

THE THEMATIC UNIT: OLD HAT OR NEW SHOES?

C. Stephen White, Greta G. Fein, Brenda H. Manning and Anne Daniel

INTRODUCTION

When kindergarten and early childhood teachers set out to formulate their own curriculum, they must inevitably answer questions that have plagued educators since the kindergarten movement began in the United States more than one hundred years ago. Questions related to what children need to know and how this knowledge is best acquired have been potent challenges to those seeking to provide meaningful experiences for young children. These questions have been at the center of debates and divisions among early educators for the last century. For those deciding what is or is not an appropriate early childhood curriculum an additional issue concerns how to apply a growing body of cognitive theory and research which provides evidence that young children are more cognitively capable than traditional developmental theory indicated. The teacher willing to offer more than standard workbooks, basal readers, and worksheets faces an awesome task of formulating an appropriate curriculum that offer young children a vision of the pleasure and surprises that comes with knowing about something important.

Like many early childhood educators across the country, we believe that the question of what and how to teach young children can be best addressed as teachers develop their own experimental curricula. A typical way for teachers to develop their own experimental curricula is through a familiar, though sometimes disparaged, technique for organizing their programs, the "thematic unit." The integration of activities around a theme is not a new approach; its antecedents can be found in the early literature describing Froebelian kindergartens. The terms unifying, correlating, and integrating have historically indicated educators' concerns with presenting learning opportunities in a cohesive and meaningful manner. For many years, "units" have been employed by early childhood teachers eager to pull together separate strands of learning. However, these units have sometimes resulted in misused, overworked, and sometimes trivialized learning. As a result, many early childhood teachers and administrators have turned to commercially published units and/or a predetermined curricula that must be taught.

A continuing problem for educators who believe that the thematic unit provides a cohesive structure for the provision of meaningful experiences to young children has been a lack of a strong, empirically elaborated psychological model of learning in the young. The authors believe that the thematic unit approach can be a highly effective means of structuring a curriculum. Additionally, recent theoretical and empirical work in developmental psychology related to how young children learn and think provides a strong rationale for using this approach in early childhood classrooms.

Our rationale comes from the theoretical and empirical work of Piaget, Vygotsky, information processing theorists and Bruner. Although Piaget and Vygotsky differ in their conceptions of the child's developing mind, we believe that when considered together, these theorists provide justification for a constructivist approach situated in a social context. Their theories offer an image of how peers and adults can contribute to learning although we consider Vygotsky to be a more vivid and powerful source through his "scaffolding strategies." We also include schema theory that was developed by information processing theorists. Schema theory, although a different but similarly generative theoretical framework, broadens a traditional developmental model to include ideas about what children learn from the countless events that contribute to their understanding of self and others. We turn to Bruner whose most recent work provides a scope and depth of theorizing about the mind. We are unable to include other theoretical frameworks because of space limitations, however, we mention other theorists such as Bahktin, primarily to indicate how we have formulated our rationale.

This article will focus upon historical perspectives of the thematic unit and recent research and theoretical work that provide reasons for using this form to organize the early childhood curriculum. The idea of thematic units can serve as a useful core for a case study in determining if history really does repeat itself. While reflecting on the history of thematic units, we have asked ourselves whether or not some educators and researchers currently view thematic units as a rather old idea with a new name. However, we do recognize that there surely are other educators and researchers who view the thematic units as a part of the history of early childhood curriculum development in which new approaches build on the best of past curricular development efforts. We, the authors of this article, favor the latter view. We do so because the changing learning theories that have served to propel curricular efforts in early childhood education over the last several decades favor different and more inclusive notions about how young children develop and acquire knowledge (Alexander et al., 1996).

In this article we address the three terms that have been used historically to designate unit-based curricula; *activity, project* and *unit*. These terms have been used interchangeably throughout the history of early childhood curriculum. We use the term used by those who originally wrote about unit-based curricula. To clarify recent trends in early childhood curriculum efforts, we provide definitions of these terms to highlight different levels of integrated curricular development and the theoretical frameworks that seem to have given them birth.

HISTORICAL PERSPECTIVES

The standard references for innovation in the kindergarten among reformers of the early 20th century in the U.S. or in Europe were to Froebel, Pestalozzi, and Rousseau (Dewey & Dewey, 1915; Ferrier, 1927; Hewes, 1995; Osborn, 1991; Pratt, 1948). Weber (1969) traced the path of kindergarten education in the United States and included certain curricular developments that can be thought of as antecedents to the current concept of the thematic unit. The Froebelian kindergarten was imported into this country in the mid-1850s with its guiding principles of unity intact. His "gifts and occupations" can be found in programs subscribing to his beliefs; teacher directed activities according to prescribed patterns. At that time, this was the officially sanctioned way of educating young children (Osborn, 1991). The priority of the search for truth was embodied in the materials and activities designed for classroom use (Froebel, 1976). Play was considered to be the avenue for children's discovery of these truths. In his book *Mother's Songs, Games, and Stories*, published in 1914, Froebel encouraged mother-child interaction, thus fostering mothers' awareness of their children's development through songs, pictures, and finger plays. Songs and stories were often unchildlike and illustrations poor, but they reinforced Froebel's guiding principles (Weber, 1984).

Cores and Correlations

Froebelian teachers began functionally using the *Mother's Songs, Games, and Stories* book by integrating class discussions and songs and verses around its illustrations. This use of the text may have been influenced by the Herbartian movement of the 1890s–1900s, which postulated that certain subjects should form the core of the curriculum, around which other subjects should be correlated. To Froebelian teachers, correlation meant that using literature as a core from which other aspects of the curriculum could be concentrated (Weber, 1969, 1984) unified the program. "Gifts and occupations" were subsequently correlated to songs and verse until, as Hill (1913) complained:

> This mistaken notion of correlation resulted in a dreary round of activities in which the teacher endeavored to repeat ... the same 'idea,' 'subject,' 'topic,' or 'thought' for the day ... (p. 265).

Susan Blow, an acknowledged Froebelian leader, decried this approach which she felt violated Froebel's original intentions for use of his materials. Ever cognizant of retaining the purity of the Froebelian practices, Blow would not give sanction to what she felt was an external imposition of "themes" upon his "complete" materials (Weber, 1969, 1984).

The disenchantment with Froebelian methods was as often personal as it was conceptual. Some indications of the depth of feeling is conveyed by Pratt's (1948) recollection of her preparation at Columbia Teacher's College in 1892:

> You taught children to dance like butterflies, when you knew they would much rather roar like lions, because lions are hard to discipline and butterflies aren't. All activity in the Kindergarten must be quiet, unexciting. All of it was designed to prepare the children for the long years of discipline ahead. Kindergarten got them ready to be bamboozled by the first grade (p. 15).

Inspired educators like Pratt began to experiment with new ideas and concepts about the relation between real-life experiences and school activities. Undeterred by lack of leadership's approval, teachers shifted gears and began selecting "great men, events, animals, and objects" as the central focus for curriculum (Weber, 1969, p. 57). Pratt demonstrated how a trip to the New York City waterfront could be used to facilitate experiences with map making and boat building (Pratt, 1924). Similarly, Lucy Sprague Mitchell's "here and now" stories and learning program were based on her belief that children should learn by exploring their immediate environment (Antler, 1987). But only a few hearty and independent souls ventured into these uncharted waters. Most often, Froebel's materials were linked to these topics in different ways, which sometimes were clever, and often inappropriate. These Froebelian materials and pedestrian topics appear to be the only curricula teachers had available. For the purposes of this chapter, the important point is that this marriage of Froebelian and Herbartian philosophies under the banner of correlation was the early application of the "thematic approach" to organizing the curriculum.

Changes in Curriculum Organization

During the first quarter of the twentieth century, allegiance to Froebelian practice began to dissipate with the arrival of new philosophies and ideas concerning education, young children, and curriculum. G. Stanley Hall's child study movement and Thorndike's behaviorism provided an impetus for a new direction for examining children, assessing their behaviors, and changing educational practices accordingly (Hewes, 1995; Osborn, 1991; Trawick-Smith, 1997; Weber, 1984). Perhaps the greatest influence for change in the direction of curriculum was the work of Dewey, whose theories provided direction and inspiration to those seeking alternatives to the Froebelian structure. Dewey's philosophy contained guiding principles rather than a prescriptive "structure" (Dewey, 1963); these principles gave support for early childhood teachers' own changing beliefs about how meaningful educational experiences can be provided. Dewey's laboratory school was a model of "Progressive Education." Those who were part of this movement (Dewey & Dewey, 1915) found these models in The Elementary School of the University of Missouri, at Columbia; Carolyn Pratt's School in New York City; Public School 45 in Indianapolis Indiana; The Francis Parker School in Chicago; and the Fairhope School in Alabama. Even so, Dewey's philosophy was not always applied in ways he approved (Weber, 1984). Varied interpretations of Dewey's thinking led to a number of different curricular applications. As Dewey's philosophy moved into more and more American classrooms, educators continued to test, adapt, and discard his ideas.

One aspect of Dewey's philosophy encouraged curricular developments that contributed to the evolutionary path of the theme or unit. Dewey believed that children's interests should be a guiding force in curriculum formulation (Dewey, 1966). According to Dewey, children's interests should be pursued in an active manner, in an atmosphere of freedom. These activities should lead the child to higher levels of thinking through experiences provided in the classroom. In elaborating a social education philosophy, Dewey envisioned classrooms as communities in which curriculum was discovered in the social life of the class (Dewey, 1963). The challenge was to view curriculum correlation not as a unification of existing separate subjects, but as a differentiation of separate studies from the unity of interests and relationships embedded in a classroom's social life (Dewey & Dewey, 1915). From children's interests would emerge activities which would by natural consequence, differentiate into separate avenues of study. In support of this idea Dewey and other reformers called upon an unrefined even romantically expressed psychological principle which today might be called intrinsic motivation.

During the early 1900s teachers had to decide how to apply this philosophy and make it work for their classrooms. Kilpatrick's essay entitled "The Project Method" written in 1918 presented a strong case for the rationale behind purposeful activity and social relations. According to Kilpatrick (1918), a project is an activity that the child approaches with understanding and purpose in a social setting. The project method or activities program as it was sometimes called, dominated the American curriculum for many years. By 1932, more than seven thousand articles on activities and projects had been published (Foster & Headley, 1948). Kilpatrick suggested four different types of projects for classroom use: (a) a project of constructive nature resulting in an external product or presentation; (b) a project showing appreciation of an aesthetic experience, such as a song, reading a story, writing a poem; (c) a problem-solving type of experience; and (d) a project leading to accumulation of particular knowledge or skills. A review of professional literature of the 1920s and 1930s by the fourth author revealed a majority of described projects to be of the first type. Children made paper boats and built block bridges; these were cooperative group projects that might or might not be supported by experiences with the real thing.

Some Tentative Definitions

As we stated earlier in this article, terms such as activity, project, and unit have been used loosely and even interchangeably. In order to clarify our own ideas about the breadth and depth of curriculum integration, we define these terms so as to draw attention to these issues. We view the activity as a basic educational event. An *activity* engages a child or group of children in a problem-solving effort planned by the teacher for a particular setting and for a short duration. An activity might engage the children in building a bus in the block corner or in experimenting with shadows cast by a goose neck lamp on a corner wall. The activity provokes children's thinking about different issues: How are the seats on a bus arranged? Or how can I make my shadow jump? It also calls upon emerging skills: counting pairs and negotiating with peers' disagreements about the seating capacity of the bus. If the activity occurs in a group setting, peer discussion and cooperation are encouraged.

A *project* connects two or more activities by a common topic. At least one activity is situated in the world outside the classroom. It is only at the project level that the relation between events in the classroom and events in the real world can be elaborated. At this level of curriculum organization, the "school bus", for example, is an object of study. Real world observations of the form, contents, size, and mechanics of the bus are rendered and represented in the classroom in words, pictures, or block structures. If shadows are the topic, shadows can be measured and monitored on the playground at different times of day; concurrently, the conditions under which shadows are created can be manipulated in the classroom. Clearly, projects take longer to plan and, depending on the children's interests, may continue for fairly lengthy periods.

Projects can also be organized at a higher, multi disciplinary level. At this level, the project becomes a *unit* and the topic becomes a theme. As a project, the school bus is studied as an object. Shadows are studied as natural phenomena. As a unit, the cultural context of study expands. The school bus or shadows are, in addition, studied as fictional, cultural, or historical entities. Children might be asked to interview parents, grandparents, and other relatives to find out how they went to school when they were children. These oral histories can become family biographies. The study of shadows might expand to "Peter Pan" and other fanciful stories. Units can be lengthy affairs and, depending on the children's interest, touch diverse domains of knowledge and skills. When families are asked to participate in unit experiences by providing information and resources only the home can provide, the unit also becomes a way to facilitate connections between the home and school.

Early Influences of Developmental Psychology

As Dewey and his followers applied philosophy to education, there were others who set out to apply psychology to education. Arnold Gesell studied early behavioral development providing detailed norms in areas of motor development, language use, adaptive behavior and personal-social behavior (Weber, 1984; Wortham, 1992). For the first time educators were asked to acknowledge a normal and desirable pattern of child growth (Wortham, 1992). Teachers needed to supply toys and equipment that would be appropriate to the child's level of development (Lazerson, 1972). How the preschool curriculum can help children achieve these developmental norms is described in the *Conduct Curriculum* developed by Patty Smith Hill at Teacher's College in New York (Hill, 1923). The conduct curriculum targeted traits and abilities similar to those studied by Gesell. Each classroom activity had the aim of improving a particular trait (manual dexterity, politeness, rhythmic movement) or habit (tying shoes, cutting on the line, standing quietly in line). One learns manual dexterity by paper cutting, and stringing beads; and rhythmicity by dancing; cooperativeness by block building with peers. If the unit encourages a broad view of curriculum, the activity encourages attention to exactly what the child is doing during the school day.

By the 1920s projects or activities were also referred to as units or units of work (Arbuthnot, 1933; Parker & Temple, 1925; Streitz, 1939). At times the terms "activity," "project," and "unit," were used in different contexts and illustrated by different examples. At other times, these terms were used interchangeably. They were not defined or deliberately distinguished from one another. Experts discussed units in somewhat broader terms than those applied to activities or projects. Units were to revolve around children's interests and their desire to pursue these interests. In keeping with Dewey's belief that "learning in school should be continuous with that out of school," (Dewey, 1966, p. 358) experts urged teachers to build units on the study of real institutions (the post office) or social activities (gardening) accompanied by concrete experiences such as a visit to the post office or planning and planting a school garden. In Caroline Pratt's City and Country School each class assumed responsibility for providing a service needed by the school as a whole such as a printshop, post office, and a school supply store (Pratt, 1924). As a result, the community would come into the classroom and the classroom would reach out to the community. Although Dewey was primarily concerned with the integration of in-school and out-of-school experiences, some writers noted in passing that units should integrate various subjects and skills (Arbuthnot, 1933).

At this time, the ideal of a unit as elaborated by early childhood educators envisioned an extraordinary depth of children's involvement in learning. After children expressed a need or interest related to a particular topic, an alert teacher could "guide" the class through a unit rich in creative opportunities and social possibilities, while integrating pertinent strands of the academic curriculum. Often, culminating activities such as performances, displays, and constructions would signify an end to the unit. A unit could continue as long as it maintained the children's interest. In the classrooms most closely aligned with this enlarged vision of Kilpatrick's project method, projects came to a natural close. That is, teachers did not push their students to end a unit in order to advance to another topic.

By the 1940s, few references to the "activities program" existed in the periodical literature. As commercial units became available, the project method as a technique for integrating academic and social experiences became altered. Teachers no longer focused on children's interests and the boundaries between home, school, and community were preserved. Eventually, the original techniques delineated by Kilpatrick and Dewey were abandoned by most teachers. Units were often assigned to specific grades regardless of need, interest, or appropriateness. For example, all kindergartners might explore "Our Home", while all second graders studied "Indians." Despite a large number of activities, the units often failed to meet Dewey's criterion of purpose. Teachers often became more concerned with the appearance of projects and displays, and/or the final product rather than the possible interrelatedness of learning within one project. In some instances, children's interests gave way to a total license for teachers to allow children to do anything within the scope of a project. These changes in interpretation and application of Dewey's philosophy began to draw criticism from advocates of Dewey's process approach as well as from Dewey himself. In addition, questions were raised by parents and teachers related to whether or not these projects insured the teaching of necessary skills.

While the first 25 years of the twentieth century had seen curricular reforms of far reaching consequences, curriculum matters in the 1940s remained relatively stable. The Progressive "project method" of an "activity curriculum" was predominately called the "unit." Foster and Headley (1948) upheld the unit as the "vehicle for bringing activity into the classroom and taking knowledge out of the classroom" (p. 127). They offered specific criteria for appropriate units: (a) the unit should grow out of children's experience; (b) the unit should be complex enough to allow for a broad range of responses from children; (c) the unit should broaden the outlook and social understanding of the group; (d) the unit should be an impetus to other units of work; and (e) the unit should offer children a degree of satisfaction.

Foster and Headley (1948) also outlined specific topics considered to be appropriate for units. These topics were aligned most closely with social studies and included specific units that are still predominant today, such as holidays and the seasons. Because of this narrow and predictable use of unit techniques, critics emerged who felt the approach was becoming trivial in its application (Weber, 1969, 1984). This criticism remains today as many early childhood educators view commercially produced and some teacher made units as irrelevant to young children's experiences. Additional criticisms of topical units included a general lack of creativity and depth, as well as the inclusion of content which children already know.

A review of periodical literature of the 1940s through the 1980s by the fourth author revealed few specific references to the unit or thematic approach as applied to young children. However, early childhood and elementary curriculum textbooks published in the last 25 years provide extensive coverage of using the unit in the curriculum. Cohen and Rudolph (1977) view units as appropriate organizers, but encourage teachers to take children further into stimulating intellectual pursuits than traditional projects allow. In addition, units can evolve from both children's interests and teacher's insights into what may be meaningful to the class. The unit, as described by Jarolimek (1977), provides a framework of learning. It serves as a means of organizing instructional materials around a broad theme that encourages integration, not fragmentation of ideas.

The most persuasive recent effort to revive the "project approach" comes from Katz and Chard (1989). These authors define a project as an "in-depth study of a particular topic that one or more children undertake" (p. 2). An example of a project is the study of the school bus in which subgroups of children sketch the motor, diagram the interior and exterior lights, measure its size using ropes, and build buses out of blocks or cardboard. By way of definition, we would add that a project consists of multiple activities connected by a common theme. The world outside the classroom is brought into the classroom, and the children go outside the classroom to see the world. Because classroom "bus" activities call forth different skills and touch different academic subjects, the project is intersituational and interdisciplinary. Further, to the degree that each activity requires children to work together, children learn about a cultural object in a socially collaborative way.

Old questions remain with regard to the value and relevance of using a thematic project or unit as the predominant organizational component of early childhood curriculum. Robinson and Spodek (1965) viewed units with reservation and emphasized the fact that often ample time is not provided for children to explore common unit topics. These topics may also be too broad for young children to assimilate under any circumstances. Units have also been criticized for an overemphasis on goals and objectives which encourage children to simply present a predetermined "chunk" of information back to the teacher.

As contemporary kindergarten and early childhood programs adapt curriculum to meet the changing needs of young children who are exposed to academic programs at earlier ages, educators continue to grapple with what curricular components comprise an appropriate program for young children (Goffin, 1994). Kindergarten and preschool teachers are expected to deal with a wider range of individual differences among young children, many of whom are difficult to educate. However, much of the curriculum of early education programs continues to remain somewhat inconsistent, often comprised of adaptations of elementary curriculum or commercially packaged activities (Goodlad, 1984). Consequently, these materials may be irrelevant to children who enter school with different backgrounds or who have specific intellectual needs (Goffin, 1994). Many early childhood programs have only begun to address the fact that many children in their programs may have difficulty in linking what they have learned out of school to school tasks (Goffin, 1994; Spodek, 1986). One solution to the lack of curriculum diversity is to develop additional curricular materials and activities. Another solution is to conceptualize an approach to curriculum that presupposes cultural connections between the child's life at home and at school. This last solution would provide an application of Vygotsky's (1962, 1978) sociocultural explanation of how young children develop cognitively.

Research related to theories of learning and cognitive development (Bransford, Nitsch & Franks, 1977; Gelman, 1978, 1979; Vygotsky, 1978) suggested that the early childhood years may be the most productive for intellectual development and that young children may be more cognitively competent than previously believed. New notions about children's development and learning encourage us to reexamine the activity-project-unit triad and connect it to a broader theoretical framework.

We propose that recent theories and research concerning the role of cognitive development in learning and teaching provide a foundation for the use of a thematic unit approach to early childhood curriculum. This body of research found in information processing studies as well as work that evolved from Piaget, Vygotsky, and more recently, Bruner provides additional evidence to support the use of thematically linked activities for teaching young children.

INSPIRATIONS FROM CONTEMPORARY PSYCHOLOGY

Despite concerted efforts to inform parents, teachers, and administrators of "developmentally appropriate practices" early childhood educators continue to express concern about the current emphasis on formal academic instruction in kindergarten and preschool classrooms (Schoonmaker & Ryan, 1996). For example, kindergarten teachers continue to feel pressure to provide an academically-oriented curriculum which fits into the existing elementary school curriculum structure (Bredekamp & Copple, 1997; Durkin, 1987; Hatch & Freeman, 1988; Walsh, 1989). Current instructional practices in kindergarten and preschool often reflect a narrowed emphasis on academic tasks which can be easily observed and assessed. This narrowed emphasis has emerged as a result of pressure from parents, administrators, and first grade teachers as well as state-mandated kindergarten curricula (Bredekamp & Copple, 1997; Walsh, 1989; White, 1988). In many classrooms, academic instruction includes the use of curricula that employ commercially prepared units that can be systematically used in limited time periods and are considered to be "teacher proof." A specific problem with the use of this type of approach is that the materials and methods used are not always intellectually engaging or demanding for many children (Bredekamp & Copple; 1997; Bredekamp & Rosegrant, 1996). Often, activities which are "watered down" versions of elementary curriculum are found to be boring and repetitive to young children (Bredekamp & Copple, 1997). These recent criticisms are reminiscent of classrooms of the early 20th century which reformers found objectionable because of ineffective teaching techniques, such as decontextualized and irrelevant drill and practice activities (Schoonmaker & Ryan, 1996).

In the last 15 to 20 years an abundance of research in early childhood education and cognitive psychology has begun to change how we view the learners' knowledge structure as well as how knowledge is acquired (Bredekamp & Copple, 1997). Current research related to young children's learning is moving away from traditional developmental and maturational theories, such as Gesell, to the study of factors that contribute to the construction of knowledge.

Activities: Piaget and a Constructivist Curriculum

Piaget's theory on cognitive learning has had a strong impact on our beliefs related to how children acquire knowledge (Schoonmaker & Ryan, 1996). His theory of the growth and development of basic structures of mind is considered to be one of the broadest and most precise (DeVries & Kohlberg, 1987; Trawick-Smith, 1997). There have been a number of applications of his theory to early childhood curriculum. In this section we will focus on the components of his theory which provide support for curricula which facilitate the construction of knowledge through the use of activities. The most consistent and innovative of these efforts is found in the work of Kamii & DeVries (1978).

According to Piaget, knowledge evolves through an interaction or dialogue with the physical and social environment, rather than by direct biological maturation or direct learning of external information from the environment (Trawick-Smith, 1997). The acquisition of knowledge occurs when children construct it from "inside the head" when interacting with the environment (Chaille & Britain, 1997). By acting on reality and transforming it, children construct knowledge as an organized whole rather than in isolated pieces. Piaget believed that children are busy trying to make sense out of everything they encounter and are capable of constructing and organizing their own knowledge in an active way (Chaille & Britain, 1997). For example, children assimilate reality as they become engaged in symbolic play. As children think about reality in their pretense play, knowledge becomes more elaborate and better organized. Children do not come to know reality simply by being exposed to it (Kamii & DeVries, 1977).

The construction of knowledge, according to Piaget, occurs through three types of knowledge: physical, social, and logical-mathematical (Trawick-Smith, 1997). Physical knowledge entails individual actions on objects which leads to knowledge of the objects themselves. By acting on objects and finding out how they react to his or her actions, the child gains further knowledge about the properties of objects through his or her senses (DeVries & Kohlberg, 1987). In logical-mathematical knowledge, actions on objects lead to more abstract thought in which relations are created by the individual. Logical-mathematical structures are constructed by the child's own mental activity. Social knowledge has people as its source and is socially derived. The social context offers possibilities for children to become aware of differences in perspectives and provides special motivation to coordinate these (DeVries & Kohlberg, 1987).

Applications of Piaget's beliefs about constructivism to early childhood curriculum have resulted in illustrative activities which consider children's interest, active involvement, and organization of knowledge. Applications which focus most specifically on Piaget's ideas related to knowledge construction are found in the curricula designed by Kamii and DeVries (1977). These curricula focus on physical knowledge activities developed to enhance the child's construction of knowledge about the nature of matter through actions on objects and observations of the reactions of objects to these actions. For example, when children are engaged in more open-ended activities, such as a "roller" activity, they are able to experiment with wooden dowel rods cut into different diameters to ride a variety of boards on a variety of rollers in different ways or hit different rollers to make music (Kamii & DeVries, 1977). In activities designed to study shadows, the children actively experiment with different relationships between lamp, screen, and object (DeVries, 1986). When we compare activities such as the roller activity, the shadow study, and other physical knowledge activities of Kamii and DeVries (1977) to earlier activities found in Hill's curricula, such as stringing beads and cutting paper dolls we can see how Piaget's theory deepens the conceptual domain of educational efforts.

We believe that Piaget's theory of knowledge construction provides a foundation for the design of early childhood activities that provide opportunities to integrate new information and knowledge into a learner's existing knowledge structure. Piaget's theory tells us that it is necessary to begin with meaningful real life experiences; questions rather than answers; and an accepting attitude toward errors.

In terms of early childhood curriculum, Piaget like Dewey, Pratt and other reformers believed that interest and active involvement were essential for learning. According to Piaget, the affective aspect that intervenes constantly in intellectual functioning is the element of interest. To provide for the constructive process, children's interest must be aroused, because without interest, children would never make the constructive effort needed to make sense out of their experience (Chaille & Britain, 1997). Active methods that permit rediscovery or reconstruction of relationships are necessary to children's overall construction of knowledge and intelligence. Children's active involvement in their own learning which permits the opportunity for errors is more beneficial than activities in which factual knowledge is obtained by way of isolated and segmented learning (Chaille & Britain, 1997). Active methods provide a broad scope to the spontaneous research of the child and require that every truth to be learned or rediscovered is, at least, reconstructed by the student and not simply provided for him/her (Kamii & DeVries, 1978).

Recent elaborations of Piagetian applications address peer collaborative efforts, a focus strongly favored by Vygotsky. From a Piagetian perspective, peer collaboration helps children become autonomous learners (DeVries & Kohlberg, 1987; DeVries & Zan, 1994). From a Vygotskian perspective, an individual's learning cannot be separated from the social and cultural milieu in which this learning originated.

Vygotsky's (1978) premise was that the social context facilitates cognitive development. If children have parents and/or teachers who are proficient language users; that is, that they model verbal problem solving and reinforce themselves verbally; then Vygotsky would predict that these children will be more advanced intellectually than those reared in a less expansive verbal environment. Adults and other more experienced members of society mediate the acquisition of knowledge and skills needed to live in a particular culture (Berk & Winsler, 1995).

Vygotsky promoted the idea that social and verbal interaction, valuable in its own right, is also a tool for educating young children (Berk & Winsler, 1995). Contemporary theoreticians in this tradition have elaborated on the role of dialogic processes in the child's appropriation of cultural knowledge (Bakhtin, 1981; Rogoff, 1990; Wertsch, 1985, 1991). In turn, the child's specific cultural knowledge is incorporated into future dialogic processes, regardless of whether these processes are interpersonal or intramental in nature (Bakhtin, 1986).

A key component in the construction of knowledge is the linkage or connectedness of elements in the knowledge base. Some researchers equate the linkage of pieces of information with conceptual level understanding. That is, conceptual knowledge is rich in relationships. Seeing relationships between units of knowledge is the basis of conceptual understanding (Prawat, 1989). The organizational structure of knowledge is provided by the connections between the elements of the knowledge base. The school bus can be understood merely as the vehicle that brings children to school; or, it can be understood as a complex object with stationary and moving parts. Even more elaboratively, it can be understood as a cultural object created to meet the changing demography of residential areas. A shadow can be understood as physical knowledge, but it can also be understood as cultural knowledge. From our perspective, instruction aimed at cultural knowledge is likely to yield richer cognitive structures and more highly motivated learning.

Schema Theory and Projects

Piaget's beliefs about the organization of knowledge and knowledge construction is commonly referred to as a basis for recent research related to schema theory. Schema theory is a kind of informal, unarticulated theory about the nature of the events, objects, or situations that we face (Rumelhart, 1980). A schema may be viewed as consisting of a system of subschemata which corresponds to the constituent parts of the ideas or concepts being represented. Schemata guide our information seeking and can represent knowledge at all levels of our experience. When one initially interprets new information, schemata are used to reinterpret the stored data in order to integrate or reconstruct the original interpretation.

The use of prior knowledge to understand new knowledge or information is a key component of schema theory. New knowledge can be acquired through a gradual process of restructuring in which either new schemata are created to augment existing schemata or old schemata are completely reorganized (Gallagher & Pearson, 1989). For example, what a person expects and knows and how deeply a person understands something influences how much she or he remembers of it.

Schema theory leads us to take children to visit a real bus, or more simply, to visit the school yard in order to see where the shadow of the flag pole is at 10:00 AM, 12:00 PM, and 3:00 PM. The Reggio Emilia curriculum provides a contemporary example of how schema theory can be applied to projects for young children based on themes. A project theme is used in the Reggio Emilia curriculum as the main cognitive support for learning, but then it depends on the children, the course of events, and the teachers to develop appropriate elaborations (Edwards et al., 1993; Staley, 1998). In a project based on "puddles in the rain", children explore the reflections they see in the water, which then leads to experimentation with mirrors and shadows. The puddle schema becomes more complex as children draw trees and people to place around a mirror to make sense of why

> when you are close to a puddle, you see everything, but if you are far way, you see less and less of these complex images (Edwards et al., 1993, p. 130).

Schema theory provides a view of cognitive development in which growth is not simply an accumulation of "pieces of knowledge." Rather growth is better viewed as a "remodeling" of the knowledge structure as a whole (Bransford et al., 1977). Implications of these ideas were in effect realized when Carolyn Pratt described the establishment of a school post office (Pratt, 1948). In this project, children built equipment, made stamped envelopes, set up a schedule of mail collections and deliveries, inaugurated a special delivery service for urgent messages, and instituted a parcel post service. This project evolved over a long period of time and included frequent trips to a branch post office. By establishing a functional post office in the school, societal and community functions became contextualized for children.

It is important that students incorporate new information into their existing world knowledge and that they avoid storing information into compartmentalized units that are primarily accessed by way of either question answering or examination (Bredekamp & Rosegrant, 1996). This compartmentalized knowledge is often learned by rote memorization for the explicit purpose of retrieval upon request rather than being incorporated with existing knowledge of other related subject areas (Bredekamp & Copple, 1997). Projects that share a common topic create classroom contexts for teachers to apply a variety of techniques to assist children in the acquisition of knowledge. In school settings, integrated approaches to instruction provide opportunities for students to learn about social studies and mathematics while they learn science (Bredekamp & Copple, 1997; Bredekamp & Rosegrant, 1996). This combination produces more effective learning than when different topics or units of information are presented in isolated parts. When topics for potential lessons are all related to a common context, students' ability to remember what they learned is much better than when the lessons are presented in a more typical, unconnected format (Bransford et al., 1985; Bredekamp & Copple, 1997; Bredekamp & Rosegrant, 1996).

Projects can also strengthen the linkages between the home and school. An integrated approach can readily facilitate a home-school connection when teachers encourage parent participation and are culturally sensitive to children's home experiences. Rather than depending upon traditional family involvement activities such as parent meetings or conferences with parents, a connection between the home and school can be embedded in the curriculum through the use of projects. This connection can come from information and resources that only the home can provide. For example, when children are asked to bring in cultural objects from the home or ask family members about their own relevant experiences, a connection to the home is automatically established.

For integration to successfully occur, the integration needs to be in the mind of the teacher as well as in the teacher's interaction with children. Social interaction plays a crucial role in learning; meanings are negotiated through exchanges with peers and adults. As we discuss in the next section, knowledge is a product of culture and learners become acculturated through interactions with members of their community.

Narrative Thought

New theories of cognition integrate themes taken from Piaget, Vygotsky, and others. Thinking is defined as an interaction between an individual and a physical or social situation (Greeno, 1989). Thinking is "situated", insofar as it occurs, as an active relation rather than as an activity in an individual's mind. Thus, thinking is inherently affected by culture; knowledge goes from culture to mind as well as from mind to culture. Even more, Bruner (1990) has argued that culture imposes upon human thought patterns that are inherent in the culture's symbolic systems especially forms of story telling or explanation. From a Vygotskian perspective, a tutor's words can support the child's use of a particular problem-solving strategy. Beyond that, the tutor also conveys the culture's judgment that the problem is worth solving, approval of the setting in which the problem is presented, and acceptance of the social resources used to find solutions. The issue, then, is not only what is said, but when and how it is said.

Bruner (1985, 1990) claimed that people do not deal with the world event by event. Rather, they frame events through larger, more inclusive structures that provide an interpretive context for experiences. People also frame experience in a narrative form. Narrative structure plays a central role in the organization of knowledge. It provides a system by which people organize their experience in, knowledge about, and transactions with the world. People have an "innate and primitive" disposition to narrative organization; this disposition allows children to quickly comprehend and use narrative forms to organize experience.

Narrative requires four conditions: agentivity, sequential order, canonicality, and perspective. Agentivity refers to an interest in human action and its outcomes. If narrative structures were used to help children integrate curriculum activities, children might be asked to tell collaborative tales about their efforts to build a school bus or to change the size of shadows. Narrative marks the ordering of events; whether realistic or not. Events are ordered in time. How does the teacher think the shadows activities would be ordered and what order do the children use? Canonicality means that narrative shows a readiness to mark the unusual; when children tell stories about their school activities they are likely to dwell upon the novel or unexpected.

We believe that ideas about narrative thought favor large and expansive thematic units. These units need not be about the "here and now;" quite the contrary. Make-believe and fantasy are intrinsic attributes of children's thinking about the physical and social world (Cuffaro, 1995). The fourth author developed a kindergarten unit called "castles and kings" that nicely illustrates the idea of an expansive unit. This thematic unit crossed several domains and multiple activities. Touching literary matters, the children read about King Arthur as a legend and Cinderella as a fairy tale. They played with a Lego castle and measured the height of play dough slapped onto the wall by a catapult made of blocks. They studied checkers as a Medieval game, made clay dragons and cucumber sandwiches, sang Lavender Blue, and wrote "royal proclamations." These and many other activities tapped children's fascination with the glitter of royalty and the violence of Medieval times.

The children experimented, played, cooked, sang, ate, danced, and shared. The activities were part of an ongoing discussion ranging from pretend characters in legends and tales to the real kings and queens of today. At least three things happen in this unit. First, children are encouraged to interpret their classroom experiences using narrative structures. Ancient castles had moats for protection (build castles out of sand or boxes) and catapults were a way for attackers to damage the castle. Second, narrative forms are used to convey ideas relevant to the curriculum. How the catapult works is connected to a larger story line; its workings mattered to these Medieval people, and these workings can be studied in the classroom. Thirdly, physical knowledge activities, such as the working of a catapult, are embedded in a larger "story" of real and fictive characters (Cuffaro, 1995). In this curriculum, the study of a physical mechanism is situated in cultural problems that human beings try to solve.

In summary, a thematic unit rests upon topically linked classroom activities and projects that cross disciplinary boundaries and engage realistic as well as make-believe thinking. When narrative theory is used to generate thematic units, the school bus goes from a self-contained physical object to a multi-connected cultural object. Several story lines might criss-cross these projects; some story lines might come from the teacher and others from the children

(Cuffaro, 1995). Story lines from the children can facilitate connections between home and school through stories about the family. In the school bus project, children could ask family members, "How did you get to school when were in first grade?" "When you were in High School?" "If you took a bus, what did it look like?" "Was it yellow?" The teacher then tries to provide an interpretive context for activities and projects and the children are encouraged to elaborate an interpretive framework to make sense of their efforts (Bakhtin, 1981; Bruner, 1985, 1990).

FROM CORRELATION TO INTEGRATION

In this article, we have presented recent theoretical work and research which provides a rationale for using *activities*, *projects*, and *thematic units* to weave cultural connections between the young child's life at home and at school. Contemporary theory points to at least 3 ways of looking at the unit of analysis in curriculum organization: activities from a Piagetian perspective; the project from a schema theory perspective; and thematic units from a narrative theory perspective.

Activities
Piaget believed that children's knowledge evolves through an interaction or dialogue with the physical and social environment. When children are interested and actively involved in their learning they construct knowledge from "inside the head" as they encounter problems and apparent solutions (Chaille & Britain, 1997). Since eliciting children's genuine interest and active involvement can be difficult, the strategies used by DeVries and Kohlberg (1987) and those found in the Reggio Emilia curriculum (Edwards et al., 1993; Staley, 1998) might be consulted. Rather than simply asking young children what they are interested in or choosing from a selected set of teacher-determined topics, DeVries and Kohlberg and the Reggio Emilia curriculum provide recommendations for asking questions and engaging children in relevant, meaningful experiences. Activities, which capitalize on interest and active involvement, can also provide structural relations among ideas and concepts and opportunities to integrate new information and knowledge into a learner's existing knowledge structure.

Projects
Schema theory advocates the consideration of a learner's prior knowledge and previous experiences when new knowledge is acquired. New information should be incorporated into existing knowledge structures by means of

meaningful experiences that are related to a common context. Parents, friends, and peers play important roles in cognitive development because they give structure to the experiences of children. Other persons may act as mediators in arranging the environment so that children encounter experiences that help them separate relevant from irrelevant information. A project based on a theme can be used as a framework of cognitive support from which learning can be elaborated by teachers, the children themselves, and the course of events.

Thematic Units
The work of Vygotsky and Bruner also demonstrates the importance of the sociolinguistic context in facilitating cognitive development and learning. From a Vygotskian perspective, adults and more experienced members of society mediate the acquisition of knowledge and skills needed to live in a particular culture. That is, through scaffolding, children advance in consciousness and control as a result of the aid provided by adults and more competent peers. The tutor scaffolds the learning task to make it possible for the child to internalize external knowledge and convert it into a tool for conscious control. Bruner believes that culture imposes upon human thought, patterns that are inherent in the culture's symbolic systems, especially forms of story telling or explanation. According to Bruner, people frame events through a narrative form and narrative structure plays a central role in the organization of knowledge. We believe that ideas about narrative thought favor large, expansive thematic units. A thematic unit rests upon topically linked classroom activities and projects that cross-disciplinary boundaries and engage realistic as well as make-believe thinking.

Each of these theories expands and enriches practices that have had a checkered or even beleaguered past. Many of these curricular approaches and ideas have gone from boom to bust, from popularity too ill repute. The past certainly does seed the future but in ways that are often unpredictable. As new theories help us to think about teaching and learning we see the past with clearer eyes and even at times figure out how to construct new ventures. Although Carolyn Pratt might not approve of Castles and Kings, and Patty Smith Hill might cringe at activities such as Rollers, we believe that contemporary theories provide support for conceptualizing an approach to curriculum that presupposes cultural connections between the child's life at home and at school. The use of thematic units in early childhood curriculum can and should contribute to knowledge construction by framing activities which consider children's interests, active involvement, and previous home and school experiences. We believe that this conceptualization of curriculum holds the potential for formulating activities, projects, and units, which promote young children's knowledge acquisition in meaningful and thoughtful ways.

REFERENCES

Alexander, P. A., Murphy, P. K., & Woods, B. S. (1996). Of Squalls and Fathoms: Navigating the seas of innovation. *Educational Researcher, 25*(3), 31–36, 39.
Antler, J. (1987). *Lucy Sprague Mitchell.* New Haven, CT: Yale University Press.
Arbuthnot, M. H. (1933). The unit of work and subject-matter growth. *Childhood Education, 9*(4), 182–188.
Bakhtin, M. M. (1981). Discourses in the novel. In: M. Holquist (Ed.), *The Dialogic Imagination.* Austin: University of Texas.
Bakhtin, M. M. (1986). *Speech genres and other late essays* (C. Emerson & M. Holquist (Eds), V. V. McGee (Trans.). Austin: The University of Texas Press. (Original work published 1979).
Berk, L. E., & Winsler, A. (1995). *Scaffolding children's learning: Vygotsky and early childhood education.* Washington, D.C.: National Association for the Education of Young Children.
Bransford, J. D., Nitsch, K. E., & Franks, J. J. (1977). Schooling and the facilitation of knowledge. In: R. C. Anderson, R. J. Spiro & W. E. Montague (Eds), *Schooling and the Acquisition of Knowledge* (pp. 31–55). Hillsdale, NJ: Erlbaum.
Bransford, J., Sherwood, R., & Hasselbring, T. (1985). The video revolution and its effect on development: Some initial thoughts. In: G. Forman & R. B. Pufall (Eds), *Constructivism in the Computer Age* (pp. 173–201). Hillsdale, NJ: Erlbaum.
Bredekamp, S., & Copple, C. (Eds) (1997). *Developmentally appropriate practice in early childhood programs.* Washington, D.C.: National Association for the Education of Young Children.
Bredekamp, S., & Rosegrant, T. (Eds) (1996). *Reaching potentials: Transforming early childhood curriculum and assessment* (Vol. 2). Washington, D.C.: National Association for the Education of Young Children.
Bruner, J. (1985). Vygotsky: A historical and conceptual perspective. In: J. Wertsch (Ed.), *Culture, Communication, and Cognition: Vygotskian Perspectives.* New York: Cambridge University Press.
Bruner, J. (1990). *Acts of meaning.* Cambridge, MA: Harvard University Press.
Chaille, C., & Britain, L. (1997). *The young child as scientist* (2nd ed.). New York: Longman.
Cohen, D. H., & Rudolph, M. (1977). *Kindergarten and early schooling.* Englewood Cliffs, NJ: Prentice-Hall.
Cuffaro, H. K. (1995). *Experimenting with the world.* New York: Teachers College Press.
DeVries, R. (1986). Children's conceptions of shadow phenomena. *Genetic, Social, and General Psychology Monographs, 112*(4), 479–530.
DeVries, R., & Kohlberg, L. (1987). *Programs of early education.* New York: Longman.
DeVries, R., & Zan, B. (1994). *Moral classrooms, moral children.* New York: Teachers College Press.
Dewey, J. (1963). *Experience and education.* New York: Collier Books. (Original work published in 1938).
Dewey, J. (1966). *Democracy and education.* New York: Free Press. (Original work published in 1917).
Dewey, J., & Dewey, E. (1915). *Schools of tomorrow.* New York: E. P. Dutton.

Durkin, D. (1987). A classroom-observation study of reading instruction in kindergarten. *Early Childhood Research Quarterly, 2*, 275–300.

Edwards, C., Gandini, L., & Forman, G. (1993). *The hundred languages of children*. Norwood, NJ: Ablex.

Ferrier, A. (1927). *The activity school*. Translated by F. D. Moore & F. C. Wooton. New York: John Day.

Foster, J. C., & Headley, N. E. (1948). *Education in the kindergarten*. New York: American Book Company.

Froebel, F. (1976). *Mother's songs, games, and stories*. New York: Arno. (Original work published in 1914).

Gallagher, M., & Pearson, P. D. (1989). *Discussion, comprehension, and knowledge acquisition in content area classrooms* (Tech. Rep. No. 480). Champaign, IL: University of Illinois, Center for the Study of Reading.

Gelman, R. (1978). Cognitive development. *Annual Review of Psychology, 29*, 297–332.

Gelman, R. (1979). Preschool thought. *American Psychologist, 34*, 900–905.

Greeno, J. G. (1989). A perspective on thinking. *American Psychologist, 44*, 134–141.

Goffin, S. G. (1994). *Curriculum models and early childhood education*. New York: Macmillan.

Goodlad, J. I. (1984). *A place called school: Prospects for the future*. New York: McGraw-Hill.

Hatch, J. A., & Freeman, E. B. (1988). Kindergarten philosophies and practices: Perspectives of teachers, principals, and supervisors. *Early Childhood Research Quarterly, 3*, 151–166.

Hewes, D. W. (1995). *Early childhood education: It's historic past and promising future*. Speech presented at the Annual Graduation Celebration: Early Childhood Education, Long Beach, CA. ERIC DOCUMENT PS 023 394.

Hill, P. S. (1913). *Reports of the committee of nineteen on the theory and practice of the kindergarten*. Boston: Houghton Mifflin.

Hill, P. S. (1923). *A conduct curriculum for the kindergarten and first grade*. New York: Scribners.

Jarolimek, J. (1977). *Social studies in elementary education*. New York: Macmillan.

Kamii, C., & DeVries, R. (1977). *Piaget for early education*. Boston: Allyn & Bacon.

Kamii, C., & DeVries, R. (1978). *Physical knowledge in preschool education*. Englewood Cliffs, NJ: Prentice Hall.

Katz, L. G., & Chard, S. C. (1989). *Engaging children's minds: The project approach*. Norwood, NJ: Ablex.

Kilpatrick, W. (1918). The project method. *Teachers College Record, 19*, 319–335.

Lazerson, M. (1972). The historical antecedents of early childhood education. In: I. J. Gordon (Ed.), *Early Childhood Education*. Chicago, IL: National Society for the Study of Education.

Osborn, D. (1991). *Early childhood education in historical perspective*. Athens, GA: Daye Press.

Parker, S. C., & Temple, A. (1925). *Unified kindergarten and first-grade teaching*. Boston: Ginn.

Pratt, C. (1924). *Experimental practice in the city and country school*. New York: E. P. Dutton.

Pratt, C. (1948). *I learn from children: An adventure in progressive education*. New York: Simon and Schuster.

Prawat, R. S. (1989). Promoting access to knowledge, strategy, and disposition in students: A research synthesis. *Review of Educational Research, 59*, 1–41.

Robinson, J. F., & Spodek, B. (1965). *New directions in the kindergarten*. New York: Teacher's College Press.

Rogoff, B. (1990). *Apprenticeship in thinking*. New York: Oxford University Press.

Rumelhart, D. E. (1980). Schemata: The building blocks of cognition. In: R. J. Spiro, B. C. Bruce & W. F. Brewer (Eds), *Theoretical Issues in Reading Comprehension* (pp. 33–58). Hillsdale, NJ: Erlbaum.
Schoonmaker, F., & Ryan, S. (1996). Does theory lead to practice? Teachers' constructs about teaching: Top-Down perspectives. In: J. Chafel & S. Reifel (Eds), *Advances in Early Education and Day Care: Theory and Practice in Early Childhood Teaching* (Vol. 8, pp. 115–151). Greenwich, CT: JAI Press.
Spodek, B. (1986). Using the knowledge base. In: B. Spodek (Ed.), *Today's Kindergarten, Exploring the Knowledge Base, Expanding the Curriculum* (pp. 137–143). New York: Teacher's College Press.
Staley, L. (1998). Beginning to implement the Reggio philosophy. *Young Children, 53*, 20–25.
Streitz, R. (1939). An evaluation of "units of work." *Childhood Education, 15*(6), 258–261.
Trawick-Smith, J. (1997). *Early childhood development.* Upper Saddle River, NJ: Prentice Hall.
Vygotsky, L. S. (1962). *Thought and language.* Cambridge, MA: MIT Press. (Original work published in 1934).
Vygotsky, L. S. (1978). *Mind in society.* Cambridge, MA: Harvard University Press.
Walsh, D. J. (1989). Changes in kindergarten Why here? Why now? *Early Childhood Research Quarterly, 4*, 377–391.
Weber, E. (1969). *The kindergarten, its encounter with educational thought in America.* New York: Teacher's College Press.
Weber, E. (1984). *Ideas influencing early childhood education.* New York: Teachers College Press.
Wertsch, J. V. (1985). *Vygotsky and the social formation of mind.* Cambridge, MA: Harvard University Press.
Wertsch, J. V. (1991). *Voices of the mind: A sociocultural approach to mental action.* Cambridge, MA: Harvard University Press.
White, C. S. (1988). Kindergarten reading instruction in Georgia. *Georgia Journal of Reading, 14*(1), 5–11.
Wortham, S. C. (1992). *Childhood 1892–1992.* Wheaton, MD: Association for Childhood Education International.

"AIR IS A KIND OF WIND": ARGUMENTATION AND THE CONSTRUCTION OF KNOWLEDGE

Sue Dockett and Bob Perry

ABSTRACT

This paper explores the use of argumentation by young children. Drawing on transcripts of children's play, as well as planned mathematical experiences for children aged five to seven years in a school setting, a social constructivist approach is employed to describe and analyze ways in which argumentation is used by some young children to construct knowledge. Implications for early childhood education are derived from these examples and from a theoretical discussion of argumentation and social constructivism.

INTRODUCTION

A suburban Sydney, Australia, Year 1 class, (children aged 6–7 years) is involved in a series of small group mathematics experiences. Two boys – Jeremy and Lazlo – draw a shape on a deflated balloon, blow up the balloon and report what happens to the shape. They are joined by a teacher.

Jeremy: I have to draw a shape, blow up the balloon and see what happens.
Jeremy blows up the balloon.
J: It's gone, cause I blew it up too much and the ink's gone, it's fade.

Teacher: *Why has it faded?*
J: *It's faded cause it goes stretches and the ink disappears. The ink stretches and leaves little dots and then it disappears. It gets smaller and smaller and smaller and it disappears.*
T: *How come this happens?*
J: *Because it was very long and once it grows they get to be little dots and then it disappears. Then it gets disappearing.*
T: *What makes it disappear?*
J: *Because it's stretching. Because it's growing bigger, cause we're blowing air into it. Air.*
T: *Does air make things grow bigger?*
J: *Yes. Because it's stretching it inside and if you stretch it inside it grows bigger on the outside as well.*

Lazlo: *When I blow up the balloon, I just write what happens.*
Lazlo is holding his blown up balloon.
L: *It gets bigger because the balloon stretches. Like when it starts off small and you blow it up and it goes pop! The balloon goes bigger and then the shape goes bigger. The shape goes bigger. Because the balloon goes like this (uses his hands to demonstrate the balloon enlarging), it stretches up like this and you keep blowing like this and the balloon stretches out. Just because you blow air into it.*
T: *Why air?*
L: *Air is, um, is a kind of a, kind of, blow up. Air is a kind of wind.*
T: *Why would wind make it blow up?*
L: *Cause wind is strong.*

In each of these transcripts, there is an attempt by the child to justify his conclusion in terms of the evidence derived from the activity and his previously constructed knowledge. The reasoning behind this justification is presented in response to both the results of the activity and the interaction with the teacher. There is also an attempt by the teacher to challenge some of those justifications, and to prompt further understanding of the activity, as well as the process of supporting claims. The process apparent in each of the transcripts is one of "argumentation", described by Krummheuer (1995, p. 29) as a

> social phenomenon, when cooperating individuals [try to] adjust their intentions and interpretations by verbally presenting the rationale of their actions.

In this paper, we consider the use of argumentation by young children and its place in an approach to learning based on social constructivism. We draw on examples of argumentation from children's play and more structured situations to consider implications for the roles of adults working with young children.

YOUNG CHILDREN AND ARGUMENTATION

Background

For many people, arguing is a feature of everyday life. Children, as well as adults, engage in the process of justifying actions, negotiating situations and implementing compromises. Early research, derived from a Piagetian perspective, attributed importance to the process of argumentation, but located the ability to argue logically within the realm of formal operations, and so considered it beyond the cognitive capabilities of young children (Inhelder & Piaget, 1958/1977). Studies which focussed on young children's use of argumentation concluded that their skills were poor in comparison with adolescents and adults (Damon & Killen, 1982; Kruger & Tomasello, 1986). For example, Berkowitz, Oser, and Althoff (1987) charted a developmental pattern of argumentative understanding and skill, noting that children between the ages of 6 and 8 years did not routinely or spontaneously provide justifications for their stated positions. In keeping with the proposed link between formal operations and argumentation, Berkowitz et al. (1987) reported that "reasoned dyadic interaction" (Stein & Miller, 1993, p. 303) was identified only in children aged over 11 years.

In contrast to this wave of research, other investigations of argumentation and conflict resolution have focussed on children's spontaneous interactions in a variety of contexts, and in interview situations where the stimulus materials have been adapted in ways that are meaningful and relevant for young children. In these situations, children have been reported to: use justifications for stated positions from about the age of 3 years (Dunn, 1988; Miller & Sperry, 1987; Shantz, 1987); expect to be provided with, and to provide for others, reasons to support assertions (Eisenberg & Garvey, 1981); and to use a range of moral justifications in familiar domains (Stein & Trabasso, 1982). Further, the work of Tesla and Dunn (1992) indicates that young children develop complex negotiating skills in the context of the family. In keeping with this view, Stein, Bernas, Calicchia, and Wright (1998, p. 11) report that children

> become skilled at demonstrating that their position has more benefits than their opponents', especially when that opponent happens to be a sibling.

The context of arguments has been raised as an issue by Stein and Miller (1993), who note that young children may not demonstrate their actual competence in spontaneous situations. While spontaneous interactions, such as those that occur within play, are often of particular interest and relevance to young children, these researchers suggest that children are aware that conflicts with friends can have serious implications for that friendship. They also note that children tend to avoid arguments if they believe they do not have sufficient knowledge to win. To reflect these cautions, investigations of children's argumentation should include play contexts, where spontaneous interactions abound, as well as some more structured situations where justification and clarification can be sought. In promoting a methodology which uses questions and probes rather than total reliance on spontaneous interactions, Stein and Miller (1993, p. 305) suggest that

> the ability to produce complex discourse spontaneously often lags behind the ability to understand and explain events when directly probed for answers.

One active area of research in children's use of argumentation involves elementary mathematics classrooms (Horn, 1999; Krummheuer, 1995; Yackel, 1998, Yackel & Cobb, 1996). For example, Yackel and Cobb (1996) have outlined the importance of creating a classroom context where children are encouraged to develop and present mathematical explanations and where the value of attempting to make sense of the explanations proffered by others is recognized. The norms established in these classrooms set up an expectation that children, as well as adults, can try out their preferred explanations by sharing them in the social context of a mathematics lesson. Further, the expectation is that arguments will be acknowledged and respected as they are discussed and tried out. In other words, children will not be ridiculed if and when they proffer an explanation that differs from other explanations of the same event. Within this context, teachers promote the discourse of argumentation.

Elements of Argumentation: What Is An Argument?

The notion of argumentation, as reported here, has developed in large part from the social interactionist approaches to mathematics learning propounded by Bauersfeld and his colleagues (Bauersfeld, 1988) and the work of Krummheuer (1995). There are several definitions and types of arguments described in the research literature, including interactive arguments (Stein et al., 1998) and evaluative arguments (Stein & Miller, 1993). Despite the common usage of the term, argumentation does not necessarily relate to a dispute or conflict. Rather, argumentation is a process of justification or support for a claim that is made. In other words, argumentation is the process of providing evidence to convince others about the validity of a claim. Sometimes argumentation occurs in the context of a dispute. Other times, it occurs when another does not seem to understand the claim being made, or to add detail to a claim. At still other times, the claim made and its justification are viable within the social context and not challenged by anyone within the context.

When, in the transcript at the beginning of this paper, Lazlo makes his claim that *the shape goes bigger because the balloon goes bigger*, he believes that he is making a rational statement and that he can justify his claim, if necessary. "... [T]hese techniques or methods of establishing the claim of a statement are called *argumentation*" (Krummheuer, 1995, p. 232). The step by step result of the argumentation process is an *argument*. It provides a justification for the claim which is acceptable to the participants of the argumentation process, at least for that moment.

Argumentation usually consists of a number of elements which together seek to justify a claim. Toulmin's (1969) model of argumentation provides a framework which identifies six elements: claim (or conclusion); evidence (data); warrants; backing; qualifiers; and counterarguments. The first step in an argument, a claim or conclusion, is a statement or stance that the arguer believes to true. In the case of Lazlo's argument, his claim that *the shape goes bigger* is his belief about what has happened. The second element of an argument is evidence, or data, that is used to support that belief. In Lazlo's example, the evidence is that *the balloon goes bigger*, hence the shape also gets bigger. Sometimes there are many reasons cited in support of a belief. The data may be drawn from multiple sources, such as observations or statements from others. In the arguments of young children, it is not uncommon for reference to be made to 'authority' figures, such as parents or teachers, as a form of support for a particular position (Perry & Dockett, 1998).

Warrants are "beliefs and assumptions that guide an arguer's choice of the kind of evidence that must be offered to support his or her claim" (Stein & Miller, 1993, p. 294). Warrants may be unstated, or hard to distinguish, as they draw on the social context for meaning. They reflect understandings within that context about what is meaningful and important and, because of this, may not be clear to others outside that context. Lazlo's warrant emphasizes the stretching of the balloon–*the balloon stretches*–and reflects the assumption that stretching explains the increase in the size of the balloon.

Warrants may by supported by backing. Backing provides additional support for the warrant. As with warrants, backing may be implicit in arguments. Lazlo seems to make the decision that his statements about the balloon getting bigger because it has stretched are not sufficient to convince the teacher. He adds some backing to his argument, stating that the balloon stretches *because you blow air into it*. When asked *Why air?*, he provides further backing, stating that *air is a kind of wind*, and *wind is strong*.

Lazlo did not provide qualifiers or counterarguments. A qualifier is some piece of information that qualifies, or limits, the conditions in which the claim is taken to be true. Counterarguments provide challenge to the claim being made. In Toulmin's model, qualifiers and counterarguments originate from the arguer. However, Stein and Miller (1993) note that if these occur at all, they are likely to come from any opponent in an argument, rather than the originator of the argument.

Toulmin's model has been used as the basis for a number of investigations of young children's use of argumentation (for example, Cobb & Bauersfeld, 1995; Cobb, Yackel & Wood, 1995; Krummheuer, 1995; Yackel, 1998). Stein and Miller (1993) have adapted a working model of argumentation from Toulmin's (1969) conceptual model. This working model emphasizes the importance of a claim and data, or a claim and a warrant in order to generate the discourse of an argument. Other elements as listed by Toulmin – the backing, qualifiers and counterarguments – are noted as adding strength and coherence to an argument, but are not considered essential to constitute discourse as an argument.

The research reported in this paper focussed on young children's use of argumentation in the early years of school. This focus derived from an interest in the nature of the curriculum presented in these years – in particular, the stated curriculum objectives of developing the skills of critical awareness–and from the stance of social constructivism, which emphasizes the importance of the social context in which learning occurs. The study set out to consider the presence, or absence, of the elements of an argument, as listed by Toulmin, in children's interactions.

The Importance of Argumentation

Why is it important to consider young children's use of argumentation? The arguments that children develop and use provide some clues about the knowledge and understandings they are constructing. Stein et al. (1998) report that the nature of knowledge about particular issues is reflected in the stance taken in an argument. The language and the logic used to justify this stance, and to negotiate with others about the issue, presents a range of insights into the knowledge constructed by different individuals.

Piaget (1965) emphasized the importance of children justifying their position and took such justification as a strong indicator of their level of understanding. An essential element of Vygotskian theory is that all higher forms of mental functioning are constructed jointly in a social context and then internalized by the participants through dialogue with others (Berk, 1994). In this way, "language is a major cultural tool that enables us to think logically and to learn new behaviors" (Bodrova & Leong, 1996, p. 95). Social experience shapes the ways in which individuals think about and interpret the world. Social experience is shared with others through language. Language then is the

> major means by which social experience is represented psychologically and it is an indispensable tool for thought (Berk, 1994, p. 30).

How children use language provides some clues as to the ways in which they are thinking and some indication of the learning that has occurred and that is likely to occur. The focus of this paper is the insight into children's learning and thinking provided by the language used in argumentation.

The aim of the study reported here is to consider younger children's use of the process of argumentation in open-ended situations. Underlying this aim is the proposition that simultaneously, young children are constructing understandings of both the process of argumentation and of the topic they are discussing. In the case of Jeremy and Lazlo, they are building up an idea of what it means to provide justification and of what constitutes appropriate justification, as well as an understanding of why the drawing disappears when the balloon is blown up.

METHOD

Observations were undertaken in two metropolitan schools in Sydney, Australia. One school was chosen for its well developed play program in Kindergarten, where three classes (approximately 70 children with an average age of 5 years) combined to engage in extended sessions of play each week. A wide range of equipment and resources was provided during these sessions and children moved freely to both the indoor and outdoor areas. While many of the planned experiences related to the current teaching/learning theme, there was no set product expected from any of the experiences. In this setting, two hours of play interaction was observed and recorded by each of two researchers.

The second school was chosen on the basis of the mathematics program that had been developed by the Year 1 teacher. This program highlighted small group experiences based on planned activities, but the ways in which children responded to these activities and were encouraged to describe and document their experiences remained open-ended and in the control of the children involved. Approximately 25 children with an average age of 6 years participated in these experiences. A further two hours of interaction in this context was observed and recorded by each of two researchers.

Children's interactions were recorded using audio tape and field notes. From transcripts, several interactions were chosen for analysis, on the basis that each provided some indication of supporting statements for a stated position or conclusion. These are not presented as representative of all children observed, but do provide some clues to answer the question of whether or not children aged 5–6 years use argumentation in the specified contexts.

Differences between this and other studies include a focus on children younger that those involved in previous studies and a focus on play as well as mathematical contexts within the classroom. The rationale for these differences rested on the belief that if children in Year 2 classrooms as described by Yackel and Cobb (1996) actively use argumentation, then the basis for that use may well be developed prior to this. Given that young children develop a great deal of conversational competence, especially when focussed on achieving their own goals (Dunn, 1988), during the preschool years, it may reasonably be expected that some of that competence relates to providing justification for a stated position or goal. The spontaneous nature of play interactions and the ability of children to focus on issues that matter to them, was behind the decision to include play interactions in the analysis of argumentation. It was expected that children would be most likely to justify their statements and positions in situations where the outcome mattered.

RESULTS

In this section, we use a diagrammatic representation of Toulmin's (1969) model, described by Krummheuer (1995), to analyze several arguments.

Jeremy's Argument

The beginning of an argument is a *conclusion* inferred from *data*. Krummheuer (1995, p. 241) represents this inference (Fig. 1).

In the transcript of Jeremy's interaction, the conclusion he makes is that his drawing of the shape (the ink) disappears, while his data appears to be that the balloon stretches when it is blown up. Diagrammatically this is represented in Fig. 2.

In both the interactions between the teacher and the two boys, some other elements of Toulmin's (1969) model of argumentation appear. The first of these

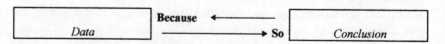

Fig. 1. Representation of the Inference "Data, so Conclusion".

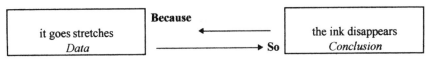

Fig. 2. Representation of Jeremy's Inference.

is the *warrant*. Utterances which act as warrants in arguments legitimize the links which have been made between the data and conclusion of the inference. That is, they establish that these links are reasonable in the current situation. Warrants are general statements, accepted by the participants, and which act as "bridges" (Krummheuer, 1995, p. 242), linking the data and conclusion. Jeremy presents two warrants arising from his observations:

(a) that there is a gradual disappearance of the shape as the balloon is inflated; and
(b) that the balloon is growing bigger as air is blown into it.

Both of these serve as links between the data and conclusion as they offer support for why it might be that stretching the balloon results in the disappearance of the ink. This gives what Krummheuer (1995) has called the core of an argument, where the warrant explains the data and links it to a conclusion. This is represented in Fig. 3.

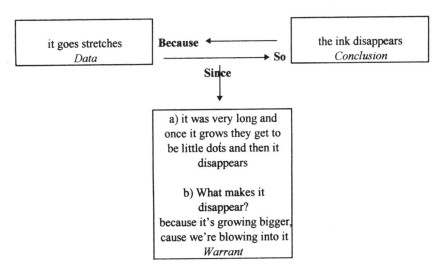

Fig. 3. Representation of Jeremy's Core Argument.

In many arguments, particularly those involving young children, it is difficult to distinguish between the data and the warrant(s). Frequently, children do not use warrants based on general principles to support their arguments. Anderson, Chinn, Chang, Waggoner and Yi (1997, p. 164) report that general principles are only used as warrants when

> a common cultural value is relevant . . . the implied warrant is surprising or confusing in some way [or when] the warrant is implicated in a dispute.

In other situations, the warrant remains implicit. Despite the difficulties in sometimes distinguishing between data and warrant, it is important that children seem aware of the need to justify a claim in order to have it accepted within the social context, and that they are becoming aware of the ways in which they might do this. Such an awareness indicates that children understand that different people have different perspectives and perceptions and that they need to explain their views in order for others to access and respond to these. In this way, an understanding of argumentation can be linked to other developments in understanding the social and cognitive world, developing at about the same time, such as children's theories of mind (Flavell, Miller & Miller, 1993). It also suggests the development of a critical, rather than an accepting, approach to events and situations.

The warrant aims to support the link between the data and the conclusion. Just as this link can be challenged, the authority of the warrant can also be questioned. To strengthen this warrant, *backing* may be used. The backing establishes the authority for each warrant:

> Standing behind our warrants . . . there will normally be other assurances, without which the warrants themselves would possess neither authority nor currency – these other things we may refer to as *backing* of the warrants (Toulmin, 1969, cited in Krummheuer, 1995, p. 243).

Jeremy's argument provides a clear example of the model (Fig. 4).

In this example, Jeremy is claiming that *the balloon goes stretches* because *it's growing bigger because we're blowing into it* so *the ink disappears*. Supporting this position is the backing that *it's stretching inside and if you stretch it inside it grows bigger on the outside as well*.

Analysis of any argument, but particularly those of young children, requires observation of the interaction as well as a consideration of what is said. In many interactions, children imply specific meanings through their actions, much as Lazlo does by indicating with his hands how the balloon gets bigger. It is also quite likely that all the elements of the model detailed above will not be present in all, or even most, of the arguments of young children. (For example, any backing which Jeremy might have for the first of his warrants

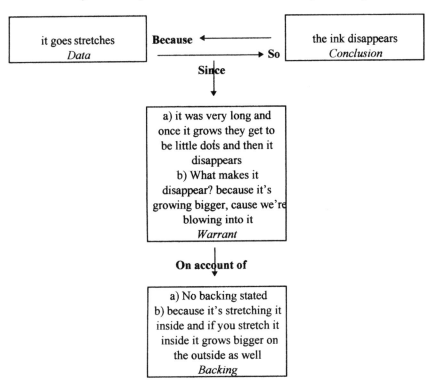

Fig. 4. Jeremy's Argument.

remains unstated or implied.) Nevertheless, it is important to recognize that in some instances, young children are becoming aware of the need to justify what is claimed and of the necessity for using some form of warrant and then backing to ensure that this justification is viable. These are complex processes that are used by young children.

Anna and Briony

The argumentation reported for Jeremy and Lazlo occurred as they interacted with an adult. While interactions with adults can be critical in promoting argumentation, interactions with peers can also contribute to children's involvement in the process. Play is one context in which argumentation among peers can develop. Such argumentation does not necessarily involve a dispute, although this may be the case, as reported elsewhere (Perry & Dockett, 1998). Play can also contribute to a collaborative process of argumentation, where children each provide some elements of the process and together justify their conclusion. This is the case in the following example:

> Anna and Briony are part of three Kindergarten classes involved in a planned play session. They are at the "weighing shop" which is equipped with balance scales, buckets and sponge blocks. A researcher(R) is observing their interaction.
>
> A: This is a weighing shop. Here's some chicken.
> B: This one here is light, this one here is the biggest. This one is both balancing because they have the same blocks in them.
> A: Yeah and they're getting heavier each time. We're putting meat in (sponge blocks).
> B: Mine is balancing because it goes like this (straight).
> R: What does it mean?
> A: It's like um ... like this (holds the beam straight). Balanced because it goes like this.
> B: But not like this (moves the beam on the balance to an angle).
> A: And if it stops and it's on the same side (that is, level), it's balancing.
> R: This side is up and this one is down. Why is that?
> A: I know ... cause that one's the heaviest. Cause it's got more blocks.
> B: And that one's lighter.

A possible analysis of this interaction, according to the Toulmin/Krummheuer model outlined above is given in Fig. 5.

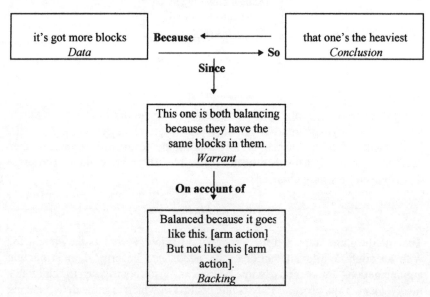

Fig. 5. Ann and Briony's Argument.

In this example, argumentation was accomplished interactively. While Anna supplied both the data and conclusion, Briony supplied the warrant and both have contributed to the backing. The backing is provided through words and actions which indicate how the beam on the balance moves when the contents of the two buckets are the same or different. Anderson et al. (1997) note that typically, arguments are developed in a socially situated activity and constructed by the interaction of several participants. This is particularly so when an issue is "engaging or controversial" (p. 140).

The play context provided the stimulus to engage in the experience and an opportunity to develop the argument collectively. However, it is noticeable that without the questions of the adult, the children could well have completed the experience without detailing or explaining the position they adopted. This does not imply that the understandings were not present: rather that they would not have been made explicit without the scaffolding provided by the adult.

Bianca and Robert

Bianca and Robert are involved in the Kindergarten play session. Bianca places some felt figures on a feltboard. She has a series of pattern cards next to her and refers to one as she places the pieces on the feltboard. Robert arrives and watches for a few minutes.

Bianca: *[looks up]* This is a robot. Playing around. Dancing, it's dancing.
          [she moves the pieces as she speaks]
R: Huh? How do you know?
B: 'Cause got these things*[points to different shapes]* and these *[points to the patterns that accompany the pieces]*.
          [she picks up a remaining piece of felt]
          I can't do it. I can't do it. I've got this one *[spare piece]*. I don't know where to put it.
R: Maybe it goes down the bottom.
B: No. I've gone right down to the bottom. Can't put it anywhere else. That's a head *[points]* with a hat on it and here's this one *[the piece still left]*. I got nowhere to put it. No. Not down the bottom.

In this instance, Robert's question of *How do you know?* prompts Bianca to refer to both the feltboard pieces she has placed on the board, as well as the instructions, to support her conclusion that *This is a robot*. Her two sources of data are interesting-reference to the pattern adds a level of authority that was not evident from the presence of the pieces themselves (Fig. 6). Bianca also makes the claim that she "can't do it" when she finds a left over piece. She supports this claim by describing her actions, thus engaging in a further use of argumentation. However, in these instances, Bianca provides neither explicit warrant nor backing, nor does Robert seek any further justification.

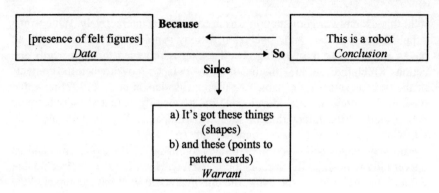

Fig. 6. Bianca's Argument.

Corrie's Argument

A similar analysis of the interactions and utterances is reflected in another transcript of play. In this instance, Corrie (from Kindergarten) had been engaged in parallel play as she painted. Other children were using the same materials as they painted, but there was no interaction. As the teacher approached, Corrie looked up, holding her painting.

Corrie: *It's a rainbow-but it went away. Cause the rain stopped. When the rain goes, it starts sprinkling and then it gets heavier, then it will slow down when the rainbow comes and then stops. It goes when the rain comes.*
Teacher: *Where?*
C: *Goes home. Back where all the rain came from. It goes to bed and it has lots of sleep. Because it gets cold. After that it goes back in the rain.*
T: *When do we see the rainbow?*
C: *It's always the same rainbow, cause it's got the same colors in it.*
T: *There's only one?*
C: *Yes. A little bit different one. Probably 'cause the colors wear out and it's a little bit different when the colors come back.*
T: *How?*
C: *Well, the rain would wash it all [color] out. Then God has to paint it again. If He ever mixes the paint up! [laughs].*

Corrie offered her explanation of why the rainbow went away. She indicated that rainbows and rain do not appear at the same time and then suggested that the rainbow went home to sleep. In what sounds like an argument that may have been used by others to convince Corrie that she should stay out of the rain,

"Air is a Kind of Wind": Argumentation and the Construction of Knowledge 241

she says that the rainbow *goes to bed and it has lots of sleep. Because it gets cold.*

Corrie then offers an explanation for why rainbows are always the same colors-in her words, *it's always the same one.* Even when they look different, she explains that they are really the same with reference to the colors being washed out by the rain and then needing to be repainted (Fig. 7).

Of particular interest in this argument is Corrie's use of a warrant – *God has to paint it again* – which has an implicit rather than stated backing. Corrie then introduces a counterargument, which acknowledges that there may be some difference in the colors of the rainbow, but then offers two explanations to negate the counterargument and lend support for her original position that we always see the same rainbow.

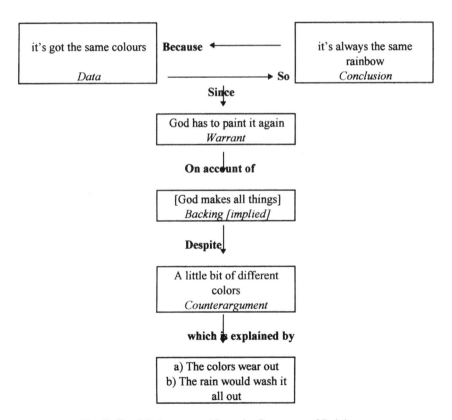

Fig. 7. Corrie's Argument About the Sameness of Rainbows.

DISCUSSION

Children's Arguments

Argumentation can occur in many contexts. While young children may not use the entire, logical process described by Toulmin (1969) when seeking to justify a claim, their interactions indicate that they are becoming aware of the need to produce evidence and of the nature of evidence that provides an acceptable link between data and conclusions. Children's arguments seem to be understood and accepted within relevant contexts. Anderson et al. (1997, p. 164), argue for a focus on conversational logic which recognizes that many arguments that do not meet the formal criteria outlined by Toulmin (1969) can be judged "logically complete and formally sound" on the basis that

> the logic of an argument can depend on the meaning as well as the form of the terms and that normally the participants in an argument are observing . . . the maxim that one ought to say no more that necessary.

One measure of the acceptability or otherwise of an argument is the response from others involved in the interaction.

From an early age, children engage in argumentation. Stein and Miller (1993, p. 329) suggest that:

> . . . in familiar contexts, children as young as 4 can bring evidence to bear on a position that they support. Moreover they can and do offer criticisms for opposing positions. The difficulties experienced by arguers often derive from their lack of knowledge about appropriate evidence to support a position.

This suggests that a number of conditions need to be fulfilled in order for young children to develop complex arguments. In particular, the context needs to be familiar, the topic of the argument needs to be one which is well understood and the process of argumentation – especially what constitutes acceptable evidence – needs to be clear.

Anderson et al. (1997), note that children's arguments can appear vague and, as described in this study, lack explicit connections, warrants or backings for warrants. However, they emphasize that what is missing in arguments can usually be inferred by other participants. Even when arguments appear to be missing elements and where connections seem vague, it is remarkable that those participating in the context generally make sense of what is being said. One way in which this may be achieved is by the conversational partners filling in the gaps themselves as the interaction progresses. In this way, the context in which the argument occurs contributes to the process. This may be because information relevant to the argument has already been discussed, or because such information is readily available in the particular context and does not need to be re-stated. In Anderson et al.'s (1997) reports of arguments involving children in Grade 4, only when the basis for the argument was unclear to others or when points were contested, did participants offer additional information in the form of warrants or backings. This suggests that disputes around particular points can provide the impetus to be more explicit about the connections made, as well as the warrants and backings used to support a position.

In analyzing children's arguments, Anderson et al. (1997) themselves call upon some general principles. One of these is the "principle of charity" (p. 146) which translates roughly as the assumption that in constructing arguments, children (indeed all of us), are rational and aiming to make sense. Hence the starting point for considering children's arguments is that they intend to make sense and to convey a position which is, at least to them, rational. A further set of conventions – maxims underlying Grice's (1975) co-operative principle which is held to underpin effective communication (Siegal, 1991) – are also used to consider children's arguments and to explain some of the gaps that are noted. The maxim of quantity, which roughly translates as "speak no more or no less than is required" (Siegal, 1991, p. 31) is one of the conversational rules that may be followed by children when they do not state what they believe to be obvious. When we assume that children intend to make sense and that they may offer only information in their arguments which they believe is necessary to convey that meaning, we are able to consider children's arguments within the context in which they occur and to derive some understanding of what knowledge children bring to the situation, both about the topic being discussed and the process of developing an argument.

Young children engage in argumentation, though it is clear that in many instances, some elements of such arguments are either vague or missing. Despite this, children's arguments do make sense to them and often, to those around them. Anderson et al. (1997), report that "the arguers understood each other most of the time" (p. 163) as participants in the argument filled in missing detail by reference to the context, previous conversations or information that had already been shared by the participants. They note that even when gaps appeared in the structure of the argument that there was an expectation on the part of the speaker that their conversational partners would fill in the gaps. The conclusion of these researchers is that

> the form of children's arguments is acceptable, as long as you take the perspective of an actively cooperative participant in the discussion (p. 166).

This is significant in that the ability to 'fill in the gaps' in an interaction is one of the characteristics of intersubjectivity, which is discussed in relation to social constructivism later in this paper.

Argumentation and Play

Play provides an ideal context for negotiation and compromise. Play provides a familiar context for many children, where they have considerable control over the nature and direction of interactions, as well as their own levels of involvement. The data collected for this study indicate that some argumentation does occur within play, however this did not seem to be a predominant focus of the observed play. Further detailed observation and analysis may reveal more. We suspect that play, for many children, is not as free and interactive as it is for others, with more dominant and popular players controlling much of the action and less dominant players having either to follow the directions offered, or leave the play (Dockett, Szarkowicz, Petrovski, Degotardi & Rovers, 1999; Main, 1996; MacNaughton, 1997). The power differentials created in some play situations may mitigate against the spontaneous use of argumentation. Nonetheless, there is potential for children to interact through play in situations which matter to them and with people who matter to them. This makes play an appropriate context for young children to construct and to test their knowledge of issues and of processes, such as argumentation.

Arguments With Friends

Several of the transcripts reported in this paper involved an adult asking questions in order to make explicit some of the assumptions children relied upon in their arguments. Peers, as well as adults, can adopt this role. The interactions of Anna and Briony, and Bianca and Robert promoted the development of arguments. These interactions were cooperative, with no evidence of dispute. When disputes do occur between friends, interactions have the potential to either promote or inhibit argumentation. On the one hand, there is evidence that play between friends generates a greater number of conflicts than play among non-friends and that, because of the importance of the friendship, there is considerable incentive to resolve these in an amicable manner (Dunn, 1993; Hartup, 1996). However, Stein et al. (1998), note that there are social and personal goals at stake in any argument and that sometimes these take precedence over the intellectual goal of developing a logical argument and evaluating evidence. The nature, or value, of the relationship may result in a greater willingness to consider other points of view:

> If arguers value their relationship with their opponent and are determined to maintain that relationship, they may take a less rigid stance on an issue .... Depending upon how important the relationship is, arguers may choose to exercise flexibility in their commitment to a stance ... arguers who are good friends may have a high degree of trust

in each other's ability to evaluate evidence. When opposing information is introduced, a close friend may be far more willing to listen to exactly what the nature of the opposition is. If a high level of trust is not present, then arguers may choose to focus on the limitations rather than the strengths of a position (Stein et al., 1998, pp. 9–10).

It may be that the stance taken in an argument between friends is deemed to be less important that the friendship, and compromise is possible. When compromise occurs, there is clear evidence of a shift in understanding among the participants, as an original position is revised, additional evidence is evaluated in relation to existing evidence and understanding, and a new, modified position adopted (Stein et al., 1998).

Argumentation and Social Constructivism

The importance of argumentation for educators relates to the window it provides onto the young child's construction of knowledge. It enables educators to consider the ways in which young children explain and connect issues and ideas as they interact within their environment, and to plan to extend, refine or complicate their understanding.

The construction of knowledge takes place in a social context and as part of a series of social interactions (Fosnot, 1995), where

mental functioning in the individual can be understood only by examining the social and cultural processes from which it derives (Wertsch & Tulviste, 1992, p. 548).

This is clearly so for argumentation. In the cases of Jeremy and Lazlo, the reported interactions are between the individual child and the teacher. In later examples, interaction between peers suggests that children can participate in the joint construction of arguments, reinforcing the importance of peer interactions. In emphasizing the social context of learning, children's interactions are considered through a social constructivism lens, where learning is regarded as "an interactive as well as a constructive activity" (Cobb, 1990, p. 209) and where

the social contexts of learning and the interactions which occur in them play a fundamental role in children's development (Buzzelli, 1995, p. 272).

As noted previously, social experiences influence the ways in which individuals interpret and make sense of the world. The role of language within social contexts is crucial to learning and development. In particular, dialogue between children and more experienced and knowledgeable others results in children internalizing the language used in interactions and applying it to their own independent experiences (Berk, 1992).

The importance of the social context of learning, and the dialogue which occurs within that context, are detailed by Buzzelli (1995) in his application of a dialogic model of the role of adult-child discourse in moral development. Because of the relevance of these elements for argumentation as a process of social construction, each is discussed below, under the broad headings of: internalization as a social process; ways in which discourse shapes thinking; and the role of the teacher.

(i) Internalization as a Social Process

Internalization is one of the central notions of the Vygotskian theory of cultural development. Through internalization, higher mental functions move from the external to the internal plane:

> Any function in the child's cultural development appears twice, or on two planes. First it appears first between people as an interpsychological category, and then within the child as an intrapsychological category (Vygotsky, 1981, p. 163).

Lawrence and Valsiner (1993, p. 151) describe internalization as

> the process by which material that is held out for the individual by social others is imported into the individual's intra-psychological domain of thinking and affective processes.

Rogoff (1990) has proposed a change of focus from internalization as a two stage process of understanding moving from the external to the internal, to a focus on appropriation, whereby

> children do not take the external and make it internal through simple importation. It is through participation in activities with others that their understanding of the social activity is changed. Children's understanding of their participation in social activity with others and of the activity itself are appropriated for use in other similar settings (Buzzelli, 1995, p. 276).

The concept of appropriation suggests that children are engaged in social experiences when they observe and interact with others. This form of engagement in social activity means that

> what is practiced in social interaction is never on the outside of a barrier, and there is no need for a separate process of internalization (Rogoff, 1990, p. 195).

Argumentation is itself a social process. Even when children do not seem to be engaged in conflict or dispute, they utilize their understanding of the social world in their arguments. The concept of appropriation suggests that as children engage in argumentation, they take on both the nature of the activity and their role within it. In other words, they are learning about argumentation, what is acceptable and appropriate within a specific context, as well as how to engage in argumentation.

(ii) Discourse Shapes Thinking

Sociocultural theory emphasizes the contexts in which interactions occur and the ways in which signs, such as language, are used in these contexts.

Communication is essential for learning and development. In relation to arguments, the interactive contexts in which they occur, and the ways in which language is used, help children build up an understanding of the "rules, standards and procedures" (Buzzelli, 1995, p. 278) of argumentation. As well as working out what sort of evidence is required to justify a claim and ways in which to deliver that information, children are also working to 'fill in the gaps' in arguments. As noted previously, not all relevant information is contained in the arguments of children, and there is a reliance on others being able to make sense of the limited information because of common experiences, understandings or assumptions.

Several of the arguments reported in this paper involve one child justifying a claim they have made. There is no dispute, so there is no need to explain a position or statement in the face of challenge from peers. However, the children apparently accept the need for justification, and in the case of Corrie, offer this with little prompting from the adult. It is almost as if she is thinking out loud as she seeks to understand her own reasoning. When Jeremy and Lazlo describe what they are doing, they seem to repeat the instructions of the teacher to guide their actions: *I have to draw a shape, blow up the balloon and see what happens.* As they proceed to make a claim and justify it, there is a move away from the words of the teacher to statements that reflect their own knowledge of what has happened and why.

In sociocultural theory, the move from external to private speech and then to inner speech, or thought, is a significant step in the move to higher order functioning (Berk & Winsler, 1995). Wertsch and Tulviste (1992, p. 550) note that

> inner speech enables humans to plan and regulate their action and derives from previous participation in verbal social interaction.

As children are asked to support claims they make, they rely not only on arguments they have heard from other children and adults, but also organize data and conclusions, warrants and backing in ways which demonstrate their own approach to understanding and solving problems. Argumentation can both support and reflect the shift children make from external, to private and then inner speech.

Children's understanding develops through discourse. In various circumstances, this discourse will occur in dialogue with adults, peers or through private speech. When communication occurs between individuals, understanding is dependent on the establishment of intersubjectivity. Through intersubjectivity, participants who approach a task with different understandings come to an agreed understanding. Intersubjectivity creates a common ground for communication as each partner adjusts to the perspective of the other (Berk, 1997, p. 248). Intersubjectivity is arrived at during interactions (Tudge & Winterhoff, 1993) and is characterized by a recognition and coordination of intentions (Trevarthen, 1988). Two elements need to occur simultaneously for intersubjectivity to develop. These are the acceptance of a "shared focus of attention and agreement on the nature of communication" (Goncü, 1993, p. 186). According to Goncü, the shared focus of attention enables participants to expand their existing knowledge. In the case of Anna and Briony, their shared focus on the 'weighing shop' enabled them to expand their knowledge of weight and weighing. The second element noted by Goncü, that of an agreed form of communication, was also in evidence in the interaction between Anna and Briony. In this case, argumentation, supported by both words and actions, was the form of communication used to present and justify a position.

When children interact with adults, adult perspectives can dominate the intersubjectivity that develops. Adults can summarize, define and revise the events of the classroom or the interaction, as they present a particular explanation for events. The teacher who finishes a learning experience by discussing "what we learnt today" seeks to establish a common focus for the events of the day by using a specific form of communication. In other words, the interpretation of the adults is often the one that becomes accepted and acceptable in interactions. This is particularly so in classrooms, where the power of the adult to shape discourse is strong. The classrooms in which the arguments reported in this paper took place were ones where the grounds for interaction had been negotiated between teachers and children. While this does not negate the power differential, it does mean that the rules were explicit and available to all.

The dialogic model emphasizes the structure of classroom interaction and the rules which guide this. The conversational rules of each classroom are shaped by the participants – the teacher and students. These rules guide the interpretation of events and the understanding of the dialogue that occurs in that context. For example, Lazlo's use of his hands to demonstrate the effect of the balloon being inflated reflect an acceptable practice in that classroom. Buzzelli (1995, p. 283) concludes that

> the structure and content of classroom discourse influence the way children interpret classroom events and the meanings they give to shared experiences in the classroom.

The process and the content of argumentation differs across social contexts. For example, the justification from Lazlo that "air is a kind of wind" to explain why air makes the balloon stretch is probably not the sort of explanation we would expect from a physicist – or even an adult non-physicist – yet it is viable for a child in Year 1. Part of the reason for this relates to the 'rationality' of the justifications provided. What is rational to a six-year-old, may not be considered rational to an adult. Participants in the process of argumentation have a role to play in creating and accepting what constitutes a rational argument within the social context. The 'logics' used by Jeremy and Lazlo have been worked out through interaction in their classroom context – they, their peers and their teacher have established a set of classroom norms that support certain actions as children undertake tasks. For example, a large part of Lazlo's argumentation is based on physical demonstration and simulation, something which he believes is acceptable within the classroom but which is unlikely to be part of a formal justification in another context. What is important here is that the child's rationality should not be dismissed because it does not match the rationality of an adult. Stein et al. (1998) emphasize that, to the arguer, the reasoning and interaction they propose, makes sense.

(iii) Roles of the Teacher
In the classroom context, much of the communication is controlled by teachers. In particular, teachers use a number of strategies to indicate to children which aspects of discourse are important and which are not. These strategies are used to define and redefine classroom experience in ways that are acceptable to the teacher. For example, teachers use markers such as "royal plurals" to indicate an expected common understanding (What do **we** think is happening here?) and strategies such as reconstructing and paraphrasing to summarize what has happened, at least according to their perspective. Buzzelli (1995, p. 285) notes that these strategies

> highlight the importance of the teacher's role in providing the words that children will use to define and interpret their experiences.

These strategies imply that the control of the classroom and the discourse within it, rests squarely with the teacher. In many instances, this may be so. However, there are also strategies teachers can employ to encourage children to assume greater levels of control. Of particular relevance to argumentation is the use of questions, particularly assistance questions which seek "to promote mental operations through which children connect pieces of information" (Buzzelli, 1995, p. 286). The questions asked by adults and reported in this paper clearly aim to help children connect pieces of information and to use them in ways that support, or challenge a conclusion they have made.

Buzzelli (1995) applied his dialogic model to moral development. However, a consideration of the elements noted above, indicates that it can be used to explain the ways in which children develop an understanding of the process and nature of argumentation, as well as acceptable content of an argument. The focus on dialogue within social contexts, and the ways discourse can shape understandings raises a number of implications for the role of teachers in classroom contexts. These are discussed in the following section.

Implications for Teachers and Teaching

One of the issues to arise from this exploration is the importance of the adult in facilitating argumentation. While there are many opportunities for differences in perspectives and opinions to be noted within classroom interactions, indications from this study are that it often requires an adult to create a situation which calls upon children to explain or justify the position they have adopted.

Such situations are created by adults, as part of the process of establishing and implementing classroom social norms (Yackel & Cobb, 1996). Where the teaching and learning environment supports and encourages a disposition of inquiry, such classroom norms may suggest that it is appropriate to question not only what happens but also the explanations offered, and that such questioning does not imply a sense of 'being wrong'. Rather, it implies that different people see and understand things in different ways. In such a context, 'errors' can stimulate discussion and inquiry, provided those who proffer statements or arguments which are not accepted by others retain a sense of self-worth. The idea of generating many possible explanations can be promoted instead of aiming to find the 'one correct answer' (Fosnot, 1995).

It also implies that there is much to be learnt from listening to and interpreting the explanations of others. There is a recognition that some explanations will be personally meaningful for the exponent, and that there are many potentially different explanations for the one event or situation. Bereiter, Scardamalia, Cassells, and Hewitt (1997, p. 334) describe a similar process as 'progressive discourse' whereby a spirit of collaboration, rather than competition, promotes a focus on understanding:

> ... students are not merely expressing opinions for the sake of being heard. There is a clear expectation of working together to figure things out.

This is similar to the position put by Fosnot (1995, pp. 29–30) when she focuses on the importance of discussion:

> Dialogue within a community engenders further thinking. The classroom needs to be seen as a 'community of discourse engaged in activity, reflection, and conversation' (Fosnot, 1989). The learners (rather than the teacher) are responsible for defending, proving, justifying, and communicating their ideas to the classroom community. Ideas are accepted as truth only insofar as they make sense to the community and thus rise to the level of 'taken-as-shared'.

A sense of collaboration and inquiry is promoted by teachers who demonstrate these qualities. Teachers shape the structure and content of classroom discourse, and through this, they shape the interpretation and meaning appropriated by children. Teachers who demonstrate their own sense of inquiry, their own ability to make mistakes, recognize competing arguments, and use negotiation and compromise, model those skills to children. They also create a discourse which questions the traditional role of the teacher as the font of knowledge in a classroom and focuses attention on the nature of the argument being constructed, rather than the person making the argument.

Collaboration doesn't just happen. Teachers need to provide opportunities for collaboration to develop and children need to be prepared to devote the time and effort to engage in successful collaboration. Jones and Nimmo (1999) note that collaboration is an active process "not a passive phenomenon; it requires risk taking with some passion about it" (p. 6). Collaboration, say these authors, occurs as a result of provocation. In other words, no one engages in collaboration without some reason for doing so. For collaboration to be successful, participants need to be willing to share their own perspectives and to consider the perspectives of others, often with the outcome that a new perspective emerges. This "wrestling with another's perspectives" (Jones & Nimmo, 1999, p. 7) is integral to collaboration.

The organization of the learning environment has an impact on the nature of interactions, and therefore, on the learning that occurs. In the observations for this study, argumentation was promoted by classroom organization. Firstly, a series of small group interactions was facilitated-both in the play session and the mathematics session-where children had access to a range of materials, and where the manipulation of materials as well as ideas, was encouraged. Underlying the organization of the environment was a commitment in time, so that children had sufficient time to engage in experiences in a meaningful way. This also meant that there was time to develop an argument as well as time for others to listen and respond to that argument.

A further element of classroom organization relates to partners in social interactions. Some children sought out friends to play with and, in the case of Anna and Briony for example, jointly constructed an argument. Other arguments were developed though the prompting of an adult. Teachers need to recognize that friendships are important to young children and that sometimes, the nature of the relationship between individuals can promote, or inhibit, the acceptance and evaluation of different perspectives (Dunn, 1993; Stein & Miller, 1993). Opportunities for children to interact with friends can be provided. Interactions with friends may encourage children to consider alternative perspectives to their own and to enter into processes of negotiation and compromise.

The nature of the curriculum within classrooms has a direct effect on children's use of argumentation. While the stated aims of curricula often promote critical and reflective thinking, the actual delivery of the same can be teacher-driven. A classroom climate that encourages inquiry and where teachers and children engage in collaborative inquiry, has the potential to encourage critical thinking. Pursuing an area of interest is one of the most powerful motivators for learning (Dockett, 2000; Tinworth, 1997). A curriculum based on child-initiated inquiry conveys a number of messages to those involved. It implies that children are knowledgeable and able to share with others some of the things they want to know. It also implies that there is value in listening to children and the questions they pose as well as the explanations they offer. It encourages teachers and children to work together in the process of learning and teaching, as well as focussing on content that is relevant and meaningful for all concerned.

As an overall summary, we believe that argumentation is promoted when teachers:

- encourage a disposition of inquiry in classrooms;
- examine and reflect upon the discourse which characterizes their classroom and the interactions within it;
- respect differences in children's thinking and understanding;
- accept 'errors' as indications of children's understandings;
- promote collaboration, rather than competition;
- demonstrate the skills of collaboration and inquiry in their own interactions;
- recognize 'mistakes' or 'errors' as rational to those making them;
- use questions to help children connect and assess their own thinking;
- organize appropriate blocks of time for experiences;
- provide a range of relevant materials and resources;
- plan a range of opportunities to interact with different people in the classroom;
- use curriculum practices which reflect stated aims of encouraging critical and reflective thinking;
- draw on child-initiated curriculum.

Such a view highlights the importance of social interaction in the construction of knowledge. It recognizes that children do not develop and learn in isolation. Rather, this occurs within a social matrix, where the "systems of social relationships and interactions" and "the cultural heritage of the society" (Nicolopoulou, 1993) come together. Through social interaction, adults and children negotiate understandings as they collaborate on issues of relevance or importance. The adult does not assume the role of the possessor of knowledge, which is to be transmitted to the children. Rather, solving a problem or explaining a situation is something that can be regarded as a community endeavor and can result in a range of possible solutions, explanations and justifications. The adult can then facilitate a process whereby the group may come to an agreed position, or consensus, about the most viable explanation or solution (Berk & Winsler, 1995).

CONCLUSION

The research question posed at the beginning of this paper asked whether or not young children engaged in argumentation. The data analyzed suggest that some young children use argumentation some of the time. Not all interactions recorded for this study involved argumentation and not all children who were observed engaged in argumentation. When argumentation was used, it often consisted of establishing a connection between data and conclusion. As indicated in the transcripts, some children did employ both a warrant and backing and they made this explicit. Others relied either on implicit warrants and backing or gave none. Only one child offered a counterargument. The same situation occurs in adult conversation as well as in the interactions of older children, with argumentation used when necessary, rather than in all interactions. The findings of this study suggest that while young children may be becoming aware of the process of argumentation, they will not all engage in the complete process, or use it on every possible occasion. Nevertheless, the developing awareness of the process, and opportunities to use it in different situations, provide an important basis for the future refinement of the process.

Argumentation plays an important role in children's construction of knowledge as they try to work out and explain experiences. As they express ideas, listen to the ideas of others and try to evaluate the reasoning behind conclusions that are made, they must grapple with many conceptual issues. The establishment of a learning environment in which the children feel confident to express their own views-with an expectation that they will be listened to-is an important element in this process.

REFERENCES

Anderson, R. C., Chinn, C., Chang, J., Waggoner, M., & Yi, H. (1997). On the logical integrity of children's arguments. *Cognition and Instruction*, *15*(2), 135–167.

Bauersfeld, H. (1988). Interaction, construction and knowledge: Alternative perspectives for mathematics education. In: T. Cooney & D. Grouws (Eds), *Effective Mathematics Teaching* (pp. 29–46). Reston, VA: National Council of Teachers of Mathematics and Lawrence Erlbaum Associates.

Bereiter, C., Scardamalia, M., Cassells, C., & Hewitt, J. (1997). Postmodernism, knowledge building and elementary science. *Elementary School Journal, 97*(4), 329–340.

Berk, L. E. (1992). Children's private speech: An overview of theory and the status of research. In: R. M. Diaz & L. E. Berk (Eds), *Private Speech: From Social Interaction to Self-Regulation* (pp. 17–53). Hillsdale, NJ: Erlbaum.

Berk, L. E. (1994). Vygotsky's theory: The importance of make-believe play. *Young Children, 50*(1), 30–39.

Berk, L. E. (1997). *Child development* (4th ed.). Needham Heights, MS: Allyn and Bacon.

Berk, L. E., & Winsler, A. (1995). *Scaffolding children's learning: Vygotsky and early childhood education.* Washington, D.C.: National Association for the Education of Young Children.

Berkowitz, M. W., Oser, F., & Althoff, W. (1987). The development of sociomoral discourse. In: W. Kurtiness & J. Gerwitz (Eds), *Moral Development Through Social Interaction* (pp. 322–352). New York: Wiley.

Bodrova, E., & Leong, D. J. (1996). *Tools of the mind: The Vygotskian approach to early childhood education.* Columbus, OH: Merrill.

Buzzelli, C. A. (1995). Teacher-child discourse in the early childhood classroom: A dialogic model of self-regulation and moral development. *Advances in Early Education and Day Care, 7,* 271–294.

Cobb, P. (1990). Multiple perspectives. In: L. Steffe & T. Wood (Eds), *Transforming Mathematics Education: International Perspectives* (pp. 200–215). Hillsdale, NJ: Erlbaum.

Cobb, P., & Bauersfeld, H. (Eds) (1995). *The emergence of mathematical meaning: Interaction in classroom cultures.* Hillsdale, NJ: Erlbaum.

Cobb, P., Yackel, E., & Wood, T. (1995). The teaching experiment classroom. In: P. Cobb & H. Bauersfeld (Eds), *The Emergence of Mathematical Meaning: Interaction in Classroom Cultures* (pp. 17–24). Hillsdale, NJ: Erlbaum.

Damon, W., & Killen, M. (1982). Peer interaction and the processes of change in children's moral reasoning. *Merrill-Palmer Quarterly, 28*(4), 347–367.

Dockett, S. (2000). Child-initiated curriculum and images of children. In: W. Schiller (Ed), *Thinking Through the Arts* (pp. 204–211). Sydney: Harwood Academic Publishers.

Dockett, S., Szarkowicz, D., Petrovski, P, Degotardi, S., & Rovers, F. (1999). 'I annoy him and he hits me': Interactions of popular, neutral and unpopular children. *Journal for Australian Research in Early Childhood Education, 6*(1), 59–71.

Dunn, J. (1988). *The beginnings of social understanding.* Oxford: Blackwell.

Dunn, J. (1993). *Young children's close relationships: Beyond attachment.* Newbury Park, CA: Sage.

Eisenberg, A. R., & Garvey, C. (1981). Children's use of verbal strategies in resolving conflicts. *Discourse Processes, 4,* 149–170.

Flavell, J. H., Miller, P. H., & Miller, S. A. (1993). *Cognitive development* (3rd ed.). Englewood Cliffs, NJ: Prentice-Hall.

Fosnot, C. (1995). Constructivism: A psychological theory of learning. In: C. Fosnot (Ed.), *Constructivism: Theory, Perspectives and Practice* (pp. 8–33). New York: Teachers College Press.

Göncü, A. (1993). Development of intersubjectivity in social pretend play. *Human Development, 36*(4), 185–198.

Grice, H. P. (1975). Logic and conversation. In: P. Cole & J. L. Morgan (Eds), *Syntax and Semantics, Vol 3: Speech Acts.* New York: Academic Press.

Hartup, W. W. (1996). The company they keep: Friendships and their developmental significance. *Child Development, 67*(1), 1–13.
Horn, I. S. (1999). Accountable argumentation as a participant structure to support learning through disagreement. Paper presented at the annual meeting of the American Educational Research Association, Montreal, Quebec, April.
Inhelder, B., & Piaget, J. (1958/1977). The growth of logical thinking from childhood to adolescence. In: H. E. Gruber & J. J. Voneceh (Eds), *The Essential Piaget*. Basic Books: New York.
Jones, E., & Nimmo, J. (1999). Collaboration, conflict and change: Thoughts on education as provocation. *Young Children, 54*(1), 5–10.
Kruger, A. C., & Tomasello, M. (1986). Transactive discussion with peers and adults. *Developmental Psychology, 22*, 681–685.
Krummheuer, G. (1995). The ethnography of argumentation. In: P. Cobb & H. Bauersfeld (Eds), *The Emergence of Mathematical Meaning: Interaction in Classroom Cultures* (pp. 220–269). Hillsdale, NJ: Erlbaum.
Lawrence, J. A., & Valsiner, J. (1993). Conceptual roots of internalization: From transmission to transformation. *Human Development, 36*(3), 150–167.
MacNaughton, G. (1997). Who's got the power? Rethinking equity strategies in early childhood. *International Journal of Early Years Education, 5*(1), 57–66.
Main, N. (1996). Sticks and stones and kings on thrones: The culture of violence in a sample of early childhood daycare settings. Paper presented at the Australian Research in Early Childhood Conference, Canberra.
Miller, P., & Sperry, L. (1987). The socialization of anger and aggression. *Merrill-Palmer Quarterly, 33*(1), 1–31.
Nicolopoulou, A. (1993). Play, cognitive development and the social world: Piaget, Vygotsky, and beyond. *Human Development, 36*(1), 1–23.
Perry, B., & Dockett, S. (1998). Play, argumentation and social constructivism. *Early Child Development and Care, 140*, 5–15.
Piaget, J. (1965). *The moral judgement of the child*. New York: Free Press.
Rogoff, B. (1990). *Apprenticeship in thinking: Cognitive development in social context*. New York: Oxford University Press.
Shantz, C. (1987). Conflicts between children. *Child Development, 58*, 283–305.
Siegal, M. (1991). *Knowing children: Experiments in conversation and cognition*. Hillsdale, NJ: Erlbaum.
Stein, N. L., Bernas, R. S., Calicchia, D. J., & Wright, A. (1998). Understanding and resolving arguments: The dynamics of negotiation. Paper presented at the annual meeting of the American Educational Research Association, San Diego, CA, April.
Stein, N. L., & Miller, C. A. (1993). The development of memory and reasoning skill in argumentative contexts: Evaluating, explaining and generating evidence. In R. Glaser (Ed.), *Advances in Instructional Psychology, 4*, 285–335.
Stein, N. L., & Trabasso, T. (1982). Children's understanding of stories: A basis for moral judgement and dilemma resolution. In: C. Brainerd & M. Pressley (Eds), *Verbal Processes in Children: Progress in Cognitive Development Research* (pp. 161–188). New York: Springer-Verlag.
Tesla, C., & Dunn, J. (1992). Getting along or getting your own way: The development of young children's use of argument in conflicts with mother and sibling. *Social Development, 1*(2), 107–121.

Tinworth, S. (1997). Whose good idea was it? Child-initiated curriculum. *Australian Journal of Early Childhood, 22*(3), 24–29.
Toulmin, S. (1969). *The uses of argument.* Cambridge, U.K.: Cambridge University Press.
Trevarthen, C. (1988). Universal cooperative motives: How infants begin to know the language and culture of their parents. In: G. Jahoda & I. M. Lewis (Eds), *Acquiring Culture: Cross-Cultural Studies in Child Development.* London: Croom Helm.
Tudge, J. R. H., & Winterhoff, P. A. (1993). Vygotsky, Piaget and Bandura: Perspectives on the relations between the social world and cognitive development. *Human Development, 36*(2), 61–81.
Vygotsky, L. S. (1981). The genesis of higher mental functions. In: J. V. Wertsch (Ed.), *The Concept of Activity in Soviet Psychology* (pp. 144–188). Armonk, NY: Sharpe.
Wertsch, J. V., & Tulviste, P. (1992). L. S. Vygotsky and contemporary developmental psychology. *Developmental Psychology, 28*(4), 548–557.
Yackel, E. (1998). A study of argumentation in a second-grade mathematics classroom. In: A. Olivier & K. Newstead (Eds), *Proceedings of the 22nd Conference of the International Group for the Psychology of Mathematics Education* (vol 4, pp. 209–216). Program Committee of the 22nd PME Conference: University of Stellenbosch, South Africa.
Yackel, E., & Cobb, P. (1996). Sociomathematical norms, argumentation and autonomy in mathematics. *Journal for Research in Mathematics Education, 27*(4), 458–477.

PART II:

RECONCEPTUALIZING PLAY

"WE DON'T PLAY THAT WAY AT PRESCHOOL": THE MORAL AND ETHICAL DIMENSIONS OF CONTROLLING CHILDREN'S PLAY

Mac H. Brown and Nancy K. Freeman

Observe children's spontaneous behavior and you will see for yourself that given time, space and freedom; they are experts at play. Scholars tell us that play helps children understand their experiences, craft their identities, expand their horizons and shape their futures. Interestingly, we know it's not just children who play. Zoologists show that all mammals can be playful. But it would be a mistake to think that an appreciation for play's universality means that there is unanimity about its definition, meaning or value. There are, in fact, almost as many definitions and theories of play as there are researchers and theorists. Scholars and practitioners from disciplines ranging from A to Z, from Anthropology to Zoology, have investigated play. It has been studied by educators, sociologists, psychologists, scholars of literature, mathematicians, psychiatrists and folklorists. These varied perspectives have contributed to our understanding of play's dimensions, meanings, value and implications, and provide the foundation for our exploration of how adults support, facilitate and control young children's play.

One explanation for the multiple constructions of play is the variety of theoretical stances assumed by researchers who have contributed to the professional conversation. Each speaks in their tradition's distinctive "ideological rhetoric" situated within their particular scholarly discourse

(Sutton-Smith, 1999). These theorists propose that human play is, or can be: (a) pleasurable, (b) spontaneous, (c) intrinsically motivated, (d) nonliteral, (e) considered unproductive and (f) can have a dark or anti-social side (Clarke, 1999; Frost, 1992; Hughes, 1995; Sutton-Smith, 1999). While bringing multiple interpretations of play's value and appropriateness in various settings, all share a desire to understand play's role in children's past, present and future.

PLAY IS HERE TO STAY

Play is at once extremely robust and remarkably fragile. It is robust in the sense that it is universal – youngsters nearly always manage to play, no matter how poor or desperate their circumstances may be. It is fragile and can be ephemeral, however, because it exists, to a large extent, in the minds of the players. Bateson (1955) observes that the play frame can be easily broken or destroyed by players themselves when they violate the rules; by outside intruders, no matter how well-intentioned; or by any distraction that pulls the players out of their fantasy.

While grounded in the modern paradigm that has (privileged) children's play for generations, this essay challenges the universality of this perspective by applying a postmodern reconceptualizer's lens to an examination of the moral and ethical implications of adults' role in children's play. A backdrop for this discussion is the UN Convention on the Rights of Children which legitimizes play as a right of childhood. The UN's stated intention is to assure all the world's children the benefits of a satisfying play life. This manifesto creates weighty responsibilities which rest upon the shoulders of adults who shape and, in more ways than they may realize, control children's play.

We begin by exploring a variety of theoretical perspectives on play; next, we situate these views within the rhetoric of early childhood education paying particular attention to the challenges created by modern families' lifestyles that place children in the care of non-familial adults for the majority of their waking hours. Finally, we consider the implications these contemporary realities create when teaching young children is examined from a moral and ethical perspective.

Our exploration of these moral dimensions of teachers' control of children's play is informed by the work of Goodlad, Soder and Sirotnik (1990), Nash (1996), Noddings (1992), Strike and Soltis (1992), Freeman (2000) and others. It is situated within the framework provided by the Code of Ethical Conduct of the National Association for the Education of Young Children (NAEYC)

(Feeney & Kipnis, 1989/1998), a treatise which reflects the voices of practitioners playing a variety of roles within the early childhood professional community.

PERSPECTIVES ON PLAY

Although children's play is universal, theorists from different theoretical traditions and historical periods offer varied interpretations of its role, value and potential. Some scholars emphasize children's abilities and others their limitations; some value play because it helps children process events in their past, others believe it prepares them for their future, and a third group keeps the focus on children's present needs and interests (Cannella, 1997).

The modernist orientation has led early childhood educators to focus on the content and consequences of play. Scholars assuming this stance attribute children's playfulness to: (a) surplus energy, (b) relaxation, (c) recapitulation, (d) cognitive functioning or (e) instinct (Clarke, 1999; Frost, 1992; Saracho & Spodek, 1998). They value play because it makes important contributions to children's cognitive and social development, their creativity and flexibility and their self actualization. Sutton-Smith (1999) describes these theories variously as foregrounding *progress, imagination* or the *self*. Piaget, Vygotsky, Bruner and Erikson, like the majority of theorists guiding contemporary practice, are situated within this discourse (Saracho & Spodek, 1998).

Alternatively postmodern scholars value the contribution satisfying play makes to children's day-to-day experiences. They focus on the processes and motives of play and appreciate its role in children's present rather than focusing on its contribution to their future (Tobin, 1997). What's more, post-moderns empower children, believing that they alone are uniquely capable of choosing personally meaningful occupations. They resist efforts to shape or control children's freely chosen activities by creating constraints about the materials, themes or time devoted to play (Gerber, 1984; Hymes, 1981; Tobin, 1997).

EXPLORING DIVERSE IDEOLOGIES

> When I was a child, I spoke as a child, I felt as a child, I thought as a child. Now that I have become a man, I have put away childish things.
> *1 Corinthians 13 (King James Version)*

Modern and postmodern educators interpret make-believe differently. Conventional wisdom applies a modern linear view to child development and looks to pretend play as an indication of children's innocence and simplicity. As moderns would have it, children become increasingly able to think abstractly as they progress through the early childhood years, so that, by the time they have reached the concrete operational stage of development, make believe play has lost its appeal. In other words, by the time children are seven or eight they are ready "to put away childish things" and to work on mastering the competencies identified in state curriculum standards and similar codifications of what they should know and be able to do. A critical alternative to this perspective is one which appreciates that fantasy play becomes increasingly complex and helps children explain the world and culture they inhabit by creating "a sophisticated, abstract system that is largely imaginary ... [which gives] structure, energize[s], and give[s] meaning to experience" (Vandenberg, 1998, p. 296). This process of making meaning through play is ephemeral and elusive. To aid our understanding we turn to what Bateson (1955) and Vandenberg (1998) call "mythmaking" or the creation of culturally sanctioned stories about reality. It is important that early childhood educators appreciate how important it is that they see and hear the important truths children communicate as play, for they are going about the business of creating the myth they will carry into adulthood and leadership roles in the twenty-first century.

Adults' views about the future also color the way they arrange for children's make-believe. Traditional teachers are apt to provide for pretend play by equipping the dramatic play center with props which encourage children to recreate the *status quo*. Those assuming a critical view, on the other hand, encourage children to experiment creating visions of the unfathomable future rather than recreating the world of today (Corsaro, 1997). They avoid leading or controlling children's play themes and offer opportunities for personal expression rather than impose parameters of acceptability. These are just two examples of how adults' values, expectations and priorities and their alignment with modern or postmodern theoretical perspectives dictate their involvement in children's play and their definition of good teaching.

THE MISMATCH BETWEEN THEORY AND PRACTICE

Play is currently in vogue, supported by mainstream professional literature which operationalizes modern child-directed play-centered constructivist theory (Bredekamp & Copple, 1997; Chaillé & Silvern, 1996; Gestwicki, 1999). Its popularity is not surprising, for learning through play has been a tradition among early childhood educators for over one hundred and fifty years. It is the centerpiece of a wide variety of accepted curriculum models including programs developed to reflect the teachings of Froebel, the MacMillan sisters, and Montessori (Saracho & Spodek, 1998). Support for play has, additionally, gained momentum as a result of efforts to fortify the field's scientific knowledge base and its standing in the academy. By linking themselves to the work of positivistic psychologists early childhood educators have come to describe play as "the work of children" and take pains to document its contributions to children's cognitive, social, emotional and physical growth and development.

However, in spite of the fact that early childhood education's professional literature advises teachers to provide children extended periods of time for self-selected playful activities, visits to typical early childhood classrooms reveal the effects of the public's hysteria about enhancing children's "readiness." As a result there is a mismatch between commonly accepted visions of best practice and current realities (Bennett, Wood & Rogers, 1997).

All too often play is forced out of the school day by principals who limit recess because they fail to appreciate its contribution to children's learning and development and by public sentiment that interprets accountability and the adoption of higher standards as a call for a pushed-down curriculum. Center and play times have been replaced by skill drills, and an overemphasis on activities intended to enhance children's emerging literacy, and large group instruction focused on narrowly defined learning outcomes prevails. Play is routinely eliminated or limited by rigidly enforced classroom routines, controlled by teachers who decide what materials and themes are appropriate and constrained by requirements that equipment and materials be used in particular ways. The fact is that even though early childhood educators say they support play, their behavior indicates that they and their superiors doubt its worth and developmental value (Hughes, 1995).

The prevailing rigid and academic mindset can be explained by the persistence of two fundamental beliefs that continue to influence Americans' thinking. The first is a puritanical suspicion of anything that is fun and not related to work (Frost, 1992), and the second is the conviction that play appears chaotic and undisciplined, undermining the school's primary purpose which is to bring order and control to children's lives. As a result of the pressures to push down the curriculum, and in spite of abundant support for play in the "best practice" literature, teachers of young children make decisions about the role of play in their classroom by judging its contribution to children's "readiness" rather than its contribution to children's emotional, social and physical well-being.

We mourn the fact that *academic accountability* has become a powerful buzzword that pressures early childhood educators to focus narrowly on academics rather than maintaining the field's historical emphasis on the whole child. We sustain hope in spite of these challenges, however, because we know that regardless of prevailing attitudes about the appropriateness of encouraging free play, whether play is in or out of favor among teachers, teacher educators or principals, and no matter what is recommended by authors of commonly-relied-upon textbooks; we know play will remain universal and important in the lives of children. The challenge is to create a theory of play which is a useful guide but doesn't inadvertently include blinders and earmuffs.

We believe this trend to censure, control and limit play highlights the moral dimensions of teaching young children; for it is critical that teachers are cognizant that decisions they make about prohibiting, permitting, condoning or encouraging play have a profound effect on children's day-to-day experiences in the out-of-home settings where they spend the majority of their waking hours.

WHAT OF THE MORAL DIMENSIONS OF PLAY?

It is our purpose to foreground the moral dimension of teachers' work and to explore the implications of creating play spaces, mediating disputes, allocating resources and permitting or forbidding aggressive, sexual or exclusionary activities. These issues are important because the moral dimensions of teaching are always present and deserve scrutiny; for teaching involves "Matters of what is right, fair, just and virtuous . . . . The teacher's conduct, at all times and in all ways, is a moral matter" (Fenstermacher, 1990, p. 133). The fact remains, however, that considerations of morality are all-too-often absent from discussions of what it means to teach well (Ayers, 1993; Freeman, 2000; Goodlad, 1990; Nash, 1996).

We assert that moral classrooms share power among all stakeholders and recognize the agency of children as well as adults by giving children input as the days, weeks and months unfold. These sensitivities prompt mindful early childhood educators to pose questions such as: "Can children decide how long, with whom or what to play? How do adults enforce their own perspectives of what is "right" when they determine what children ought or ought not do? How should teachers equip and organize classrooms? and How do adults respond when their version of what is appropriate is at odds with children's interests and inclinations?" (It is critical that teachers make decisions like those posed above about if and how they provide children opportunities for play only after they reflect upon their personal values and the professional values that guide their work.) It is imperative that classroom policies be deliberately and carefully considered and not just be routinely enforced mandates created because "that's the way it's always been done" or "because they echo what the principal, parents or supervisor expect teachers me to do."

WHAT DOES ETHICS HAVE TO DO WITH IT?

Acknowledging the moral dimensions of teaching leads to an examination of the ethics of working with young children and their families, for ethics is "the study of right and wrong, duty and obligation . . . involv[ing] critical reflection on morality" (Feeney & Freeman, 1999, p. 99). These ethical deliberations are most productive when put into the context of the NAEYC Code of Ethical Conduct which was first adopted in 1989, before the current popularization of postmodern thinking within the early childhood community. The Code comes from a modern, rather than postmodern perspective of young children and early childhood education; but serves, none-the-less, as a useful resource for those striving to create humane and ethical early childhood environments. It is our conviction, moreover, that the professional ethics of the field take on an unexamined dimension when considered from the postmodern perspective which recognizes the independence and capabilities of children and imbues them with legitimate rights as individual human beings and full members of society (Dahlberg, Moss & Pence, 1999).

The Code fits comfortably in a postmodern framework because it views ethical decision making not as the application of arbitrary absolutes but as, instead, a tool for decision making situated in particular circumstances involving individuals who bring their unique perspectives and experiences to the classroom. It speaks to the truth that "The foolproof – universal and unshakably founded – ethical code will never be found . . . an ethics that is universal and 'objectively founded,' is a practical impossibility; and also an *oxymoron*, a contradiction in terms" (Bauman, 1993, p. 10). Decisions of informed practitioners are not dictated by this Code; but are instead guided by it, as they put it within the framework of the personal values and morality they bring to their work. It legitimizes the tradition that early childhood educators are expected to take responsibility for our own choices, including moral choices. They can not "abandon this responsibility in favor of conformity to universal rules and absolute truths . . . . Postmodern ethics means each of us, from childhood, must take responsibility for making difficult decisions. We are our own moral agents, bearing responsibility for making – constructing – moral choices" (Dahlberg, Moss & Pence, 1999, p. 56).

The Code brings these considerations to the fore. It reminds those who care for and educate young children that their values are reflected in how they plan for, interact with and evaluate young children. It reminds practitioners that they should be guided not just by their personal idiosyncratic beliefs but also by the ethical standards adopted by their learned society which are intended to help them answer the difficult question, "What should a good teacher do?" The sections to follow explore a variety of seemingly routine decisions teachers and caregivers frequently make which upon deliberation, we realize, have important moral and ethical dimensions which should not be overlooked.

THE MORAL DIMENSION CONTROLLING PLAY MATERIALS AND THEMES

Visits to a variety of child care settings illustrate the moral dimension of how teachers of young children equip their classrooms. Sometimes toys reflect a *laissez faire* attitude; anything that keeps children busy and doesn't hurt anyone is allowed. This means that Barbie and Ken, Star Wars figures and the latest superheros are all welcome, and guns are likely to be the only toys that are explicitly "outlawed." In these classrooms children are permitted to bring toys from home as long as they are prepared to "share them with all their friends."

Other programs take a traditional approach, insisting that classrooms should include only classic equipment like unit blocks, baby dolls, cars, trucks and fire engines. Teachers make an effort to provide open-ended materials that rely on children's imaginations, rather than toys which tend to lead play in predetermined directions. These teachers' selections are apt to support broad themes like "house" and to discourage Barbie dolls and action figures. In these classrooms children are strongly discouraged from bringing figures from the latest Disney release, the latest superhero or other heavily-promoted commercial toys from home, for teachers insist there are plenty of materials in their classrooms to keep all children productively occupied during the school day (Corbett, 1994).

EXAMINING ASSUMPTIONS ABOUT MATERIALS AND THEMES

These two approaches to equipping children's classrooms are likely to be unexamined practices that mirror teachers' experience and training. Think about what might happen, however, if a teacher with traditional values were to assume a critical interpretive perspective. She might reverse her policy strictly banning Barbie or Star Wars, deciding to give children access to these materials some would judge inappropriate for classrooms. She could justify her decision by observing that these toys hold an undeniable fascination for today's children, and that youngsters all-too-often have limited opportunities to play freely with self selected playmates. If she believes providing children opportunities to use materials of their own choosing outweighs her preference for traditional classroom activities she can justify her willingness to follow children's leads and allow these materials that are virtually impossible-to-escape artifacts of modern society (Corbett, 1994). The teacher who embraces this challenge has to consider how to welcome Barbie and Star Wars into her classroom. She has to be willing to let children exhaust their curiosity if these toys are novelties at school, and, after they have explored the prefabricated themes and story lines, she can then encourage children to put these materials into perspective and to integrate these commercial toys into spontaneous play themes that come from children's own experiences and imaginations.

In addition to deciding what toys to encourage in the classroom, teachers also have responsibilities to examine their assumptions and consider the impact of their decisions about whether to allow children to use traditional materials in non-traditional ways or to explore themes forbidden by polite society. These issues emerge when blocks become guns in the hands of classroom commandoes, when boys cross-dress in the dress up corner and when children explore aggressive, violent or sexual themes (Tobin, 1997).

Examining teachers' beliefs, values and policies about gun play offers a particularly useful way to illustrate the implications of imposing adult sensibilities on young children without giving these positions careful consideration. We know, for example, that guns are likely to be the toy most-universally prohibited in classrooms, but the view that guns are not acceptable is clearly *not true* in large segments of society. For proof, just look at the news stories that report the strongly-stated opinions of the 3.5 million members of the National Rifle Association or consider the experiences of children who see their parents go off to work as police officers or members of our armed forces, where weapons are important tools of the trade. Other children observe or even accompany adults when they hunt, for pleasure or out of necessity.[1] The value and acceptability of gun play is clearly related to one's cultural perspective – individuals' occupations, hobbies or life styles. Who are we to perpetuate stereotypes that put guns only in the hands of criminals, bullies or terrorists?

Gun play also comes to preschool as a result of the violence children have experienced themselves or seen on TV. Children who have witnessed shootings in their homes or neighborhoods, for example, are likely to use guns in ways that are therapeutically cathartic, to work through the fears and anxiety created by these traumatic experiences. It has been noted, moreover, that children of battered women use guns and other weapons defensively rather than aggressively.[2] It is important for teachers to consider the possibility that they may be unaware of children's personal circumstances, and that limiting gun play may be depriving youngsters of an activity which may help them make sense of, and gain control over, difficult experiences.

Images children see on television are also likely to bring gun play into the classroom. Should adults' reactions be different when the play involves an original plot as compared to play that simply re-enacts scenes children have seen on Saturday morning cartoons? How do teachers react when play depicts violence shown on the evening news, as it did in the wake of the Oklahoma City bombing or the Columbine High School shootings, stories that exposed children to violence and fear as well as the grief, pain and suffering that came in its aftermath (Levin, 1998)?

Teachers who use their unchallenged power over children to impose their own biases, often cloaked as the "politically correct" position, their "professional" opinions, or simply their view of "what's good for" children are likely to ban gun play without giving their decision much thought. They are likely to privilege their culturally determined stance and enforce their belief that guns don't belong at preschool. It can be argued however, that outlawing gun play is more harmful than welcoming it in the classroom. Coming to terms with violence and aggression, fear and grief during early childhood in a physically and psychologically safe play space may lead children to construct a real understanding of power and mastery. It may help them become adept at peaceful means of conflict resolution, a disposition sorely needed in a society where the nightly news reminds us that twelve children die from gunshot wounds every day (Clinton, April 11, 2000).

We believe that forcing underground the overt expressions of sexuality, racial and gender identity, aggression, power and mastery that are part of these evocative play themes makes these important matters part of the null curriculum. When that happens teachers and caregivers can neither deflect attention away from a child who feels overwhelmed or threatened, nor can they smooth entry for a child who is struggling to figure out how to join the fray. What's more, when this play is hidden from view children miss the opportunity to seek adults' support or feedback to help them make sense of it or the feelings it elicits, are denied the opportunity to visit and revisit the stressful issues as they achieve increasingly mature perspectives and are likely to be stuck forever with their five-year-old understandings of the world and their place in it (Gardner, 1991). We caution teachers to be aware that many opportunities to help children learn about these evocative and invariably fascinating topics are lost when gun play and other often-forbidden behaviors are relegated to behind the playhouse where they are secret and hidden-from-view.

Teachers who take carefully thought out, professional and ethically defensible stances to condone behaviors that others would prohibit might refer to the Code, Sections I-1.2 "to base program practices upon current knowledge in the field ... and upon particular knowledge of each child", I-1.4 "to appreciate the special vulnerability of children" and I-1.5 "to create and maintain safe and healthy settings that foster children's social, [and] emotional ... development and that respect their dignity and their contributions" to add the weight of the wisdom of their professional colleagues to their own. In our view theirs would be morally defensible decisions which empower children and give them a rich set of experiences from which to construct their understandings of these issues as well as to become self-directed and autonomous thinkers.

THE MORAL DIMENSIONS OF PRIVACY IN PUBLIC AND PRIVATE SPACES

How do we respond to children's need for privacy and do we consider private spaces a right or a privilege? We know that the UN has taken a stance that children are entitled to play but what behaviors are condoned and should teachers and other "responsible" adults provide private play spaces or only public, adult-monitored ones? Do we construct play to be limited to "healthy" or "appropriate" activities? If so, then we must declare all play spaces public so they can be monitored and controlled. Conversely, if we construct play as being genuinely "free" then truly private play becomes acceptable and teachers are given license to, with a well-thought-out rationale, "look the other way" in the face of activities which explore the margins of propriety (Tobin, 1997; Sutton-Smith, 1997).

We believe the moral dimensions of teachers' decisions about private play vary, depending on the setting, age and schedules of children in their care. Those who teach in public schools are working with children who are required to be there about six hours a day. Their classrooms are dominated by standards, schedules and lesson plans. Teachers may find it impossible to provide private spaces indoors, in part because their administrators insist they face legal mandates requiring children to be under constant adult supervision while they are in the classroom. Even teachers working in these conditions are probably aware, however, that children are expert at keeping some of their interactions private and that they regularly sustain clandestine games, particularly on the playground, where they remain under adults' watchful eyes but out of their earshot (King, 1987 as cited in Johnson, Christie & Yawkey, 1999). Teachers can decide either to ignore and implicitly sanction such behavior, making no attempt to interfere with children's personal affairs, or they can try to control all of children's interactions, investing a significant amount of effort and attention in an endeavor that will be, no matter how hard they try, unlikely to succeed. When they truly value privacy and children's rights to self-determination, they are likely to acknowledge children's needs for privacy and to support, as much as they are able, children's efforts to create and sustain a culture of childhood out of adults' reach. They could find justification for this decision by referring to the last of the Code's Core Values, which commits them to "helping children ... achieve their full potential in ... relationships based on trust, respect, and positive regard" – they trust children to know what they need to realize their potential to the fullest.

While teachers in public school settings face one kind of challenge, those working in childcare and after school programs face another. That is because the hurried childhood Elkind saw on the horizon is now a reality (Elkind, 1988). Parents' demanding work schedules often require them to create comprehensive child care arrangements placing their children in out-of-home settings for 10–12 hours a day. Opportunities for informal, spontaneous, unstructured play with self-selected neighborhood play mates have been eroded by the schedules imposed by parents' workday responsibilities. Children more often than not find themselves in the care of a series of non-familial adults who routinely make decisions about which toys are appropriate, which play themes are acceptable, and how and when adults should direct, interrupt or interfere. Truly free play, Elkind's anecdote to stress, where children choose their play partners and themes in the private spaces of childhood, is becoming increasingly difficult to find.

Teachers working in less rigidly regulated childcare and after school programs are likely to have more autonomy than those in public schools. In addition to increased freedom they are likely to have increased responsibilities, for they are probably working with children who spend most of their waking hours in public spaces. If they embrace values of agency, autonomy and self-actualization they would be justified in giving children increased access to play materials, time and freely chosen play themes. We contend these children, particularly, should be encouraged to create private spaces indoors and out. We assert that teachers are justified in turning a blind eye and a deaf ear to some "antisocial" or "naughty" behavior, and that they would do well to make an effort to resist the tendency to settle the conflicts between children that inevitably accompany private play. We believe that to do otherwise, caregiving adults are overstepping their authority and infringing on children's freedom in unethical and immoral ways[3].

POWER AND PLAY – LESSONS IN DEMOCRACY: ANOTHER PERSPECTIVE ON MORALITY IN THE CLASSROOM

A third way play reflects teachers' personal and professional values is demonstrated by how they schedule play in their classrooms. Free play is most likely to be integrated into classroom routines during children's preschool and kindergarten years, when it is more apt to be valued as an important tool which helps children reach developmental benchmarks. Time for play decreases in the primary grades and, as children approach nine or ten, play is most likely to be eliminated as academic content becomes the priority (Frost, 1992) and its contribution to the curriculum is less apparent. Corsaro (1997) observes that play continues to offer valuable opportunities which contribute to children's mastery of developmental milestones, particularly when they negotiate through power struggles among children and between children and adults. He appreciates that it is through these kinds of experiences that children develop an appreciation for their own agency and that children do not outgrow the need to negotiate, compromise and cooperate when they leave preschool behind.

Piaget (1952), Vygotsky (1976), and Erikson (1963) all cite children's interactions with peers, including power struggles, as being critical to resolving the development crises young children face. They help us see peer interactions and play as fundamental to the child's development of democracy's ideals and principles. These observations make it clear that play can become a metaphor for the kind of society in which we all live. Do we provide opportunities and support for autonomous action and thought or do we control, limit and restrict children's thoughts and actions (Kamii, 1982)? If democracy is the goal, then the ethical stance must include opportunities for free play where all children learn to exercise power in their lives and discover how to share their newly-won power with each other.

The balance between freedom and control must be on the minds of educators who strive for democracy but recognize "the paradox of educating children for participation in a democratic society within the inherently undemocratic context of . . . school" (Chevalier, 1998, p. 48). The interplay between adults' responsibility for children's in-school experiences and their desire to respect the internally constructed and mutually created control exerted by groups of children, often during free-play, requires adults to take a moral stance and search for the best balance. A teacher who resists pressures to intrude in children's play and maintains, as much as possible in a school setting, a hands-off policy, offers children time, resources and freedom to pursue meaningful themes to their logical conclusions. That teacher would likely allow the rambunctious rough and tumble game of chase to continue until it loses its appeal and would not intrude in the innocent "I'll show you mine if you show me yours" explorations in the bathroom which would, before long, have satisfied childrens' natural curiosity. Neither a totally *laissez faire* approach which might let some children abuse their power, for example, nor a totally Orwellian police state, where the children never learn to use their power and acquire self-restraint, is desirable.

Once more we realize how seemingly routine decisions have ethical and far reaching effects, and appreciate that they require teachers to have courage in their convictions. We come to the realization that a decision to give children control of their own activities could be justified, however, by relying not just on personal values but also on the vision of respect for children and their personal autonomy supported by our profession's Code of Ethics.

This examination of the moral dimensions of play in school and child care takes a postmodern view of the child as a capable agent in his own right. It illustrates why it is important that teachers anticipate their reactions to children's freely chosen activities and be able to identify the values that guide their classroom's policies. The decisions they make about play materials, themes, and privacy and scheduling all have important moral dimensions which may not have occurred to many who care for and educate young children. Reflective early childhood professionals who are willing to condone, or even encourage, often-forbidden play are wise to think carefully about the rationale that guides their decision. Particularly if they are challenging the prevailing canon of early childhood best practices they need to be ready to justify their position by articulating teachers' moral obligations to support children's self actualization. Teachers who take this less traveled road are likely to find support for their views by referring to the NAEYC Code of Ethics. They need to be able to clarify, however, how their own values, when combined with this statement of professional values and ideals, guides their well reasoned and intentional decisions.

Examining the moral and ethical dimensions of controlling preschoolers play means that we are engaged in the process of deconstructing early childhood education. As we see children's play as serious behavior to be respected and nurtured rather than controlled, we are reconceptualizing our expectations for children and play. In other words, we are engaged in the process of creating a new set of "myths" about childhood that may eventually become the new canon that will guide us well into this century.

NOTES

1. One author consults in a community where it is not uncommon for preschool boys to accompany their fathers when they go deer hunting.
2. These two examples were provided by Dr. Sandra Frick-Helms, Ph.D., Adjunct Professor, University of South Carolina School of Medicine.
3. Teachers and caregivers must never lose sight of their responsibilities to always, without exception, assure the safety of the children in their care. The challenge is to balance privacy WITH sufficient supervision to avoid injury, predation or victimization of any child.

REFERENCES

Ayers, W. (1993). *To become a teacher.* New York: Teachers College.
Bateson, G. (1976). A theory of play and fantasy. In: J. S. Bruner, A. Jolly & K. Sylva (Eds), *Play – Its Role in Development and Evolution* (pp. 119–129). NY: Basic Books Inc., Publishers. Originally published in *Psychiatric Research Reports*, No. 2, pp. 39–51, 1955.
Bauman, Z. (1993). *Postmodern ethics.* Cambridge, MA: Blackwell Publishers.
Bennett, N., Wood, L., & Rogers, S. (1997). *Teaching through play: Teachers' thinking and classroom practice.* Buckingham: U.K.: Open University Press.
Bredekamp, S., & Copple, C. (Eds) (1997). *Developmentally appropriate practice in early childhood programs* (Rev. Ed.). Washington, D.C.: NAEYC.
Cannella, G. S. (1997). *Reconstructing early childhood education: Social justice & revolution.* New York: Peter Lang Publishing.
Chaillé, C., & Silvern, S. B. (1996). Understanding through play. *Childhood Education, 72*(5), 274–277.
Chevalier, M. (1998). Paradoxes of social control: Children's perspectives and actions. *Journal of Research in Childhood Education, 13,* 48–55.
Clarke, L. J. (1999). Development reflected in chase games. In: S. Reifel (Ed.), *Play and Culture Studies* (Vol. 2, pp. 73–82). Stamford, CT: Ablex Publishing.
Clinton, W. J. (2000). Remarks by the President at bill signing of Responsible Gun Safety Act of 2000 Maryland State Capitol Annapolis, Maryland.
Corbett, S. M. (1994). Teaching in the Twilight Zone: A child-sensitive approach to politically incorrect activities. *Young Children, 74*(4), 54–58.
Corsaro, W. A. (1997). *The sociology of childhood.* Thousand Oaks, CA: Pine Forge Press.
Dahlberg, G., Moss, P., & Pence, A. (1999). *Beyond quality in early childhood education and care: Postmodern perspectives.* Philadelphia, PA: Falmer Press.
Elkind, D. (1988). *The hurried child: Growing up too fast too soon* (Rev. ed.). Reading, MA: Addison-Wessley Publishing Company.
Erikson, E. H. (1963). *Childhood and society.* NY: Norton.
Feeney, S., & Freeman, N. K. (1999). *Ethics and the early childhood educator: Using the NAEYC Code.* Washington, D.C.: NAEYC.
Feeney, S., & Kipnis, K. (1989/1998). *Code of professional ethics and statement of commitment* (Rev. ed.). Washington, D.C.: NAEYC.

Fenstermacher, G. D. (1990). Some moral considerations on teaching as a profession. In: J. I. Goodlad, R. Soder & K. A. Sirotnik (Eds), *The Moral Dimensions of Teaching* (pp. 130–154). San Francisco: Jossey-Bass.

Freeman, N. K. (2000). Professional ethics: A cornerstone of teachers' preservice curriculum. *ACTION in Teacher Education, 22*(3), 12–18.

Frost, J. L. (1992). *Play and playscapes.* Albany, NY: Delmar Publishers.

Gardner, H. (1991). *Unschooled mind: How children think and how schools should teach.* New York: Basic Books.

Gerber, M. (1984). Caring for infants with respect: The RIE approach. *Zero to Three, 4*(3), 1–3.

Gestwicki, C. (1999). *Developmentally appropriate practice: Curriculum and development in early education.* Albany, NY: Delmar.

Goodlad, J. I., Soder, R., & Sirotnik, K. A. (Eds) (1990). *The moral dimensions of teaching.* San Francisco: Jossey-Bass.

Hymes, J. (1981). *Teaching the child under six.* Columbus, OH: Merrill.

Johnson, J. E., Christie, J. F., & Yawkey, T. D. (1999). *Play and early childhood development* (2nd ed.). New York: Longman.

Hughes, F. P. (1995). *Children, play, & development* (2nd ed.). Boston, MA: Allyn and Bacon.

Kamii, C. (1982). *Number in preschool and kindergarten.* Washington, D.C.: NAEYC.

Levin, D. (1998). *Remote control childhood? Combating the hazards of media culture.* Washington, D.C.: NAEYC.

Nash, R. J. (1996). *Real world ethics.* New York: Teachers College Press.

Noddings, N. (1992). *The challenge to care in schools: An alternative approach to education.* New York: Teachers College.

Piaget, J. (1952). *The origins of intelligence in children* (M. Cook, trans.). NY: New American Library.

Saracho, O. N., & Spodek, B. (1998). A historical overview of theories of play. In: O. N. Saracho & B. Spodek, (Eds), *Multiple Perspectives on Play un Early Childhood Education* (pp. 1–10). Albany, NY: State University of New York Press.

Strike, K. A., & Soltis, J. F. (Eds). (1992). *The ethics of teaching* (2nd ed.). New York: Teachers College Press.

Sutton-Smith, B. (1999). The rhetorics of adult and child play theories. In: S. Reifel (Ed.), *Advances in Early Education and Day Care: Foundations, Adult Dynamics, Teacher Education and Play* (Vol. 10, pp. 149–162). Stamford, CT: Jai Press.

Tobin, J. J. (Ed.) (1997). *Making a place for pleasure in early childhood education.* New Haven, CT: Yale University Press.

Vandenberg, B. (1998). Real and not real: A vital developmental dichotomy. In: O. Saracho & R. Spode (Eds), *Multiple Perspectives on Play in Early Childhood Education* (pp. 295–305). Albany, NY: SUNY Press.

Vygotsky, L. S. (1976). Play and its role in the mental development of the child. In: J. S. Bruner, A. Jolly & K. Sylva (Eds), *Play – Its Role in Development and Evolution* (pp. 535–554). New York: Basic Books. Originally published in *Soviet Psychology, 12*(6), 62–76, 1966.

THE DANGEROUSLY RADICAL CONCEPT OF FREE PLAY

David Kuschner

In the introduction to the 1920 edition of a slim book entitled, *A study of the kindergarten problem in the public kindergartens of Santa Barbara, California, 1898–1899* (Burk & Burk, 1920), noted nursery and kindergarten educator Patty Smith Hill wrote the following.

> As far back as 1898–1899, Dr. Frederic Burk carried out one of the most significant and scientific experiments ever made in the kindergarten field. At the time, the record of it met with a chilly reception . . . it was stamped with the vehement disapproval of the greatest kindergarten authorities of the day, and pronounced heretical and dangerously radical. It was almost a forbidden book in some circles . . . (Hill, 1920, p. vi).

Why would a book of only ninety-eight pages, a book which was essentially about kindergarten methods and curriculum, be called "dangerously radical" and why would it engender "vehement disapproval?" Shapiro (1983), in his study of the history of the early kindergarten movement writes that the Santa Barbara study was the "first intensive investigation of the kindergarten classroom by a person trained in child study and sympathetic to early childhood education" (p. 123). Why then, was the book "met with a chilly reception" and considered by some to be a "forbidden book?" As surprising as it may seem given our 21st century beliefs, the radical idea was that young children should be given the opportunity to engage in free play.

To understand the impact of the Burk study, it is necessary to consider the historical context in which it occurred. At the time of the study, the kindergarten movement in this country had been growing for better than thirty years. Free kindergarten associations were established in cities from coast to coast, the International Kindergarten Union had been organized in 1892, working kindergartens were displayed at international expositions, and training schools were established to meet the growing need for kindergarten teachers. An emphasis on children's play was one of the core characteristics of the early kindergarten movement and all of this had been set in motion by the introduction of the ideas of Friedrich Froebel to this country in the late 1850s.

FROEBEL PLAY

> Into his dilapidated powder mill he has succeeded in gathering a little band of village children, not without some hesitation on the part of parents, it would seem, who naturally wondered what this curious business meant – an old gray-haired man spending his time playing with little children (Snider, 1900, p. 296).

The old gray-haired man was Friedrich Froebel (1782–1852), and this 1837 gathering of children in the powder mill in Blankenburg, Germany is now considered to be the first kindergarten.[1] This program for young children was, "In its view of children and in its educational purpose . . . different from any school that had been designed before" (Spodek, Saracho & Davis, 1991, p. 19).

During Froebel's time, schools, essentially for children seven years and older, emphasized subject matter and basic skills, focusing on reading, writing, memorization, and recitation (Morrison, 1997; Paciorek & Munro, 1996; Ross, 1976). Froebel's kindergarten, by contrast, was the first program in the history of early childhood education to emphasize the activity of children (Spodek & Saracho, 1988). According to Weber (1969), "It was daringly new to suggest that a major portion of a child's time in school be used to manipulate and construct objects" (p. 7). The focus on activity and manipulation of objects led Froebel to emphasize play as a major component of the kindergarten curriculum, believing that, "Play is the highest phase of child development . . . . The play of childhood are the germinal leaves of all later development" (Froebel, 1887, pp. 54–55). According to Gutek (1997), "Froebel's exaltation of the role of play was a strikingly different approach from that of many conventional educators up to the nineteenth century" and Froebel should be credited with being "one of the pioneers in legitimizing the concept of play in Western educational history" (p. 249).

Understanding Froebel's emphasis on children's play requires a brief discussion of the philosophical underpinnings of his educational theory and methods. The foundation of Froebel's educational system was a philosophical belief in the unity of all things, animate and inanimate alike, and in the idea that man contained within himself a divine spirit, that, "mankind was the physical embodiment of God's reason" (Shapiro, 1983, p. 20).

Froebel lived during a time when the philosophical school of thought called *idealism* captured the attention of much of Europe (Weber, 1969). Linked to such philosophers as Fichte, Schelling, Kant, and Hegel, idealism posited the existence of an Absolute Idea, embodying both absolute truth and absolute beauty. The concept of an absolute idea manifested itself in Kant's theory of *a priori* categories of mind, such as space and time. Kant believed that these mental structures exist before – *a priori* – sensory experience. Like Kant, Fichte also argued for the existence of knowledge prior to experience, knowledge that had its origins in some divine source (Gutek, 1997).

If there was an absolute idea, one that had divine origins, then for Froebel it logically followed that all things in the world must be interconnected by virtue of this relationship to God. Goals for today's kindergartens might include literacy development, school work habits, and learning to cooperate with other children, but for Froebel there was something more transcendent; there was the need to bring children to the understanding that everything in the universe stands in relation to God. This theme of connectedness led Froebel to emphasize the concept of *Gliedganzes*. The concept of *Gliedganzes* refers to part-whole, or member-whole, relationships where each individual entity, say a person or a plant, is a whole unto itself and at the same time is a member of a larger whole. Man, for example, is an individual while also a member of a larger social group.

For Froebel, the goal of education in the kindergarten, "was the unfoldment of the child's powers in order to achieve this divine unity" (Weber, 1969, p. 8) and for Froebel, play was the vehicle through which children would attain the understanding of their relationship to God, Gliedganzes, and unity. Froebel offered the following definition for play: "Play is the self-active representation of the inner – representation of the inner from inner necessity and impulse" (Froebel, 1887, p. 55). Inner activity was primary for Froebel (Weber, 1969); he believed that there was always something more in the mind than what was taken in through the senses. Here the influences of Idealism, Kant, Fichte, and the concept of the Absolute Idea are evident. Froebel also believed in the necessity of self-activity because he saw this as, in some sense, the essential destiny of the child since the child was imbued with the creative force of God. He called this the 'eternal divine principle,' and wrote that it, "demands and requires free self-activity and self-determination on the part of man, the being created for freedom in the image of God" (Froebel, 1887, p. 11).

This self-activity, then, would be the means through which the child's development and inner life would unfold. According to Froebel, self-activity needed to be more than simply the object observation of Pestalozzi and more than the absorption of verbal instruction; self-activity needed to be the active, physical handling of concrete materials. Froebel maintained that children learn best through the physical handling of objects rather than the language of adults. His suggestion to adults on this matter was direct: "Do not ... tell him words much more than he could find himself without your words" (Froebel, 1887, p. 86). According to Froebel, recitation as a form of expression needed to be replaced by play as a form of self-expression (Brosterman, 1997).

But to Froebel's mind, it wasn't enough for a child to simply represent the inner through play and self-activity. For Froebel, the purpose of education,

> consists of leading man, as a thinking, intelligent being, growing into self-consciousness, to a pure and unsullied, conscious and free representation of the inner law of Divine Unity, and in teaching him ways and means thereto (Froebel, 1887, p. 2).

To this end, Froebel believed that the child's activity needed to be reflected in the external world, that, "The deepest craving of this inner life is to behold itself mirrored in some external object" (p. 238). Froebel wasn't arguing for the object observation of Pestalozzi; the child wouldn't just be absorbing abstractions from the shapes and forms of the outer world but would be seeing in the shapes and forms, "the expression of his spirit, of the laws and activities of his own mind" (Froebel, 1887, p. 108). Froebel was suggesting an interaction between the child and the material world, an interaction in which the child would be, "making the internal external, the external internal, and ... [would be] perceiving the harmony and accord of both" (Froebel, 1900, p. 174).

In order for children to perceive this harmony between the inner and the outer, the play of children, according to Froebel, needed guidance and direction (Morrison, 1995). To facilitate the child's 'making of the internal external and the external internal,' Froebel created a specially designed set of educational materials, perhaps the first of its kind (Brosterman, 1997), known as the *gifts and occupations*.[2] The significance of Froebel's genius, according to Brosterman, was to recognize that there were certain aspects of a child's life and activity that would benefit from planned and thoughtful direction. The design of the gifts and occupations, and the prescriptions for their use would provide that direction – and would ultimately lead to Burk's 'dangerously radical' study.

Gifts and occupations. Froebel took more than a decade to design these special educational materials. Generally thought to be twenty in number, they have come to be known as the *gifts and occupations*, although Froebel himself didn't focus on the division of the materials into two categories, simply calling all of them, *gifts* (Brosterman, 1997). Briefly, the gifts and occupations included balls, various size building blocks, wooden tablets, materials to represent lines and circles, materials to represent points (e.g,. beans, seeds, pebbles), clay, paper-folding, parquetry, sewing cards, and beads for stringing. In general, the sequence of the materials moved from solids to points.[3]

Many of the materials reflected aspects of Froebel's early experience in mathematics, architecture, and crystallography (Weber, 1969). Embedded within the materials, according to Froebel's design, were universal mathematical truths and the sense of perspective and proportion so much a part of architecture (Gutek, 1997). By the design of the gifts and occupations, Froebel "brilliantly deconstructed all physical and conceptual realms and conveyed vivid impression of the continuity of nature" (Brosterman, 1997, p. 36). This deconstruction relates to the symbolism embedded in the Froebelian materials. A child's activity would result in the production of particular concrete, tangible forms, forms that symbolized and it was what those forms symbolized that was important. According to Froebel (1887), "In the forms he [the child] fashions he does not see outer forms from which he is to take in and understand; but he sees in them the expression of his spirit, of the laws and activities of his own mind" (p. 108). The child, in other words, would have his consciousness formed by seeing in the outer, concrete forms, symbolic representations of the emerging inner ideas. A look at one particular material in the sequence, the 3rd Gift, provides a good illustration of Froebel's emphasis on the materials symbolically representing certain concepts.

The 3rd Gift, along with gifts 4, 5, and 6 are collectively referred to as the building gifts. The building gifts are the first ones in the sequence that were to be used with preschool age children and the ones we might recognize today since they are the forerunners of the common building blocks (Brosterman, 1997).[4]

Froebel (1900) called the 3rd Gift the, "the first divisible plaything in child life" (p. 120), and two important Froebelian concepts were embedded in the 3rd Gift and in the building gifts in general. One concept concerns the relationship of parts and wholes. The philosophical idea that Froebel 'designed' into the building gifts was that any part is a complete entity unto itself (the child, for example) but also a member of a larger whole (i.e. society). This was important because in the 3rd Gift, "the aim of his [Froebel] educational method is secured when the child finds himself at the same time a separate individual and a member of a higher unity" (Harrison & Woodson, 1905, p. 86). The 3rd Gift was the first of the gifts to embody in symbolic form the part-whole relationship that was so important to Froebel; the first gift to represent the Gliedganzes.

The second general concept symbolized by the building gifts relates to Froebel's ideas about children's development. The internal unfolding of development needed to be mirrored in external objects and experience.

Froebel's focus on play with concrete materials is based on the belief that, "all knowledge and comprehension of life are connected with making the internal external, the external internal, and perceiving the harmony and accord of both" (Froebel, 1900, p. 174). As the child separates and recombines the cube and its parts, he is experiencing, according to Froebel, this relationship between the external (the surface appearance of the whole cube) and the internal (the individual parts that make up the cube).

> Thus, after comprehending the *outside* of the object, the child likes also to investigate its *inside*; after a perception of the whole, to see it separated into its *parts*; if he obtained a glimpse of the *first*, if he has attained the *second*, he would like from the parts again to *create the whole* (Froebel, 1900, p. 118).

The building gifts were also designed to symbolically represent, and therefore foster within the child, what Froebel called the three forms or realms of knowledge: forms of life, knowledge, and beauty. As children played with the building gifts they would at times construct representations of things in their environment, for example, tables and chairs, houses, and mountains. These are the forms of life. The mathematical relationships inherent in the design of the blocks (halves and quarters) are an example of the forms of knowledge that can be represented through the building gifts. These mathematical relationships were important to Froebel because they were essential for the development of child's understanding of his surroundings. The particular design of the building gifts was necessary because if the building blocks were irregular or arbitrary in shapes and sizes, they would

> leave only vague, unsatisfactory impressions on the child's mind and would not assist him much in the intelligent investigation of the world of nature as all the divisions and subdivisions of crystals, plants, and animals are according to mathematical laws (Harrison & Woodson, 1905, p. 96).

Forms of beauty are represented by the symmetrical designs that can be produced by the positioning and arranging of the blocks themselves. Froebel believed that these symmetrical designs were pleasing to the child and a key aspect of their creation was the transformation of one design to another, not destroying the first but using the same parts to create a new whole; thus, again, symbolizing the philosophical idea that all things were related.

The symbolism embedded in the gifts and occupations would be pointed out by teachers through words, songs, and rhymes, and it was assumed that children would understand the representations (Krogh, 1994). Froebel believed that children would be able to naturally understand the metaphysical truths embedded in the materials (Weber, 1969), that there was a "connection between the toy and the child's soul" (Almqvist, 1994, pp. 48–49).

The Dangerously Radical Concept of Free Play 281

A result of this emphasis on the symbolic representation of truths is that while the child might initially use the gifts in some self-directed manner, the gifts, "were never available for entirely 'Free Play.' Always tethered in some fashion to the forms of the three realms [life, knowledge, beauty], their use was subordinate to the greater whole" (Brosterman, 1997, p. 37). The gifts, according to Gutek (1997), were designed, "to ensure that this development follows the correct pathway, which is both God's and nature's plan of development" (p. 248). As Snider (1900) writes in an early biography of Froebel, the "games, gifts, occupations, are to train the child, not merely amuse him" (p. 290).

This emphasis on the symbolic truths embedded in the materials led Froebel to some fairly prescriptive descriptions of how the materials were to be used and, perhaps inevitably, to some contradictions within his own writings. These descriptions, of what Froebel called "sufficing instructions" (Froebel, 1900, p. 20), were intended to "render prominent the laws of mental growth, proceeding from and leading to the use of the play and its different representations" (p. 21). Accompanying these instructions were detailed diagrams that represented various constructions and designs that could (or perhaps, should) be made with the particular gifts. It may be open for question as to whether Froebel intended these diagrams to be illustrative of *possible activity* or prescriptive of *necessary activity*, but the following description taken from *Pedagogics of the Kindergarten* (Froebel, 1900) does suggest a prescriptive sense to Froebel's 'sufficing instructions.' (Plates and figures refer to diagrams Froebel provided for the reader.)

> We begin with the arrangement of Plate V, Fig. 1; we move one after another of the four outer cubes which now stand with surface against surface, round to the left or right . . . so that the edges come to end in (i.e. touch) edges (see Plate V. Fig. 2); then further, edges touching surfaces (Plate V, Fig. 3), and finally surfaces touching edges (Fig. 4, Plate V); and we have thus attained and represented before the child's eye what we wished: we have made manifest and clear the inner unity of shape in the manifoldness of the movements in and through the change of shapes . . . (p. 134).

Froebel suggested that the diagrams be used in the following ways. First, they can help the parent or teacher understand what children are doing with a particular gift. This purpose emphasizes observation of children's play with the diagram as a guide for that observation. The second use focuses on helping children understand the meaning that they have produced with the particular gift: here the child is being exposed to the symbolism embedded in the materials. Finally, the diagram can be something the child may imitate. Now the child doesn't initiate the play but is guided to the reproduction of certain patterns and configurations, again with an emphasis on 'seeing' the symbolism inherent in the materials (Froebel, 1900, p. 236). Froebel believed that this direction was necessary because "unstructured play represented a potential danger ... a child left to his own devices may not learn much" (Morrison, 1995, pp. 65–66).

The emphasis on the symbolism in the materials led Froebel to include imitation and dictation in his educational system – at least as interpreted by his followers in this country. It was through imitation that children could make a connection between their inner impulses and the ideas and concrete experiences of the outer world (Weber, 1969). Harrison and Woodson (1905) argue that for the teacher, "There comes the time ... when her superior knowledge of the Play-Gift, as well as her knowledge of what facts it is best of the child to learn, necessitates her definite control of his play" (p. 92). They go on to suggest that this control can take the form of imitation.

> This [taking control] may be done by joining in the play and letting him imitate the form she [the teacher] makes .... By this kind of play she introduces him to an orderly use of all the parts of the Gift in each new object made, and also leads him to build one form out of another without destroying the forms, thus guiding him into transforming rather than destroying (p. 93).

Confusion as to the nature of play and self-activity in Froebel's system is understandable given the contradictory statements that can be found within Froebel's own writings. In discussing the use of the 3rd Gift, for example, Froebel (1900) writes that the child must be allowed "the greatest possible freedom of invention; the experience of the adult only accompanies and explains" (p. 130). But later in the very same paragraph, Froebel writes, "It is here quite essential to remark that all eight cubes always belong to each design – that is, they must stand in some relation to the whole" (p. 130). Similarly, when describing a child's work with the forms of beauty, Froebel (1900) writes, "The greatest freedom is always given in respect to the choice of starting point" (p. 226), but once the child is engaged in the activity, "how essential to the welfare of the child ... to restrict the freedom of change by limiting it to fixed members and by determining it to a definite direction and goal" (p. 234). Here lies the seeming contradiction: freedom of invention, yes, but only to the degree that it serves the greater goal of the child experiencing the symbolism believed to be inherent in the materials. Again, the use of the materials "was subordinate to the greater whole" (Brosterman, 1997, p. 37). And if the child does not perceive the 'greater whole,' the mother or kindergartner "has many opportunities of correcting the child's perceptions by his representations ... as the building gifts afford a means of clearing the perceptions of the child" (Froebel, 1900, p. 222).

This, then, was the situation when Frederic Burk and his colleagues offered the results of their 'dangerously radical' study: Kindergartens had been in existence for over forty years, there were free kindergartens (and some public school kindergartens) in most major cities, dedicated women and men were committed to the education of young children in these kindergarten programs, and the curriculum of the kindergartens was heavily influenced by the ideas of Froebel. There was by no means, however, total uniformity of opinion about what constituted, in today's terms, 'appropriate kindergarten practice,' and the Santa Barbara study was an important part of the growing opposition to the Froebelian educational system.

A STUDY OF THE KINDERGARTEN "PROBLEM"

Frederic Burk, superintendent of the Santa Barbara schools at the time of the study, had been one of G. Stanley Hall's doctoral students at Clark University in Worcester, Massachusetts. As the 'father of the child study movement,' Hall's work as a psychologist set the stage for an approach to the study of children that has influenced child development research in this country for most of the twentieth century (Thomas, 1996), and he is credited with conducting the first large scale research project using children as the subjects of the study (Charlesworth, 1987). Hall's research methodology consisted of two approaches that are a part of the child development tradition: observing groups of children and summarizing the results of these observations in the form of averages or norms for different age levels (Thomas, 1996). With the emergence of the child study movement during the last quarter of the nineteenth century, the young child was no longer of interest just to the teacher; she could now be a subject for the laboratory (Lazerson, 1972). Children were to be understood and educational methods designed not through introspection and philosophical intuition – the methods of Froebel – but through direct study of groups of children.

Hall and the child study movement provided what was perhaps the main challenge to the Froebelian kindergarten. Hall believed that Froebel should be given credit for, among other things, his emphasis on play, his belief in the inherent goodness of man, and his focus on childhood as a separate stage of development. Hall's criticisms of Frobelian psychology and practice fell into two broad categories, one dealing with the appropriateness of some of the Froebelian activities and the second related to who determined and controlled what the children did in their play. From this perspective, Hall believed that the Burk study was extremely important because "it showed how real children differ from the manikin children of the Froebelian metaphysicians" (Hall, 1911, p. 29).

Free Play Versus Froebel Play

For Frederic Burk, the study was needed because, "the fundamental weakness of the prevailing kindergarten consists in its gross neglect of instincts which properly belong to its period, and attempts prematurely to develop instincts which do not bud until the adolescent period" (Burk. F., 1920, p. xi). The purpose of the study was to identify these instincts by observing their emergence during children's free play. During the course of the three month study, the kindergarten teachers in four classrooms observed the play of the children in their rooms. Two of the classes had children ages five and six who were in their second year of kindergarten, and two of the classes had children four and five years of age, in their first year of kindergarten. The study consisted of two parts. First, the spontaneous play of the children was observed during two twenty minute periods. In the second part of the study, the children were offered the gifts and occupations as materials for their play. Instead of their activity with the materials being guided by adults, as was the standard practice in Froebelian kindergartens, "Every day for half an hour the kindergarten materials, the gifts and occupations, were spread on a table and each child chose one thing he cared to play with for that time" (Burk, C. F., 1920a, p. 68). This part of the study focused on two aspects of children's play with the Froebelian materials: material choice and material use. The researchers were interested in which materials the children voluntarily chose for their play, and once that choice was made, how they then used the material. The researchers wanted to know if the children would use the Froebelian materials in the manner prescribed by Froebel – if they were not directed to do so by adults.

Results of the Study

In terms of the children's spontaneous play, the teachers made note of over 100 different kinds of play and organized these plays into three categories: plays of physical action, representative plays, and traditional games (e.g. "London Bridge is Falling Down"). In terms of frequency of occurrence, plays of physical action and representative plays were the most predominant while the traditional games were observed to occur at a fairly insignificant level.

The researchers' analysis of the physical play of the children supported Hall's belief that the Froebelian kindergarten emphasized the wrong kind of physical activity. The teachers observed that the children engaged in relatively few physical plays that required fine motor skills. This particular finding supported Hall's own beliefs about young children's development and what he suggested was one of the problems with Froebelian practice. Hall believed that Froebel's educational system did not adequately consider the child's physical development and that more attention should be given to the strengthening of the body and care should be taken not to put undue strain on the eyes and hands. Hall faulted Froebel's emphasis on the perceived spiritual needs of children and argued that kindergarten needed to "recognize that the child's bodily needs are as great as perhaps and paramount in importance to, the needs of the soul" (Hall, 1911, p. 7). Hall's own research led him to conclude that the large, fundamental muscles, those that control the limbs and trunk, developed before the accessory muscles which are needed for the fine motor handling of Froebel's gifts and occupations (Hall, 1911). The gifts and occupations, according to Hall, did not engage children in large motor activity just at the time they most needed it for development, and this overemphasis on sedentary activities might result in fatigue and nervousness (Weber, 1969). Hall couched these criticisms in terms of concerns for children's health and wrote that, "one of the most heinous offenses of the modern kindergarten" (Hall, 1911, p. 18) was its neglect of health issues. According to the Santa Barbara study, during spontaneous play children tended to engage in physical activity which utilized the large muscle groups, and

> The bearing this fact has on kindergarten is obvious. If we take the cue from children's natural play we must bid Godspeed to the already departing 'fine' work and the 'accurate' work of the kindergarten .... It is tolerably evident ... that no system of calisthenics can take the place of or even compete with the set of exercises in which Dame Instinct instructs the kindergarten child. Her "system" is quite complete as most systems furnished by the logical adult – and not half so stupid either (Burk, C. F., 1920b, p. 30).

In terms of representative plays, the researchers identified three categories: being an animal, making things, and representations of adult occupations. They concluded that the children were imitating what they had seen and observed, that "Their analogies are analogies of external features based on sense impressions. There is not particular evidence of any appetite for symbolism" (Burk, C. F., 1920b, p. 34). C. F. Burk goes on to dispute the claim by Susan Blow, perhaps Froebel's staunchest defender, that the symbolism of the kindergarten would "stir the child with faraway presentiments of his ideal nature, his spiritual relationships, and his divine destiny" (p. 34). The researchers argued that

> there is certainly nothing in the analogizing found in children's natural play to sanction the kindergartner in any strained attempts to arouse spiritual "adumbrations" in the child

through the symbolic games of the orthodox kindergarten. Rather than worry about mysterious "presentiments," may it not be safer to give the child what his healthy imitative, constructive, and dramatic instincts clearly and simply demand, again trusting Dame Instinct to utilize her material to the best advantage? (p. 34–35).

As an example, C. F. Burk (1920b) focuses on the kindergarten circle. The symbol of the circle, often painted on the floor of the classroom and utilized to represent unity and community, was a major staple of the Froebelian kindergarten. Based on the results of the study, she concluded that "The sacred circle of kindergarten paraphernalia does not seem to be based on any natural penchant of children of the kindergarten age for the traditional circle games, for these seldom appear in their undirected or unindicated play" (p. 36). She also suggested that, "The kindergarten circle may stir presentiments of universal unity, but in equal probability the child takes it for what it is worth to him – a toe-line whereon, forsooth, he must march" (p. 35).

In terms of material choice, i.e. which Froebel materials children chose to use, the teachers found that here too the children like to be active, and that they do "not like to sit at a table and pretend to be doing something with something – a stupid occupation for an active child" (Burk, C. F., 1920a, p. 74). Specifically, the researchers found that the second gift (sphere, cylinder, and cube) was rarely selected by the children, and that "two materials far outstripping the rest in interest are clay and sewing cards" (p. 71).[5] They also found differences in the materials chosen by the younger children when compared with the materials chosen by the older kindergarten children, thus indicating to the researchers that there is a developmental progression in both interests and ability. This was one more argument for letting the unfolding development of the children guide their activity.

In terms of material use, the teachers/researchers recorded the children's use of the materials according to the following categories: no order or aimless use; form arrangement; color arrangement; design; and representation. The teachers found that aimless play occurred primarily with the beads, parquetry, and tileboard materials. There was very little aimless play with clay or blocks; furthermore, the blocks and clay were used almost exclusively for representation. In fact, representation was the largest component of purposeful play in all of the classrooms. In Froebelian terms, the representation itself was largely made up of natural and life forms rather than the forms of beauty or geometric designs (Burk, C. F., 1920a, p. 76). The younger children created representations of eleven different kinds of animals while the older children produced twenty-one different animal representations. As far as object representations, both the younger and older children produced representations of over eighty different objects.

Similar to their observations of the children's spontaneous play, the Santa Barbara teachers concluded that the children were creating representations based on their own interests and past experiences. When given a choice, the children were creating symbols to *represent* their own ideas rather than *reproducing* the symbolism supposedly inherent in the Froebelian materials. According to the researchers, "The child prefers to imitate from memory, with delightful freedom, rather than to confine himself to the narrow restraint of symmetrical proportions" (Burk, C. F., 1920a, p. 76).

Based on the findings of the study, the Santa Barbara researchers offered this assessment of the state of kindergartens at the end of the century.

> The kindergarten is loaded down with an unsifted mass of material which has been chosen by the adult mind as suitable for the logical development of the child and which has been used as the basis of dictation exercises, arranged in formal sequence. The child has not been particularly consulted either in the choice of material, or in the use to be of it (Burk, C. F., 1920a, p. 69).

In particular, the Santa Barbara researchers took issue with the adult imposed symbolism that was found in the Froebelian kindergarten. As might be expected, Frederic Burk's mentor, G. S. Hall also had problems with the symbolism of Froebel's materials. He didn't accept the way in which Froebel translated the philosophy of idealism into kindergarten practice, a translation Hall referred to as Froebel's "weird and bizarre version of this metaphysical ferment" (Hall, 1911, p. 2). Hall believed that with the gifts and occupations, Froebel was attempting to create "a perfect grammar of play and an alphabet of industries" (Hall, 1911, p. 17). With this predetermined 'grammar' and 'alphabet' children would then be led to physical representations of Froebel's beliefs. But for Hall, these "meanings seen or claimed exist solely for the teacher and not at all for the child" (Hall, 1911, p. 18). Activities and materials, according to Hall, should be based on children's interests and should be appropriate to their stage of physical and mental development. This was the value of child study and Hall believed that the kindergarten movement hadn't contributed anything of significance to the child study movement and that the lack of contribution was at least partly due to the kindergarten's misplaced emphasis on "its cult of attenuated symbolism" (Hall, 1911, p. 26).

Once the study was completed, Burk and his teachers felt so strongly about the importance of what they had observed during this study that they chose to continue providing free play time for children after the study had ended, although, "it was such a radical departure from current kindergarten programs that the periods were labeled 'recess'" (Weber, 1969, p. 59). About this 'free play' or 'recess period' C. F. Burk (1920a) writes,

During no other hour of the day is there such close absorption in their play, such deep interest, such concentration, such unconsciousness of the doings of other children. This is the hour when each child is most thoroughly bent on self-expression, on spontaneous doing, on self-activity (p. 82).

Reactions to the Study

As might be expected, this attack on the symbolism of Froebel by Hall and the Santa Barbara group was decried by the more conservative followers of Froebel. Susan Blow, not surprisingly, strongly objected to the conclusions of the Santa Barbara study. Besides questioning some of the methods of the study, Blow also criticized the conception of imitation and representation presented by the Santa Barbara study, citing the "fatal blindness of the experimenting kindergartners to the meaning of imitation and their ceaseless surrender to the momentary caprices of little children" (Blow, 1908, p. 176). Blow believed that three principles characterized the imitative process: (1) what a child imitates he tends to become, (2) what he imitates he will notice, and (3) what he imitates he begins to understand. For Blow, then, selection of what a child imitates is critical. The following passage is quoted at length because it captures Blow's criticism of the free play program's belief in allowing children to represent, or imitate, whatever they choose.

> Blind to the fact that nothing less than the child's personality is at stake in the method and matter of his imitations, the free play revolutionizers of Froebelian games are perfectly willing to have children transform themselves into sneaking foxes and writhing rattlesnakes. Ignoring the reaction of imitation upon selective interest they eye with equal favor the representation of garden-planting, or human burial .... In short, through a specific application of the fatal heresy which underlies the whole free-play programme, they shift upon children the entire responsibility of selecting what they will represent and how they will represent it. By this act of abdication they reduce the kindergarten games to a hotchpotch not only devoid of educational value, but absolutely perverting in its reaction upon intellect, emotion, and will (Blow, 1908, pp. 176–177).

In summary, for Blow, "The gist of contention between the free-play reformers and the traditional kindergarten is, whether the immediate interests of little children are sufficient index of what is contributory to their development" (Blow, 1908, p. 179). Her answer, clearly, was in the negative.

DISCUSSION

It has not been uncommon in the literature of early childhood education for questions to be raised about the Froebelian conception of play. Weber (1969) suggested that Froebel's ideas regarding self-activity needed to be reassessed

The Dangerously Radical Concept of Free Play

because they were "excessively teacher directed" (p. 5), while Krogh (1994) concluded that his "close-ended, prescribed, teacher directed activities might be enjoyable but could not be described as play" (p. 21). This position was echoed by Spodek and Saracho (1988) when they concluded that, "Even though Froebel . . . uses the word freedom extensively in his writing, the activities he designed were highly prescriptive. Because the children had to do what they were told with the materials, the activities cannot really be termed play" (p. 9).

Play, according to Reynolds and Jones (1997), can be defined as, "Children's self-chosen process of recreating experience in order to understand it" (p. 3). The symbolic, or representative, play of Froebel was more about the *re-production* of experience than the *re-creation* of it. The Froebelian materials and pedagogy that accompanied them were designed to lead children to the re-production of particular symbolic representations. These representations would serve as 'templates' for the child's developing understanding of what Froebel believed to be the eternal truths of unity and wholeness. If the design of the materials contained representations of these truths, then whether or not children experienced the representations could not be left to chance. Froebel introduced dictation and imitation into his educational system so that the child could be led systematically to these re-productions and to the perceptions of the symbolism within them. Dewey (1944) summarized this aspect of Froebel's system when he wrote,

> After the scheme of symbolism has been settled upon, some definite technique must be invented by which the inner meaning of the sensible symbols used may be brought home to children. Adults being the formulators of the symbolism are naturally the authors and controllers of the technique. The result was that Froebel's love of abstract symbolism often got the better of his sympathetic insight; and there was substituted for development as arbitrary and externally imposed a scheme of dictation as the history of instruction has ever seen (p. 59).

The Santa Barbara teachers, on the other, emphasized how children would spontaneously create representations based on their own experience, rather than based on the perception of externally determined symbolism. These representations were symbolic re-creations of the children's understandings and the meanings their experiences held for them. For Susan Blow, the danger in this perspective was that children might in fact symbolically represent, for example, 'writhing rattlesnakes' and 'human burials,' and these representations would not lead the children to the perception and understanding of the universal truths identified by Froebel as necessary for their development. Dewey (1972) captures this distinction quite well when he writes that, "The symbolism of the child must be taken as genuine, as intrinsic, as having meaning for the child himself and not simply for us" (pp. 219–220).[6]

Brian Sutton-Smith (1986) suggests that although conceptions of play have changed through history, these changing conceptualizations almost always exhibit various "bipolar characteristics" (p. 3). One of the bipolar pairs, originating with Aristotle and Plato but perhaps still with us today, is the idea that there exists *good* play and *bad* play. For Plato and Aristotle, *good* play were those activities considered to be noble pastimes. According to Sutton-Smith, today,

> this separation between *good* play and *bad* play is presumably between play that socializes, encourages learning, or is cognitive, and those other kinds of play that are a danger or nuisance in the household or classroom and for which children are sent outside (p. 7).

When the adult, predetermined aspects of 'good play' are designed into materials, as was the case with Froebel's materials and is perhaps the case with certain contemporary 'educational' toys and computer software, it follows that an effort would be needed to ensure that the children achieve these results when playing with the materials, so that "we have attained and represented before the child's eyes what we wished" (Froebel, 1900, p. 134). If, as Susan Blow wrote, "organized materials organizes the mind" (Blow, 1913, p. 135), then it would follow that allowing children to choose what to represent and how to represent it is "the fatal heresy which underlies the whole free-play programme" (Blow, 1908, p. 176–177). *Good* play leads to the orderly attainment of the desired learning goals.

Conversely, according to Sutton-Smith, the adjective, 'bad,' is used to refer to play that might originate from the children's interests and needs but, as opposed to representing order and achievement, is more suggestive, from the adult perspective at least, of "disorder and subversion" (Sutton-Smith, 1986, p. 7). In Sutton-Smith's view, the goal assumed by many adults is to emphasize *good play* and "The major technique used to obscure the differences between good and bad play has been the gradual control of children so that under our supervision they can only produce *good* play" (Sutton-Smith, 1986, p. 11). This need for adult supervision was expressed by Susan Blow (1908) when she wrote the following concerning the Santa Barbara study: "There is not a single word in the diary to suggest any effort was made to lead the children to build better, draw better, model better, or do anything better than they were originally able to do" (p. 156). For Blow, Froebel play was good play; the play of the Santa Barbara study, free play, was just the opposite. Blow might have taken to heart a second metaphoric distinction Sutton-Smith (1988) makes concerning society's view of children's play, a distinction between *sacred play* and *festival play*.

> When in 1695 John Locke discouraged his readers from allowing their children to play in the streets and urged that they were better off inside the house with their alphabet blocks ... he initiated what was to become a new distinction. Now there was to be educational play. In the course of the next few centuries, and in the hand of Pestalozzi and Froebel and their successors, this was to become a new kind of sacralized play .... What we are left with then in modern life is the notion ... of two kinds of play, an educational one that is somewhat sacred ... and another more festive kind that we hope to confine to the playground, but that often has a tendency to sneak into the school and upset our lessons ...
> (Sutton-Smith, 1988, p. 45).

While looking back to the time of Froebel's influence on the early kindergarten movement might be of historical interest and significance, one hundred years after the first publication of the Santa Barbara study there may still exist a tension among educators as to the nature of play and its value in the curriculum. According to Klugman and Fasoli (1995), "Teachers are finding it increasingly difficult to justify play in early childhood programs because of a back-to-basics emphasis on abstract, pencil-and-paper tasks for children at the expense of more interactive, child-focused play activities" (p. 195). Seefeldt (1995) writes that "Instead of children's play being the foundation for learning, the kindergarten and primary curriculum now revolves around scope and sequence charts and drill and practice of basic skills" (p. 187). Finally, in words that echo the debate of a hundred years earlier, Glickman (1984) suggests that "where preschool and elementary school meet, at the kindergarten level, the raging controversy of purpose and curriculum as it relates to play is most evident" (p. 256).

The Santa Barbara teachers concluded that the play which originated in the spontaneous interests of the children, play that might be active and full of 'sneaking foxes and writhing rattlesnakes' does in fact belong in the classroom. One hundred years after the Santa Barbara study, the question remains as to whether or not the 'kindergarten problem' is still be with us and whether the concept of free play is still a somewhat 'dangerously radical' idea.

NOTES

1. As Hewes (1997) points out, the original Froebelian kindergarten was not just for five year olds; children from ages two through six were included in Froebel's original program.
2. Detailed descriptions of the gifts and occupations can be found in Froebel's, *The education of man* (Froebel, 1887) and Brosterman's, *Inventing kindergarten* (Brosterman, 1997).
3. It is important to note the age range of interest for Froebel. The first two gifts in the series were designed to be used by mothers with their infants.
4. The 3rd Gift was made up of eight one-inch cubes, forming a two-inch cube ($2 \times 2 \times 2$). The 4th Gift consisted of eight brick-shaped blocks ($2 \times 1 \times \frac{1}{2}$) forming a two-

inch cube. The 5th Gift contained twenty-seven one-inch cubes, three bisected and three quadrisected diagonally, forming a three-inch cube; the 6th Gift was made up of twenty-seven brick-shaped blocks, three bisected longitudinally and six bisected transversely, forming a three-inch cube (Froebel, 1887, pp. 285–287).

5. The use of the second gift with four, five, and six year old children might have been a misinterpretation of Froebelian theory. The second gift was more rightly used with infants (Hewes, 1997).

6. It is interesting to note that similar criticisms have been leveled at the methods and materials created by Maria Montessori. Piaget, for example, stated that, "Montessori's idea of focusing on activity is excellent, but the materials are disastrous .... With standardized material, one doesn't dare try to change it. And yet the really important thing is for the child to construct his own material" (Evans, 1973, p. 52).

REFERENCES

Almqvist, B. (1994). Educational toys, creative toys. In: J. H. Goldstein (Ed.), *Toys, Play and Child Development* (pp. 46–66). NY: Cambridge University Press.

Blow, S. E. (1908). *Educational issues in the kindergarten*. NY: D. Appleton and Company.

Blow, S. E. (1913). First Report. In: The Committee of Nineteen (Ed.), *The Kindergarten: Reports of the Committee of Nineteen on the Theory and Practice of the Kindergarten* (pp. 1–230). Boston: Houghton Mifflin Company.

Brosterman, N. (1997). *Inventing kindergarten*. NY: Harry N. Abrams, Publishers.

Burk, C. F. (1920a). Children's spontaneous choices and use of kindergarten materials. In: F. Burk & C. F. Burk (Eds), *A Study of the Kindergarten Problem in the Public Kindergartens of Santa Barbara, California, 1898–1899* (2nd ed., pp. 63–83). NY: Teachers College, Columbia University.

Burk, C. F. (1920b). Play – a study of kindergarten children. In: F. Burk & C. F. Burk (Eds), *A Study of the Kindergarten Problem in the Public Kindergartens of Santa Barbara, California, 1898–1899* (2nd ed., pp. 26–39). NY: Teachers College, Columbia University.

Burk, F. (1920). Preface to the first edition. In: F. Burk & C. F. Burk (Eds), *a Study of the Kindergarten Problem in the Public Kindergartens of Santa Barbara, California, 1898–1899* (2nd ed., pp. ix-xii). NY: Teachers College, Columbia University.

Burk, F., & Burk, C. F. (1920). *A study of the kindergarten problem in the public kindergartens of Santa Barbara, California, 1898–1899* (2nd ed.). NY: Teachers College, Columbia University.

Charlesworth, R. (1987). *Understanding child development* (2nd ed.). Albany, NY: Delmar Publishers Inc.

Dewey, J. (1944). *Democracy and education*. NY: Macmillan.

Dewey, J. (1972). Early essays: 1895–1898 (The early works, 1882–1898, Volume 5). Carbondale and Edwardsville, IL: Southern Illinois Press.

Evans, R. I. (1973). *Jean Piaget: The man and his ideas*. NY: E. P. Dutton & Co., Inc.

Froebel, F. (1887). *The education of man* (W. A. Hailmann, Trans.). NY: D. Appleton and Company.

Froebel, F. (1900). *Pedagogics of the kindergarten – or, his ideas concerning the play and playthings of the child* (Josephine Jarvis, Trans.). NY: D. Appleton and Company.

Glickman, C. D. (1984). Play in public school settings: A philosophical question. In: T. D. Yawkey & A. D. Pellegrini (Eds), *Child's play: Developmental and Applied* (pp. 255–271). Hillsdale, NJ: Lawrence Erlbaum Associates.

Gutek, G. L. (1997). *Historical and philosophical foundations of education: A biographical introduction* (2nd ed.). Upper Saddle River, NJ: Merrill/Prentice Hall.

Hall, G. S. (1911). *Educational problems* (Vol. 1). NY: D. Appleton and Company.

Harrison, E., & Woodson, B. (1905). *The kindergarten building gifts – with hints on programmaking* (2nd ed.). St. Louis, MO: Sigma Publishing Company.

Hewes, D. (1997). Fallacies, phantasies, and egregious prevarications in ECE history. Paper presented at the Annual Conference of the National Association for the Education of Young Children, Anaheim, CA. (Eric Document Reproduction Service No. Ed 414 058)

Hill, P. S. (1920). Introduction to the Second Edition. In: F. Burk & C. F. Burk (Eds), *A Study of the Kindergarten Problem in the Public Kindergartens of Santa Barbara, California, 1898–1899* (pp. v-viii). NY: Teachers College, Columbia University.

Klugman, E., & Fasoli, L. (1995). Taking the high road toward a definition of play. In: E. Klugman (Ed.), *Play, Policy & Practice* (pp. 195–201). St. Paul, MN: Redleaf Press.

Krogh, S. L. (1994). *Educating young children: Infancy to grade three.* NY: McGraw-Hill, Inc.

Lazerson, M. (1972). The historical antecedents of early childhood education. In: I. J. Gordon (Ed.), *early Childhood Education: The Seventy-First Yearbook of the National Society for the Study of Education* (pp. 33–53). Chicago: The National Society for the Study of Education.

Morrison, G. S. (1995). *Early childhood education today* (6th ed.). Columbus, OH: Merrill.

Paciorek, K. M., & Munro, J. H. (Eds) (1996). *Sources: Notable selections in early childhood education.* Guilford, CT: Dushkin Publishing Group/Brown & Benchmark Publishers.

Piaget, J. (1962). *Play, dreams and imitation in childhood.* NY: W. W. Norton & Co., Inc.

Reynolds, G., & Jones, E. (1997). *Master players: Learning from children at play.* NY: Teachers College Press.

Ross, E. D. (1976). *The kindergarten crusade: The establishment of preschool education in the United States.* Athens, OH: Ohio University Press.

Seefeldt, C. (1995). Playing with policy: A serious undertaking. In: E. Klugman (Ed.), *Play, Policy & Practice* (pp. 185–194). St. Paul, MN: Redleaf Press.

Shapiro, M. (1983). *Child's garden.* University Park, PA: Penn State University Press.

Snider, D. J. (1900). *The life of Friedrich Froebel: Founder of the kindergarten.* Chicago: Sigma Publishing Co.

Spodek, B., & Saracho, O. N. (1988). The challenge of educational play. In: D. Bergen (Ed.), *Play as a Medium for Learning and Development* (pp. 9–22). Portsmouth, NH: Heinemann.

Spodek, B., Saracho, O. N., & Davis, M. D. (1991). *Foundations of early childhood education: Teaching three-, four-, and five-year-old children* (2nd ed.). Boston, MA: Allyn & Bacon.

Sutton-Smith, B. (1986). The spirit of play. In: G. Fein & M. Rivkin (Eds), *The Young Child at Play: Review of Research* (Vol. 4, pp. 3–15). Washington, D.C.: NAEYC.

Sutton-Smith, B. (1988). The struggle between sacred play and festive play. In: D. Bergen (Ed.), *Play as a Medium for Learning and Development: A Handbook of Theory and Practice* (pp. 45–47). Portsmouth, NH: Heinemann.

Thomas, R. M. (1996). *Comparing Theories of Child Development* (4th ed.). Pacific Grove, CA: Brooks/Cole.

Weber, E. (1969). *The Kindergarten: Its Encounter with educational thought in America.* NY: Teachers College Press.

PLAY AND DIVERSE CULTURES: IMPLICATIONS FOR EARLY CHILDHOOD EDUCATION

Jaipaul L. Roopnarine and James E. Johnson

The purpose of this position paper is multifold. In part, this paper is a narrative research review; sections are included which selectively summarize the current literature on children's play across cultures – internationally as well as within the United States. Attention is given to parent-child, child-child, and teacher-child relations. Research studies are reviewed to cull descriptive knowledge about the range, diversity, and typicality of various play forms of children from different cultures. These studies are scrutinized for information about the importance of children's ages and gender and the social and physical environments within and across cultures as factors affecting play behavior broadly defined. In another way, this paper is a concept or think piece. In considering the extant empirical and theoretical evidence from the literature, we critically analyze the strengths and weaknesses of our current knowledge base in this area of diverse cultures and children's play in order to suggest new avenues of empirical inquiry and to draw out implications for practice and policy within the field of early childhood education. In this latter vein, attention is given to curriculum development and children's learning and to staff and parent development.

The organization of this position paper is as follows: philosophical assumptions; culturally sensitive models of development and conceptual issues; a review of parent-child play activities, children's play, and teacher-child relations in select cultural groups; and implications for early childhood education practices and policies. We begin, however, with brief definitions of the terms play and diverse cultures as they will be used in this paper.

For our purposes here *play* is defined broadly with reference both to the play text proper and to the play context including social and physical environmental factors; moreover, associated or concurrent behaviors such as social interaction doing projects or activities, and exploratory behaviors are included in this broad definition. If a more particular meaning of the term play is intended, this will be indicated. Our broad inclusive treatment of play, then, covers a variety of play behaviors such as games, behaviors of chance or risk, sports, expressive dance, music and the like (drama), make-believe, physical and sensory activity, and cooperative constructive/constructional or more work-like behaviors. All fit our broad definition of play: ludic and epistemic play, natural and instrumental play, real and school play, educational and recreational play. We feel a more open and inclusive treatment regarding the definition of play is needed in any cross-cultural examination of the construct.

By *diverse cultures* we are making reference to regional ethnic and national groups around the world including developing and more technologically and economically and politically advanced countries. Diverse cultures also can refer to groups who came to the United States voluntarily and those who were brought here involuntarily (Ogbu, 1991).

Philosophical Assumptions

An important feature of postmodern thinking, which is compatible with discourse on culture within the field of early childhood education, is the *contextuality thesis*. According to this thesis, states of knowing which guide our policies and practices are local, historically contextual, and particular. People as *knowers* are historically located which means that they live in particular places, times, and societies. Hence, their beliefs and attitudes which influence behaviors are themselves influenced by ideas current in their cultures. It is assumed that the ideas of individuals are not determined by those prevailing in their cultures only. Obviously, individuals may accept, reject, or challenge prevailing social norms. Thus, ideas about culture are contested, emergent, and processural (Rosaldo, 1989). Furthermore, individuals' ways of thinking stand in reciprocal relation to cultural states of knowing. They are two sides of the same coin and are always open to change due to the mutual pressures inherent in their relationship.

The relevance of the contextuality thesis to discussions of culture and play within the field of early childhood education presupposes an acceptance of a fundamental philosophical distinction between metaphysical and epistemological knowledge. Although *metaphysical* knowledge (cast in terms such as "is logically necessary, is the cause of, is self-contradictory") is knowledge which itself can be viewed as universal, general, and independent of any specific historical context (e.g. truths as found in the natural or mathematical world), *epistemological* knowledge (cast in terms such as "is known, is believed, is considered") is knowledge possessed by persons and is thereby local and particular. Metaphysical knowledge can be true or false, while epistemological knowledge is unassailable or assailable to varying degrees (Yandell, 1997). Adherents of positivistic and post-positivistic paradigms within the philosophy of science relax or ignore this distinction, but those aligned with the constructivistic or emancipatory paradigms take this distinction to heart (Mertens, 1998). Veridicality of constructs, then, is nonsensical in accord with the contextuality thesis. Communication and shared states of knowing are perhaps best depicted as slightly overlapping spheres (i.e. Venn diagrams), with each sphere a knower or communicator. The zone(s) of interpenetration between or among persons which result from their communication (shared reality or realities) is or are very small in relation to the unshared regions. This is because personal backgrounds, values, and experiences are not made explicit in the communication process. If such is the case for knowers within the same culture, a fortiori the unshared regions are large when knowers of different cultures communicate and attempt to share states of knowing.

What is shared are the overlapping areas of knowledge, beliefs, and practices connecting knowers within and between cultures. Sharing is possible due to the fact that knowers encase their knowledge in containers with permeable membranes. However, during the act of sharing knowledge, individuals from varying contexts never partake the same morsel of 'truth'. What is shared is not identical. Comprehension of what is shared is filtered back through and assimilated by each person's own beliefs and interpretive schemes. Even with greatest empathy and extended altruism one never knows others' states of knowing as others cannot know one's own state of knowing. Humility is required for fruitful discourse and action to occur on the human stage whether the players are all from the same culture or from diverse ones. Admittedly, room for misunderstanding and error is much greater in the latter case.

To this end, co-construction of shared realities can come about through honest or authentic attempts at communication even though each participant's representation of this shared reality is not identical. Assumed is a human developmental interest in communication and mutual understanding by peoples of varied backgrounds and cultures (Heath, 1997). Overcoming physical and social oppression needed to advance civilization and promote collective cultural evolution requires nothing less. People need to stay in touch, "up close" and "over time" in order to commence, build and to validate and utilize shared understandings and collaborative endeavors. As might be gleaned from this paper, the tremendous power and promise of children's play can serve as a useful "social bridge" between and among teachers, parents, and children of diverse cultures to help foster universal multicultural early childhood education in the 21st century.

Culturally-Sensitive Theoretical Models

With few exceptions (e.g. Vygotsky, 1978), the major theories of human development and play have skirted any serious treatment of the role of the ecocultural system in the acquisition and molding of cognitive and social skills in diverse groups of children. This state of affairs has changed a bit as child development researchers have benefited from the foresight and thinking of anthropologists and ethnic scholars who have advanced culturally sensitive models of human development. We discuss a few of these models here before proposing a few conceptual issues in the study of children's play in diverse cultural settings.

Among the integrative frameworks that have been advanced for studying human development and culture are: the cultural-ecological perspective (Whiting, 1980), the developmental niche-an extension of the cultural-ecological perspective (Harkness & Super, 1996), the cultural ecology model (Ogbu, 1991), and the culturally different models (Boykin, 1978; Coll, Lamberty, Jenkins, McAdoo, Crnic, Wasik & Garcia, 1996). Others (Draper & Harpending, 1988; see also Hewlett, 1992) have proposed biologically-based models to examine cultural variations in parental investment in childrearing.

With the realization that the ecocultural system (along with geography, history, and climate) constitutes a major force in human development, collectively the ecocultural frameworks place greater emphasis on the distal and proximal ecologies of families and children. For example, the cultural-ecological model considers child development and behaviors, including play behaviors, as related to: (a) the distal influences of cultural, economic, historical, political, societal, and technical factors which are outside children's direct experiences but nonetheless exert a powerful influence on development, (b) by the physical opportunities available to children within the immediate settings in which they live and play (e.g. terrain and the climate with seasonal variations, natural objects and culturally defined human artifacts for play, danger from roadways), and (c) the social network ties with same-age and mixed-age peers and adults, and the resulting experiences, interactions, and relationships that children have with these individuals (Bloch & Adler, 1994).

Building on the integrative-cultural orientation, the developmental niche model (Super & Harkness, 1997) proposes that peer interaction and play behavior, for instance, cannot be understood apart from the settings in which they occur. These settings combine material and psychological aspects of the actors. Three major dimensions are included in any analysis: (a) physical and social settings; (b) inner psychologies of participants (or the mental representations of caregivers concerning child development, socialization, and education); and (c) cultural customs concerning care and educational practice. Thus, play behaviors are not only influenced by the immediate social and physical context, but also by the beliefs, attitudes, and values (inner psychology) of the parents, teachers, and caregivers concerning the importance of play for early development (Roopnarine, Shin, Jung & Hossain, in press). These adult representations or cultural working models about play and development and learning in children help determine the kinds of play settings and experiences children will encounter. Consequently, cultural customs/beliefs about care and education must be factored into every analysis of children's play. Yet other discussions within cultural psychology, child development, and education have shown that considerations of history or time (expressed in years, months, or generations) assist our understanding of patterns of adaptation that result after different ethnic/cultural groups migrate within and external to their culture of origin, are involuntarily brought to a different culture (e.g. African-Americans; Asian Indians as indentured servants), or are forced to become disenfranchised (e.g. Native Indians) (Greenfield & Cocking, 1994; Gibson & Ogbu, 1991). Different analyses of the experiences of disenfranchised and immigrant groups in the United States paint a confusing picture. It appears that initially new immigrants voluntarily accept language and cultural differences and are willing to adapt to the demands of the formal educational system and to society at large (Ogbu, 1991) but lose their idealism after they have been within the dominant culture for two or three generations. While some (Slaughter-Defoe, Nakagawa, Takanishi & Johnson, 1990) have proposed an examination of changes in immigrant family adaptation in terms of culturally continuous (living for more than two generations in the broad socio-ecological context) and culturally discontinuous contexts (e.g. refugee, migrants, international families), the process of assimilation to a new culture is a very complex endeavor (Fuligni, in press).

Along a multivariate continuum, families may engage in segmented assimilation, overacculturation, underacculturation, accommodation without assimilation and a host of other processes as they negotiate their way through a host culture (Roopnarine & Shin, in press). It is often challenging for families to integrate child training techniques, games, crafts, songs, toys, and play activities from the natal culture with the play behavior repertoire and accompanying play attitudes and values of the new culture in the process of eschewing a transcultural identity (enculturation).

Relatedly, in our quest to define a multicultural democracy we believe it important to ask some basic questions. Do we focus on the center or the boundaries of our cultural heritages, our sameness or our differences, the whole or the parts, in working through our present national identity crisis? Conceivably, stressing the whole can produce stifling conformity and stagnation in our nation's cultural evolution. On the other hand, stressing the parts can lead to separatism and societal fragmentation. In the face of these concerns, we cannot deny the pernicious tendency within the play literature to focus on the deficits of non-white non-European children relative to white middle class norms. Often, there has been a flagrant disregard for factors within children's lives (e.g. discrimination, prejudice, poverty, racism) that may impede growth and development. In the present context, culturally different models (see Coll et al., 1996) argue that the cultural practices and family socialization patterns of some ethnic groups in the United States should not be compared or held against White middle-class norms and values. Instead, they view within group cultural practices as legitimate and functional for child rearing and child training and highlight the unique experiences and daily social and cognitive transactions of families and children as essential in understanding their growth and development.

Rounding out this section, the final model we call attention to is more biologically rooted than those discussed so far. The biosocial perspective (Draper & Harpending, 1988) has been utilized to catalog parents' investment in children along a continuum of father-absent to father-present societies. Essentially, this view purports two types of reproductive strategies: high male parental investment in which the male provisions for the offspring and mating partner (DADS), and low parental investment in which the male is more interested in mating than parenting, thus showing little responsibility in caring or provisioning for the offspring and mate (CADS). Paternal investment, of which involvement is a component, is useful in our discussions of parent-child activities given that, above other activities, fathers seem to avail themselves as play partners to young children across cultures.

Conceptual Issues

Using the culturally-sensitive conceptual frameworks in human development outlined above as a guide, we contend that *existing play theories may be inadequate in guiding research on diverse groups of children because they appear insensitive to considerations of factors within the ecocultural system that may influence growth and development.* We assert seven propositions that may assist us in further delineating the intersections between culture and play.

(1) *Play as Indicative of Competence.* To date, the validity of using traditional play behaviors as a yardstick of early competence in diverse cultural groups is not well substantiated. While different modes of play behaviors have been shown to be associated with better cognitive and social outcomes in some cultural groups (see Lilliard, 1993; Pellegrini, 1985; Saltz, Dixon & Johnson, 1977), it is not clear whether play alone, as traditionally measured, accurately represents the social and cognitive competencies of children across diverse cultures. Play-based programs for culturally heterogeneous groups of children may be better informed by broader considerations of children's early social and cognitive activities.

(2) *Culturally-relevant Early Childhood Practice.* There is a need to draw links between culturally-relevant early childhood play experiences and materials and self-esteem, ethnic/cultural identity, and cultural pride. We do not know to what degree culturally compatible play experiences reduce disequilibration in the child's organization of knowledge about self and other, lead to a greater integration of information about self and culture, and make collaboration or joint ventures with others in the early childhood environment more feasible (Bruner, 1977; Verba, 1993).

(3) *Appropriate Childrearing/Child Training and Child Development Knowledge.* Most of the child development literature is biased in favor of white middle-class children. Knowledge about childrearing and child training techniques and norms for the emergence and acquisition of childhood development and play skills in diverse groups are not fully established. Consequently, there is inadequate information about the different learning styles of children, and family goals, expectations, and practices in the socialization of children in a large segment of the world's population This lack of knowledge can only hamper interpretations of children's mental activity – culturally laden play themes and language use during play, and to inefficient and less productive educational planning and learning.

(4) *Parental Belief Structures.* It is essential that we establish symmetries and asymmetries between parental psychology about the value of play in intellectual and social development and culture-specific developmental

expectations of children from diverse cultural backgrounds and the values about play proposed by the early childhood education community. Beliefs or ideas about play meeting children's developmental and educational needs may not be uniform for parents of children from non-European ethnic and cultural groups, children of immigrants and immigrant children, and limited-English proficient children (Roopnarine et al., in press).

(5) *Categorical Comparisons.* Because of the increasing allure of cross-national and ethnic comparisons of children's play to utilize "culture" or "ethnicity" as a categorical or independent variable, we argue that this approach may provide little in the way of aiding our understanding of the value individual cultures place on specific modes of thinking and behaving. That is, these comparisons are quick to interpret group differences as being synonymous with cultural differences without delving into culturally-contextualized constructions of the meaning of play for the acquisition of skills relative to the social and cognitive demands of individual cultures (e.g. technological versus social intelligence). Inevitably, this can lead to increments in the negative stereotyping that has plagued race/culture comparative interpretations of the play behaviors of non-White, non-dominant groups and of children in diverse societies.

(6) *Multiple Contexts.* The contexts for children's play vary widely and play may be intertwined with work (as in some African societies), spiritual observances and celebrations (e.g. Holi among East Indian children), and may occur on beaches (Marquesans), in courtyards and compounds, urban playgrounds and streets (see Roopnarine, Johnson & Hooper, 1994). These diverse contexts offer different opportunities and materials for play and different insights into children's cognitive representations. Straightforward comparisons of play across two or more settings and decontextualized explanations of children's play behavior may grossly underestimate the meaning of children's play or may mistakeningly suggest an impoverishment of play.

(7) *Social Position.* Race, gender, ethnicity, social class and other social position variables are inextricably tied to children's everyday life experiences. They directly/indirectly affect the nature of developmental processes, social and cognitive outcomes, and ultimately children's life chances. Under certain circumstances, they may account for more variation in children's play than culture itself. Theoretical considerations of structural to process to outcome models may be more fruitful in teasing out the value of different modes of play in children's early social and cognitive development.

PLAY IN DIVERSE CULTURAL SETTINGS

Acknowledging that there are several excellent reviews on the play behaviors of White North American and European children (see Harris & Kavanaugh, 1993; Howes et al., 1992; Hughes, 1995; Johnson, Christie & Yawkey, 1999; Reifel, 2001; Rubin, Fein & Vandenburg, 1983), we direct our attention to the play of children from other ethnic groups in the United States and diverse cultures around the world. As noted in a previous section, the play norms for non-European groups in the United States and in cultures around the world have not been fully determined (see Segoe, 1971; Roopnarine et al., 1994) nor have they been sufficiently explored. Furthermore, much has been written about the methodological and conceptual problems that have plagued the research on play in diverse groups of children (see McLoyd, 1982).

Parent-Child Play During Early Childhood

Despite claims about the ubiquity of parent-child play across cultures, there is tremendous variation in the style and amount of parent-child play and tacit knowledge about the importance of play for early development across cultures. It is suggested, nonetheless, that parent-child play and social activities provide the nexus for examining developmental processes in diverse cultural settings (see volume by Macdonald, 1993) In this segment of the paper, we draw heavily upon the cross-cultural play literature we have reviewed elsewhere (Johnson, et al., 1999; Roopnarine, Lasker, Sacks & Stores, 1998; Roopnarine, Shin, Suppal & Donovan, 2000; Roopnarine, Ahmeduzzaman, Hossain & Riegraf, 1993) to provide a synopsis of play across cultures. Because the terrain is vast, we focus on specific areas of play where there is a concentrated body of work and attempt to include studies conducted on non-white ethnic/cultural groups.

Parent-Child Physical Play

From the western social science literature, it has been determined that: (a) parent-child physical play begins early in infancy, peaking in the preschool years (1–4), and is uncommon after 10 years of age (Macdonald & Parke, 1986), (b) there is a negative relationship between physical play and age of parent, (c) there are strong gender differences in parental engagement in physical games with infants; limb movement games, bouncing and lifting games are more characteristic of the play of fathers, whereas conventional games (e.g. pat-a-cake and peek-a-boo) are more characteristic of the play of mothers, (d) boys are more likely to be the recipients of physical play than girls, and (e) father-child physical play is correlated with popularity in the peer group.

Although the cross-cultural literature on parent-child rough play is comparatively small, it is nonetheless instructive. During the infancy period, physical play (rough-housing, tossing, tickling, bouncing, poking) has been observed at relatively low frequencies in New Delhi, India (on average less than one incident per 1 hour of observation) (Roopnarine, Talukder, Jain, Joshi & Srivastav, 1991), in Chi Chi, Dongkong, Taipei, and Kaushung, Taiwan (on average less than one incident per 1 hour of observation) (Sun & Roopnarine, 1996), among the AKA in the Central African Republic (9 times in 264 hours of observations) (Hewlett, 1987), and is rarely reported by parents in the play of Malaysian parents in Kuching, Sarawak (Roopnarine, Lu & Ahmeduzzaman, 1989) and families in Kingston, Jamaica (Roopnarine, Brown, Snell-White, Riegraf, Hossain & Webb, 1995). Among older children, however, it accounted for about a third of the play activities of mothers and fathers and preschoolers residing in Chaing Mai Province in northern Thailand but was still infrequent (on average less than one incident across families in two hours of observations) (Tulananda & Roopnarine, 2001).

Despite the low frequencies of rough stimulating play across these cultures, in India and Taiwan, fathers were more likely to engage in rough play with infants than mothers. There was no significant tendency for parents to engage in more physical play with boys than girls. It is the case, however, that fathers seem more available to children as play partners than caregivers. For instance, fathers were far more likely to engage in play with children than feed them in Jamaican (twice as likely), Indian (thrice as likely), and Taiwanese families (twice as likely). This tendency appears robust across several other cultures as well, at least during infancy.

In our recent observations conducted on preschoolers in northern Thailand, we attempted to assess the associations between children's social competence in preschool (cooperativeness, independence, and interactions) as measured by the Preschool-Kindergarten Childhood Behavior Scales and parent-child rough play. There were no significant associations between parent-child physical play and children's social competence in preschool.

From this body of cross-cultural work, it is not clear that rough play is a central part of early parent-child activities in a number of cultures, possibly undermining broad definitions about the functions of rough play for childhood competence and its presumed biological underpinning. So far, rough play seems more prevalent among white fathers in the United States than in other groups of fathers around the world. Similarly, the benefits of rough play for the development of social competence in children has only been demonstrated within a narrow cultural frame.

Parent-Child Games
The developmental significance of different parent-child games for children's intellectual and social development has been discussed extensively (see volume by Macdonald, 1993). Similarly, the diverse games parents and children engage in have been described in some detail (see Sutton-Smith, 1976). Bearing this in mind, we focus on the universality and stylistic differences in parent-child games.

Broadly speaking, parent-child games, particularly during the first two years of life, reveal some salient features: games that involve face-to-face encounters may be more characteristic of the play of parents and children in some cultures than in others (Roopnarine et al., 1998) and reveal profound postural differences in parental engagement across cultures (for example, Japanese mothers loom in and out, tap the infant to create visual displays, and hold and touch the infant continuously whereas white U.S. mothers use their voice and are more responsive to the infant's vocalizations during face-to-face encounters) (Fogel, Nwokak & Karns, 1993); in some developing societies (e.g. India), early parent-child games are embedded in close physical contact or may involve specific texts (Igbo and Sinhala mothers) that include songs, lullabies, poetry, or rhymes (Roopnarine, Hossain, Gill & Brophy, 1994) and games that involve touching, clapping, tickling, and swinging have been observed among Mexican, Filipino, Chinese, and White parents residing in the United States (Van Hoorn, 1987); verbal interactions and object use vary tremendously during parent-child playful encounters – in some cultures, mothers find talking to the baby during play "silly" (e.g. Gusii) and thus may rely on movement to songs and action during social activities; a majority of parent-child games are passed down from parents, relatives, and friends to children (see Van Hoorn, 1987); and parents use a variety of instructional and socially-oriented texts during parent-child games across cultures but variations may exist depending on whether the society stresses social or technological intelligence.

To illustrate these stylistic differences further, let us consider a parent-infant game, peek-a-boo. Different forms of this game have been observed widely (e.g. in India, South Africa, Japan, Malaysia, Iran, Korea, and a host of western societies) though it can hardly be argued that it is universal. In depth analysis (Fernad & O'Neill, 1993) of peek-a-boo reveals distinct acoustic signals as parents reappear after hiding the face. While the acoustic call in most cultures signals the mothers reappearance, there are significant variations in the acoustic cues. Whereas in some, vowel elongation had a rough and gravely quality (e.g.

South Africa), in others it was smoother, and the calls may range from nonsense syllables in some to more meaningful use of words in others (e.g. Tamil, Brazilian Portuguese).

In our own work (Roopnarine et al., 1994) on the play of East Indian children, we were able to document distinct features in parent-infant games. Both behavioral observations and an analysis of parent-child games in India revealed that the games are marked by physical closeness involving holding, rocking, and face-to-face social activities. For example, 87% of the games we analyzed contained elaborate tactile stimulation in the absence of objects where the baby was held close to the body or massaged, in 21% of the same games the mother sang to the baby, and 52% involved face to face interactions. A closer examination of the games also revealed the rich use of language in the form of songs and lullabies.

Presumably, if we consider parent-child games as mirroring socio-cultural action and thought (Vygotsky, 1978), then mothers and fathers are in a unique position to engage the child in an "uptake" (Bruner, 1983) of the rudiments of salient cultural information and skills.

Parent-Child Pretend Play

Most would agree that during parent-child pretend episodes there are opportunities to teach communication skills, there are didactic social exchanges, demonstration, encouragement, and parents do permit child-initiated fantasy in sensitive ways (Fiese, 1990; Howes et al., 1992; O'Connell & Bretherton, 1984; Tamis-Lemonda & Bornstein, 1991; Youngblade & Dunn, 1995). As you may have already guessed, though, data on parent-child pretend activities in diverse cultural groups has received scant attention.

The studies (see Chin & Reifel, 2001; Haight, Wang, Fung, Williams & Mintz, 1999) on the participation of parents in fantasy play with children is fragmentary. For instance, parent-child pretend play has been observed among preschoolers during mother-child and father-child social interactions in Thailand (Tulananda & Roopnarine, 2001), in mother-toddler play in comparative work on Japanese and American mothers (Tamis-Lemonda, Bornstein, Cyphers, Toda & Ogino, 1992), in questionnaire assessments of the types of play Korean-American and Taiwanese parents engage in with children (Farver, Kim & Lee, 1995; Pan, 1994), and during Mexican and Anglo-American mother-child activities (Farver, 1993).

Taken together, these studies indicate that some mothers took advantage of opportunities during pretend activities to teach their children communication skills (Tamis-Lemonda et al., 1992), but others were less inclined to participate in pretend activities with children (Farver, 1993). Even if we agree that there is evidence of varying levels of parent-child pretense, these findings raise additional questions about parental knowledge about play meeting the educational and social needs of young children rather than the inability of parents to engage in meaningful social activities with their offspring (Roopnarine et al., in press).

Child-Child and Teacher-Child Play
From the foregoing, you may have gathered that cross-cultural play studies have moved well beyond the issue of ascertaining the presence or absence of different modes of play and testing the universality of Piagetian play stages. Today play researchers are more inclined to describe the myriad of ways in which children play, the enormous variation in the content and style of children's play, and how play affects and is affected by cultural context. (Roopnarine et al., 2000). Note for example, research (deMarrais, Nelson & Baker, 1994) conducted on Yup'ik Eskimo girls living in villages along the Kuskokwim River in southwestern Alaska. Participant observations and detailed interviews revealed the importance of storyknifing in the mud for enculturation and development; mudknifing with older peers helped 6–12 year-old Yup'ik girls to incorporate cultural knowledge about kinship patterns, gender roles, and community norms and values, as well as providing them with opportunities to consolidate their feelings about their culture and their own identities, habits, and attitudes.

Rich descriptions of play enactments have been cataloged in East Indian (Roopnarine et al., 1994), Marquesan (Martini, 1994), Senegalese (Bloch & Adler, 1994), Guyanese (Taharally, 1991), and Thai cultures (Tulananda & Roopnarine (2001), and for children in Singapore (Lim & Honig, 1997). While most of the studies recorded different rates of play, the observations conducted in Singapore (Lim & Honig, 1997) revealed some interesting patterns. Parallel and functional play happened more at home and cooperative and associative play occurred more at the child care center, and girls engaged in more sociodramatic play than did boys. No information on teacher involvement in play was given in any of these studies.

Perhaps in an effort to establish play norms in other cultures, a group of studies has compared the play activities of preschoolers in other societies to those of preschoolers in the United States (Pan, 1994; Tobin, Yu & Davidson, 1989). Accordingly, Pan (1994) compared the play of kindergartners in Taipei with data collected about a decade earlier on North American children (Rubin, Watson & Jambor, 1978). In addition, she was interested in finding out whether associations found between cognitive and role-taking and different play forms in Taiwan were similar to those calculated for North American children (Rubin & Maioni, 1975), and whether maternal attitudes towards play were similar in Taiwan and the United States (Johnson, 1986). She reported that constructive play was commonplace in both countries, but interactive dramatic play was more prevalent in the North American sample (twice as much in preschool and thrice as much at the kindergarten level), while parallel constructive and interactive games with rules were more prevalent in the Taiwanese sample. In the North American samples, interactive dramatic play was positively correlated with intelligence and role-taking and in the Taiwanese sample interactive games with rules was positively and significantly correlated with mental age (0.42) and with role-taking (0.28). The Taiwanese mothers did not endorse games with rules but did prefer constructive play and academic activities to a greater extent relative to the North American mothers.

Pan's findings are consistent with those of the *Preschool in Three Cultures* study (Tobin et al., 1989). Using multivocal visual ethnographic techniques, these researchers compared children in early childhood settings in the United States, Japan, and China. It was reported that the United States fell between Japan and China on dimensions of play and academic values for young children. A case in point, in response to the question "Why should a society have preschool?", 70% of the Japanese, 42% of the United States, and only 25% of the Chinese sample gave 'opportunities for playing with other children' in their top three reasons. Academic goals were within the top three choices of 67% of the Chinese sample, 51% of the United States sample, and only 2% of the Japanese sample. The Japanese, it seems, fear educational burn-out before the educational 'rat race' is over and prefer to emphasize kodomo rashii kodomo ("child-like children") during the early preschool years, while the Chinese favor preschool as a place for serious learning. Rather than 'play is the child's work', 'work (academics) is a form of child's play'. In the United States, the picture appears mixed with many teachers and parents of young children torn between wanting to give young children a head start but worrying about inducing stress and achievement anxiety by hurrying them along. In line with the preceding, parents and teachers in the United States seem confused or uninformed about the value of play for giving young children an academic head start in a developmentally appropriate way.

More recently, empirical spotlight has been shed on the play of immigrant children attending preschools or day care programs in the United States. In an attempt to tease out cultural differences in play activities and behaviors of children of immigrants, Farver and Shin (1997) compared the classroom activities of Korean- and Anglo-American three-to-five-years-olds. From direct observations of children within their respective preschool activity settings (which were markedly divergent), parental reports of home play, and teacher ratings of social competence, it was determined that Korean-American children engaged in social pretend play at less than half the rate as their Anglo-American counterparts, unoccupied states were coded twice as often and parallel play almost thrice as often in the former relative to the latter sample of preschoolers. By contrast, Korean-American children were more cooperative than Anglo-American children. Teachers rated Korean-American children as more hesitant and less sociable than the Anglo-American children.

What could account for these differences? While the cooperative tendencies of the Korean-American children may be attributed to the cultural value of interdependence and field sensitivity, the educational focus and the preschool environment may have accounted for the other differences. Although each setting focused on school readiness, the way this was carried out was quite different. In the Korean-American activity setting the nature of the activities performed included memorization games, tasks requiring persistence and effort and passive learning centered on academic achievement. In the Anglo-American setting, the aim was to foster social and cognitive skills in the youngsters by means of active involvement in imaginative and other more convergent types of problem solving situations where individual expression was valued more than group harmony. Parents of the Anglo-American children prized play, especially pretend play as a tool for cognitive and language growth and school readiness and reported that their children did a lot of it at home. By contrast, the Korean-American parents did not share these beliefs.

The gist of the play literature just reviewed is that all young children it would seem can engage in diverse play activities regardless of their cultural and social-class membership (Schwartzman, 1978; Sutton-Smith & Heath, 1981). Play deficiencies that have been attributed to certain groups of children in the past may have been due to ethnocentric or class bias and to ignorance about using appropriate research tools. As we have shown, there is considerable variation and overlap in play expression among children across cultures (Mertens, 1998). It appears that educational forms of play take on a more homogeneous flavor whereas recreational or expressive play retain more cultural specificity (Lasater & Johnson, 1994). Further, children are more likely to engage in particular forms of play that are supported by the adults of the culture. Consider for a moment the game of soccer, a favorite sport in most countries of the world. Preschool children as young as two years of age engage in this activity even though it is a game with rules which, according to Piagetian theory, makes it more developmentally advanced than symbolic play. By the same token, sociodramatic play is not seen in many of the same cultures even though it requires similar levels of cooperation. These examples do not point to the cognitive limitations or strengths of the players but to differences in cultural emphasis and encouragement.

IMPLICATIONS FOR PRACTICES AND POLICIES IN ECE

Early childhood education as a field is increasingly realizing the inadequacy of traditional child development knowledge as a foundation for practice and for policy (see Jipson, 1991; Katz, 1996; Lubeck, 1996; Scott & Bowman, 1996). Our field must be informed by other disciplines such as cultural studies and cultural psychology (Schweder, 1990) as well as keep abreast of latest theoretical revisions within the parent discipline of human development proper (e.g. Meacham, 1996). The theoretical and empirical evidence reviewed here concerning children's play in diverse cultures underscores the importance of understanding the ecocutural system of children and their families for defining and articulating guidelines for practices and policies in early childhood education. This importance is seen in the areas of teacher dispositions and roles and curriculum and learning (Reifel, 2001).

Teacher Dispositions and Roles
As we begin to realize how inextricably tied we are to other cultural groups, an important goal is to develop a global perspective aimed at improving the quality of life for all children. Nearer to home, within our pluralistic democracy we must model positive attitudes for our children by showing not only tolerance and acceptance, but also respect for and enjoyment of group differences. It is in these domains that early childhood education teachers have an important responsibility to nurture positive attitudes toward children from diverse backgrounds. This process, however, does not primarily entail being a walking storehouse of facts about other lands and other peoples. Having a cultural perspective means that we realize and appreciate in an intellectual and affective way the special qualities of each unique child in unison with the cultural forces working within the child's life space. Having a cultural perspective means that we are always open to new learning about cultural diversity and about the changing milieu of culture itself. Ethnosensitive teachers realize that they must seek shared knowledge and forge partnerships with parents in working to build bridges between the natal and school cultures. The following practical tips are offered:
(1) For those whom play is not viewed as educationally beneficial (i.e. having a play-based program and not a traditionally academic one) teachers may have to provide assurances and information to parents that suggest that a

Play and Diverse Cultures: Implications for Early Childhood Education 311

play-based curriculum is not antithetical but an integral and necessary step to the long-term educational goals of their children. Through newsletters (ideally in different languages if needed and devoid of jargon), parent meetings, conferences and the like, teachers can explain why play is necessary for children and that educational play and related behavior (such as doing projects) are often quite different from what goes on at home.

(2) Encourage, as needed, parents to adjust their "inner psychologies" or working models about schooling. For example, parents may need to grapple with the idea that schooling and socialization are evolving processes. Thus what worked for them or worked in their culture of origin may also be undergoing some metamorphosis.

(3) Remain cognizant of the content of each child's own real-life experiences and provide play opportunities in line with children's backgrounds. Teachers can learn directly from children by carefully observing and listening to them play at school, and by communicating with parents. Additionally, parents can be invited to help teachers set up culturally appropriate play activities in school that resonate with children's experiences outside the classroom. An enriched, culturally appropriate program can facilitate play, peer relations and friendships among diverse children, teachers, and parents.

(4) Welcoming input from parents and creating working relationships with them also foster, as an added bonus, greater multi-ethnic perspective taking skills (Hyun & Marshall, 1997). Over and beyond getting accustomed to the idea of many life styles, languages, and cultures, teachers need to promote empathy and perspective-taking ability, both in themselves and in children, in order that all might successfully navigate together the diverse cultural landscape.

In short, teachers need detailed, up-close knowledge of the "whole child" and the family in its cultural context. This usually translates into a better understanding of intra-group variation, which as an antidote to stereotyping is a critical ingredient of any curriculum which strives to be "anti-bias" (Derman-Sparks, 1989).

Curriculum and Learning
Curricular adaptation is exemplified in two anecdotes: one from Curry (1971) involving a group of Navajo children and the other from Monighan-Nourot (1995) on an immigrant child from Laos. In the first, apparently the Navajo children were not familiar with all the props available for dramatic play in the housekeeping corner of their middle-class-oriented preschool center. They did not use the domestic corner as it was usually set up for free play. One day, but

not by teacher design, the toys were left against the wall after cleaning. The Navajo children then vigorously engaged in sociodramatic play. Apparently the props were in the position where children recognized them as being in a spot similar to where items were placed in their circular hogan homes. In the second, Sysavath, a little girl newly arrived from Laos, entered her preschool classroom with a baby carrier on her back. She had brought in her Asian doll. During the entire school day the 'baby' was taken along by Sysavath as she participated in program activities with the teachers and the other children – from easel painting, to water table activities, to blocks and so on. The opportunity for Sysavath to mother her doll this way helped her acclimate to the program and find a sense of belonging and mastery. It gave her peer group recognition, too. Sysavath's mother was asked for replicas of the cloth baby carrier so others may share in this new play in the classroom.

Play and curriculum can come together in numerous ways including in the form of free play, structured free play, and play tutoring. As these anecdotes suggest, implementing play-based education that is multicultural, responsive teachers should always seek to support the self image and confidence of all children.

Free Play

There are several steps to enrich free play periods for all children in a group setting:

(1) Free play should have available objects and symbols in the environment that show children that a broad range of cultures are respected and validated. Parents can be consulted and invited to share materials and accessories useful for free play: items such as jewelry, costumes, scarves, clothing, props for foods, music, etc. At the same time, different kinds of toys, artifacts, books, photographs, magazine pictures that are multicultural should always be present – although specific items are changed through rotation on a regular basis.
(2) Games and small manipulatives, puzzles, and other activities that are the staple of any program can be adapted with words, numbers, and pictures signifying the content and themes of various sociocultural groups.
(3) Teachers must be ready to intervene whenever needed to foster multicultural competencies and attitudes during free play. For example, if the teacher sees children who are playing restaurant express a stereotype like "all Chinese like rice and all Mexicans like beans", s/he would suggest that some Mexicans do not like beans more than rice, and that some Chinese like beans and so forth (Boutte, Scoy, & Hendley, 1996).

On different occasions, teacher interventions may be required to counteract in children or other teachers any biased behaviors or words as well as any expressions of the superficial "tourist approach" that fails to recognize intracultural variation and individual differences.

Structured free play. There are several things that can improve structured free play for children. We state a few of them briefly:

(1) For structured play, multicultural theme boxes can be employed (Boutte et al., 1996). Theme or prop boxes contain toys and other objects related to a particular theme such as "Beauty Shop" or "Barber Shop" or "Bakery." For example, multicultural theme box for "bakery" would contain simulations of baked goods representing different cultures (e.g. tortillas, baklava, pitas, challah, etc.) with accompanying cookbooks and magazine recipes and pictures, and different kinds of cooking and eating utensils from different countries.
(2) Travel posters and folders could also be on display to augment play atmosphere related to specific cultures or subcultures.
(3) For constructive play teachers can set up classification activities with various sociocultural content. Family pictures, art work, stories, important family objects can be brought in from home and compared and contrasted. Educational constructive play can include matching or memory games using culturally specific items (e.g. types of dwellings or buildings in a culture, clothing, musical instruments).
(4) Favorite culturally related playthings and games can be a lively stimulus for group discussion, and even the start of an extended project on toys and play activities from different lands and eras.

Play becomes richer and more cooperative when children are respected and encouraged to reflect on their own and each other's backgrounds.

Play Tutoring

Play tutoring also has a place in the multicultural curriculum – whether the teacher is interested in implementing Smilansky's sociodramatic play training techniques, or the story-telling and enactment techniques that are part of Vivian Paley's narrative curriculum (Wiltz & Fein, 1996), or some other variation of it, like thematic-fantasy play training (Saltz & Johnson, 1974). Depending on the needs and interests of the children, the teacher can select a sociodramatic play theme with a cultural focus (such as role enactment of a bus trip to an ethnic festival, or a meal at an ethnic restaurant). A story or fairy-tale from a certain culture or a favorite activity or festival of a specific social group could be chosen for group dramatization. A teacher can prompt and direct role play of families in different cultures. Furthermore, stories children create themselves can be recorded by the teacher a la the Paley technique, and later employed in discussion and enactment in a large group to foster early childhood education that is multicultural.

These curricular adaptations in free play, structured free play, or in play tutoring sessions can come to exemplify what Fromberg (1995) means by play being a "dynamic representation" of multiculturalism in early childhood education. Put differently, early childhood education that is multicultural integrates play throughout the program much as it does multiculturalism itself – neither is an "add-on" but an essential dimension of the inner workings of the overall curriculum. In this effort, it is essential that adults negotiate and share power with children (Stremmel, 1997). In the anecdotes in the beginning of this section, the teachers behaved in a culturally responsive manner by accommodating to the play needs of Sysavath and the group of Navajo children. When children's knowledge, culture, and life experiences are taken seriously, their play behaviors can reveal quite a bit about themselves and their backgrounds and cultures.

CONCLUSION

We hope that this paper can serve as a catalyst for more vigorous discussions regarding play and culture and implications for early childhood education practice. Universally accepted or highly controversial, play holds tremendous promise for understanding the growth and development and the educational needs of children from diverse backgrounds. Happily, more researchers are abandoning the practice of focusing on play deficits in favor of examining play from a resilient-adaptive framework. This shift in conceptual orientation is an important first step if we are to fairly and adequately meet the needs of all children.

REFERENCES

Boutte, G., Van Scoy, I., & Hendley, S. (1996). Multicultural and nonsexist prop boxes. *Young Children*, (November), 34–39.
Bloch, M., & Adler, L. (1994). African children's play and the emergence of the sexual division of labor. In: J. Roopnarine, J. Johnson & F. Hooper (Eds), *Children's Play in Diverse Cultures* (pp. 148–178). Albany: SUNY Press.
Boykin, A. W. (1978). Psychological/behavioral verve in academic/task performance: A pretheoretical consideration. *Journal of Negro Education, 47*, 343–354.
Bruner, J. (1997). Celebrating divergence: Piaget and Vygotsky. *Human Development, 40*, 63–73.
Bruner, J. (1983). *Child's talk: Learning to use language*. New York: Norton.

Chin, J., & Reifel, S. (2001). Material scafffolding of Taiwanese play: Qualitative patterns. In: S. Reifel (Ed.), *Theory in Context and Out. Play and Culture Studies* (Vol. 3, pp. 263–289). Westport, CT: Ablex.

Coll, C. G., Lamberty, G., Jenkins, R., McAdoo, H. P., Cmic, K., Wasik, B. H., & Garcia, H. V. (1996). An integrative model for the study of developmental competencies in minority children. *Child Development, 67*, 1891–1914.

Curry, N., (1971). Consideration of current basic issues in play. In: N. Curry & S. Arnaud (Eds), *Play: The Child Strives Toward Self-Realization*. Washington, D.C.: NAEYC.

Derman-Sparks, L. (1989) *Anti-bias curriculum: Tools for empowering young children*. Washington, D.C.: National Association for the Education of Young Children.

deMarrias, K., Nelson, P., & Baker, T. (1994). Meaning in Mud:Yup'ik Eskimo girls at play. In: J. Roopnarine, J. Johnson & F. Hooper (Eds), *Children's Play in Diverse Cultures*. Albany, NY: SUNY Press.

Draper, P., & Harpending, H. (1987). Parent investment and the child's environment. In: J. Lancaster, A. S. Rossi & L. Sherrod (Eds), *Parenting Across the Life-Span: Biosocial Perspectives*. New York: Aldine de Gruyter.

Farver, J. (1993). Cultural differences in scaffolding pretend play: A comparison of American and Mexican mother-child and sibling-child pairs. In: K. Macdonald (Ed.), *Parent-child Play: Descriptions and Implications*. Albany: State University of New York Press.

Farver, J. A. M., Kim, Y. K., & Lee, Y. (1995). Cultural difference in Korean- and Anglo-American preschoolers' social interaction and play behaviors. *Child Development, 66*, 1088–1099.

Farver, J. A. M., & Shin, Y. L. (1997). Social pretend play in Korean- and Anglo-American preschoolers. *Child Development, 68*, 544–556.

Fernald, A., & O'Neill, D. (1993). Peek-a-boo across cultures. In: K. Macdonald (Ed.), *Parent-Child Play* (pp. 259–285). Albany, NY: SUNY Press.

Fiese, B. H. (1990). Playful relationships: A contextual analysis of mother-toddler interaction and symbolic play. *Child Development, 61*, 1648–1656.

Fogel, A., Nwokak, E., & Karns, J. (1993). Parent-infant games as dynamic social systems. In: K. Macdonald (Ed.), *Parent-Child Play* (pp. 43–70). Albany, NY: SUNY Press.

Fromberg, D. (1995). Politics, pretend play, and pedagogy in early childhood preservice and inservice education. In: E. Klugman (Ed.), *Play, Policy, and Practice*. St. Paul, MN: Redleaf Press.

Fuligni, A. (in press). The adaptation and acculturation of children from immigrant families. In: U. Gielen & J. L. Roopnarine (Eds), *Childhood and Adolescence in Cross-cultural Perspective*. Greenwood Press.

Gibson, M. & Ogbu, J. (Eds) (1991). *Minority status and schooling: A comparative study of immigrant and involuntary minorities*. NY: Garland Publishing.

Greenfield, P., & Cocking, R. (Eds), *Cross-Cultural Roots of Minority Child Development*. Hillsdale, NJ: Erlbaum.

Haight, W., Wang, Z., Fung, H., Williams, K., & Mintz, J. (1999). Universal, developmental, and variable aspects of young children's play: A cross-culotural comparison of pretending at home. *Child Development, 70*, 1477–1488.

Haight, W. (1998). The pragmatics of caregiver-child pretending at home: Understanding culturally specific socialization practices. In: A. Goncu (Ed.), *Children's Engagement in the World: Sociocultural Perspectives* (pp. 128–147). New York: Cambridge University Press.

Harkness, S., & Super, C. (Eds.) (1996). *Parents' cultural belief systems: Their origins, expressions, and consequences*. New York: Guilford.

Harris, P., & Kavanaugh, R. (1993). Young children's understanding of pretense. *Monographs of the Society for Research in Child Development*, 58(231).

Heath, S. B. (1997). Culture: Contested realm in research on children and youth. *Applied Developmental Science, L*, 113–123.

Hewlett, B. (Ed.) (1992). *Father-child relations: Cultural and biosocial perspectives*. New York: Aldine deGruyter.

Hewlett, B. (1987). Patterns of parental holding among AKA pygmies. In: M. Lamb (Ed.), *The Father's Role: Cross-Cultural Perspectives*. Hillsdale, NJ: Erlbaum.

Howes, C. (1985). Sharing fantasy: Social pretend play in toddlers. *Child Development*, 56, 1253–1268.

Howes, C., Ungerer, O. A., & Matheson, C. C. (1992). *The collaborative construction of pretend*. Albany, NY: State University of New York Press.

Hughes, F. P. (1995). *Children, play, and development*. Boston, MA: Allyn and Bacon.

Hyun, E., & Marshall, D. (1997). Theory of multiple/multiethnic perspective-taking ability for teachers' developmentally and culturally appropriate practice (DCAP). *Journal of Research in Childhood Education*, 11(2), 188–198.

Jipson, J. (1991). Developmentally appropriate practices: Culture, curriculum, connections. *Early Education and Development*, 2, 120–136.

Johnson, J. E. (1976). Relations of divergent thinking and intelligence test scores with social and non-social make-believe play of preschool children. *Child Development*, 47, 1200–1203.

Johnson, J. (1986). Attitudes toward play and beliefs about development. In: B. Mergen (Ed.), *Cultural Dimensions of Play, Games, and Sport. The Association for the Study of Play* (Vol. 10, pp. 98–102). Champaign, IL: Human Kinetic Publishers

Johnson, J. E., Christie, J. F., & Yawkey, T. D. (1999). *Play and early childhood development*. Glenview, IL: Scott, Foresman & Company.

Katz, L. (1996). Child development knowledge and teacher preparation: Confronting assumptions. *Early Childhood Research Ouarterly*, 11(2), 135–146.

Lasater, C., & Johnson, J. (1994). Culture, play, and early childhood education. In: J. Roopnarine, J. Johnson & F. Hooper (Eds), *Children's Play in Diverse Cultures* (pp. 210–228), Albany, NY: SUNY Press.

Lilliard, A. (1993). Pretend play skills and the child's theory of mind. *Child Development*, 64, 348–371.

Lim, S. E., & Honig, A. S. (1997). Singapore preschoolers' play in relation to social class, sex, and setting. *Early Child Development and Care*, 135, 35–39.

Lubbeck, S. (1996). Deconstructing "Child Development Knowledge" and "Teacher preparation," *Early Childhood Research Quarterly*, 11(2), 147–168.

Macdonald, K. (Ed.) (1993). *Parent-child play*. Albany: SUNY Press.

Macdonald, K., & Parke, R. (1986). Parent-child physical play: The effects of sex and age of parents. *Sex Roles*, 7, 367–368.

Martini, M. (1994). Peer interactions in Polynesia: A view from the Marquesas. In: J. Roopnarine, J. Johnson & F. Hooper (Eds), *Children's Play in Diverse Cultures* (pp.73–103). Albany: SUNY Press.

McLoyd, V. C. (1982). Social class differences in sociodramatic play: A critical review. *Developmental Review*, 2, 1–30.

Meacham, J. (1996). Mind, society, and racism. *Human Development*, 39, 301–306.

Mertens, D. M. (1998). Research methods in education and psychology: Integrating: diversity with quantitative and qualitative approaches (pp. 6–21). Thousands Oaks, CA:Sage.

Monighan-Nourot, P. (1995). Play across curriculum and culture: Strengthening early primary education in California. In: E. Klugman (Ed.), *Play, Policy, and Practice* (pp. 1–20). St. Paul, MN: Redleaf Press.

O'Connell, B., & Bretherton, I. (1984). Toddlers' play alone and with mothers. In: I. Bretherton (Ed.), *Symbolic Play*. New York: Academic Press.

Ogbu, T. (1991). Immigrant and involuntary minorities in comparative perspective. In: M. Gibson & J. Ogbu (Eds), *Minority Status and Schooling: A Comparative Study of Immigrant and Involuntary Minorities* (pp. 3–33). NY: Garland Publishing.

Ogbu, J. (1981). Origins of human competence: A cultural-ecological perspective. *Child Development*, 52, 413–429.

Pan, W. H. L. (1994). Children's play in Taiwan. In: J. Roopnarine, J. Johnson & F. Hooper (Eds), *Children's play in diverse cultures* (pp. 31–50). Albany: SUNY Press.

Pellegrini, A. D. (1985). The relations between symbolic play and literate behavior: A review and critique of the empirical literature. *Review of Educational Research*, 55, 107–121.

Piaget, J. (1962). *Play, dreams, and imitation in childhood*. New York: Norton.

Reifel, S. (Ed.) (2001). *Theory in context and out. Play and culture studies* (Vol. 3). Westport, CT: Ablex.

Roopnarine, J. L., Brown, J., Snell-White, P., Riegraf, N., Webb, W., & Hossain, Z. (1995). Father involvement in childcare and household work in common-law dual-earner and single-earner Jamaican families. *Journal of Applied Developmental Psychology*, 16, 35–52.

Roopnarine, J., Johnson, J., & Hooper, F. (Eds) (1994). *Children's Play in diverse cultures*. Albany: SUNY Press.

Roopnarine, J. L., Ahmeduzzaman, M., Hossain, Z., & Riegraf, N. (1992). Parent-infant rough play: Its cultural specificity. *Early Education and Development*, 4, 298–311.

Roopnarine, T., Hossain, Z., Gill, P., & Brophy, H. (1994). Play in the East Indian context. In: J Roopnarine, J. Johnson & F. Hooper (Eds), *Children's Play in Diverse Cultures* (pp. 9–30). Albany, New York: SUNY Press.

Roopnarine, J. L., Lu, M., & Ahmeduzzaman, M. (1989). Parental reports of early patterns of caregiving, play, and discipline in India and Malaysia. *Early Child Development and Care*, 50, 109–120.

Roopnarine, J. L., Talukder, E., Jain, D., Joshi, P., & Srivastav, P. (1990). Characteristics of holding, patterns of play, and social behaviors between parents and infants in New Delhi, India. *Developmental Psychology*, 26. 667–673.

Roopnarine, J. L., Lasker, J., Sacks, M., & Stores, M. (1998). The cultural contexts of children's play. In: O. Saracho & B. Spodek (Eds), *Multiple Perspectives on Play in Early Childhood Education* (pp. 194–219). Albany, NY: SUNY Press.

Roopnarine, J. L., Shin, M., Supppal, P., & Donovan, B. (2000). Sociocultural contexts of dramatic play: Implications for early education. In: J. Christie & K. Roskos (Eds), *Literacy and Play in the Early Years: Cognitive, Ecological, and Sociocultural Perspectives*. Mahwah, NJ: Erlbaum.

Roopnarine, J. L., & Shin, M. (in press). Caribbean immigrants from English-speaking countries: Sociohistorical forces, migratory patterns, and psychological issues in family functioning. In: L. Adler & U. Gielen (Eds), *Migration, Immigration, and Emigration*. Westport, CT: Greenwood Press.

Roopnarine, J. L., Shin, M., Jung, K., & Hossain, Z. (in press). Play and early development and education: The instantiation of parental belief systems. In: O. Soracho & B. Spodek (Eds), *Contemporary Issues in Early Childhood Education*. New Age Publishers.

Rosaldo, R. (1989). *Culture and truth: The remaking of social analysis.* Boston, MA: Beacon Press.
Rubin, K., Fein,. G., & Vandenburg, B. (1983). Play. In: P. H. Mussen (Ed.), *Handbook of Child Psychology. Vol. 4: Socialization, Personality and Social Development* (pp. 393–474). New York: Wiley.
Rubin, K. H., & Maioni, T. (1975). Play preference and its relationship to egocentrism, popularity, and classification skills in preschoolers. *Merrill-Palmer Quarterly, 21,* 171–179.
Rubin, K. H., Watson, K., & Jambor, T. (1978), Free play behavior in preschool and kindergarten children. *Child Development, 49,* 534–536.
Saltz, E., Dixon, D., & Johnson, J. (1977). Training disadvantaged preschoolers in various fantasy activities: Effects on cognitive functioning and impulse control. *Child Development, 48,* 367–380.
Saltz, E., & Johnson, J. (1974). Training for thematic-fantasy play in culturally disadvantaged children: Preliminary results. *Journal of Educational Psychology, 66,* 623–630.
Schwartzman, H. (1978). *Transformations: The anthropology of children's play.* New York: Plenum.
Shweder, R. (1990) "Cultural psychology-what is it?" In: J. Stigler, R. Schweder & G. Herdt (Eds), *Cultural Psychology: Essays On Comparative Human Development.* Cambridge: Cambridge University Press.
Slaughter-Defoe, D., Nakagawa, K., Takanishi, R., & Johnson, D. (1990). Toward cultural/ ecological perspectives toward schooling and achievement in African- and Asian-American children. *Child Development, 61,* 363–383.
Scott, F., & Bowman, B. (1996). Child development knowledge: A slippery base for practice. *Early Childhood Research Quarterly, 11*(2), 169–184.
Segoe, M. V. (1971). A comparison of children's play in six modern cultures. *Journal of School Psychology, 9,* 61–72.
Stremmel, A. (1997). Diversity and the multicultural perspective. In: G. Hart, D. Burts & R. Charlesworth (Eds), *Integrated Curriculum and Developmentallv Appropriate Practices: Birth to Age Eight.* Albany, NY: SUNY.
Sun, L., & Roopnarine, J. L. (1996). Mother-infant and father-infant interaction and involvement in childcare and household labor among Taiwanese families. *Infant Behavior and Development, 14,* 121–129.
Super, C., & Harkness, S. (1997). The cultural structuring of child development. In: J. Berry, P. Dasen & T. Saraswathi (Eds), *Handbook of Cross-cultural Psychology: Basic Processes and Human Development* (pp. 1–39). Needham, MA: Allyn & Bacon.
Sutton-Smith, B., & Heath, S. B. (1981). Paradigms of pretense. *Quarterly Newsletter of the Laboratory of Comparative Human Cognition, 3,* 41–45.
Sutton-Smith, B. (Ed.) (1976). *A children's game anthology.* New York: Arno Press.
Taharally, L. C. (1991). Fantasy play, language, and cognitive ability of four-year-old children in Guyana, South America. *Child Study Journal, 21,* 37–56.
Tamis-Lemonda, C. S., & Bornstein, M. H. (1991). Individual variation, correspondence, stability, and change in mother and toddler play. *Infant Behavior and Development, 14,* 143–162.
Tamis-Lemonda, C. S., Bornstein, M. H., Cyphers, L., Toda, S., & Ogino, M. (1992). Language and play at one year: A comparison of toddlers and mothers in the United States and Japan. *International Journal of Behavioral Development, 15.*
Tobin, J. J., Wu, D. Y., & Davidson, D. H. (1989). *Preschool in three cultures: Japan, China, and the United States.* New Haven, CT: Yale University Press.

Tulananda, O., & Roopnarine, J. (2001). Mothers' and fathers' interactions with preschoolers in the home in Northern Thailand: Relationships to Teachers' assessments of children's social skills. *Journal of Family Psychology.*

Van Hoorn, J. (1987). Games that mothers and babies play. In: P. Monighan-Nourot, B. Scales, J. Van Hoorn & M. Almy (Eds), *Looking at Children's Play: A Bridge Between Theory and Practice* (pp. 38–62). New York: Teachers College Press.

Verba, M. (1993). Cooperative formats in pretend play among young children. *Cognition and Instruction, 11,* 265–280.

Vygotsky, L. S. (1978). *Mind in Society.* Cambridge: Harvard University Press.

Wiltz, N., & Fein, G., (1996), Evolution of a narrative curriculum: The contributions of Vivian Gussin Paley. *Young Children,* 61–68.

Yandell, K. (1997). Modernism, post-modernism, and the minimalist canons of common grace. *Christian Scholar's Review, 27,* 15–26.

Youngblade, L., M., & Dunn, T. (1995). Individual differences in young children's pretend play with mother and sibling: Links to relationships and understanding of other people's feelings and beliefs. *Child Development, 66,* 1472–1492.

UNDER THE LENS: THE PLAY-LITERACY RELATIONSHIP IN THEORY AND PRACTICE

Kathleen A. Roskos and James F. Christie

Over the past several decades the study of relationships between play activity and early literacy development has made rapid progress both scientifically and pedagogically. It's now widely recognized that play and literacy share common ground as early forming meaning-making systems on many levels, including language use and narration. Children make sense of their lives through play interactions, and these same processes apply to their early grasp of written language. In this article, we offer an overview of current theorizing about the play-literacy interface. We think our endeavor can contribute to a deepening understanding of the significance of play-literacy relationships in children's play lives and their literacy acquisition. We look to three developmental theories to furnish accounts of how play and literacy processes intersect in the developing minds of children. We then examine what these theoretical "ideas" imply for professional practice in real classrooms and in preparing early childhood educators for their role as literacy teachers.

THEORETICAL IDEAS

Piaget – Individual Actions in the Play World

Out of planks of action on objects in the physical world, the young child builds intelligence, at first sensori-motor in nature but increasingly mental as the full

Early Education and Care, and Reconceptualizing Play, Volume 11, pages 321–337.
Copyright © 2001 by Elsevier Science Ltd.
All rights of reproduction in any form reserved.
ISBN: 0-7623-0810-9

power of the mind is realized in abstract logical thought by adolescence. This is the Piagetian view – the "view from above" that explains cognitive development as an orderly, stage-like progression from reflexive responses to the facile manipulation of mental representations (Thelen & Smith, 1995).

Piaget (1963) stipulated that in order for learning and development to take place, there must be adaptation. Adaptation requires a balance between two complementary processes: assimilation, in which the child incorporates new information into existing cognitive structures (often 'bending' reality in the process); and accommodation, in which the child modifies existing cognitive structures to match, imitate, or otherwise conform with the reality of the physical world. Piaget (1962) viewed play as an imbalanced state in which assimilation dominates over accommodation. Imbalanced states in themselves do not engender genuine learning or development and are in this sense non-adaptive intellectually. Children do not learn new skills when they play. Play does, however, offer children valuable opportunities to practice and consolidate recently acquired skills. Although play is not the 'cutting edge' of cognitive development according to Piagetian theory, it makes an important contribution through this practice/consolidation role.

Through the lens of Piagetian theory, the interface between play and literacy involves shared common mental processes associated with perceptual discrimination skills, representational abilities, and narrative competencies that not only advance play activity, but also have an important role in learning to read and write. For example, when children engage in pretend play, they use a variety of make-believe transformations in which objects, action, and words stand for other objects, actions, and situations. Piagetian theory stipulates that these playful transformations enable children to practice using and interpreting symbols, providing mental resources that will later come in handy in dealing with the second-order symbolism of written language. Similarly, when children plan and act out stories during dramatic play, they have an opportunity to consolidate their growing knowledge about narrative story structure, building a foundation for comprehending and writing stories.

Research has provided some support for these cognitive links between play and literacy. Pellegrini, Galda, Dresden and Cox (1991), for example, found that children's use of symbolic transformations in play at age $3\frac{1}{2}$ predicted children's emergent writing, but not emergent reading, at age 5. Other investigators examined play as a context for story production and recall (e.g. Silvern, Taylor, Williamson, Surbeck & Kelley, 1986). They found that dramatic play prompted children to use the "building blocks of story" (e.g. setting, character, and plot), helping them better remember (and therefore comprehend) stories they had been read.

By far the richest area of Piagetian-based research, however, has focused on literacy-enriched play centers that contain materials that create a environmental "press" for play-related reading and writing activities – pencils, markers, note pads, signs, menus, etc. The basic rationale behind these studies is that, through interacting with literacy objects during play, children will have opportunities to consolidate their emerging concepts about the functions and structure of print and to practice emergent forms of reading and writing. Data showed that these print-enriched play settings result in large increases in emergent reading and writing activity during play (Christie & Enz, 1992a; Morrow, 1990; Neuman & Roskos, 1990, 1992). While attempts were also made to link children's play in print-enriched setting with gains in literacy development, the results were mixed. For example, Neuman and Roskos (1990) found that adding literacy materials to play centers resulted in a significant gain in preschoolers' scores on Marie Clay's Concepts About Print test, whereas Christie and Enz (1992a) failed to find connections between the two variables.

These early Piaget-inspired studies made important contributions to our knowledge of play-literacy relationships. They brought attention to the fact that two outwardly dissimilar acts, play and literacy, apparently shared common mental processes. In addition, these studies brought to the limelight the literacy-enriched play center – an effective strategy for encouraging meaningful literacy interactions in early childhood classrooms.

At the same time, the Piagetian perspective on play and literacy has some significant shortcomings. At a very basic level, development probably does not progress lockstep in a linear set of stages as stipulated by Piaget, but rather is more shifting and fluid. For example, the recursive nature of literacy development has been revealed in research on developmental trends in emergent writing (Sulzby & Teale, 1991). A second, problem involves the general focus of the perspective. By focusing primarily on person-object interactions, the Piagetian perspective gives short shrift to the important influence of social interaction and cultural factors on development. It is this second limitation that has given rise to our next theoretical perspective – Vygotsky's theory.

Vygotsky – Social Assistance in the Play World

Play has several important roles in Vygotsky's theory of development. At the most basic level, Vygotsky (1976) believed that make-believe play has a key role in abstract thought, enabling children to think about meanings independently of the objects they represent. At a second level, play can provide a context for socially-assisted learning about literacy. Vygotsky (1978) distinguished between two levels of development: "actual development" (independent performance) and "potential development" (assisted performance), with the *zone of proximal development* defined as the distance between the two. If temporary assistance or scaffolding is provided by an adult or more competent peer, the child is able to engage in activities that he or she could not do alone. This, in turn, extends the child's knowledge and skills to higher levels.

At a third level, Vygotsky viewed play is a 'self-help' tool. He maintained that, when children engage in play, they can create their own scaffold, stretching themselves in such areas as self-control, cooperation with others, memory, language use, and literacy (Bodrova & Leong, 1996). Vygotsky saw play as a kind of magnifying glass revealing potential new abilities before these same abilities became actualized in other situations. As a result, children engaged in play often seem ahead of themselves developmentally. This raised the possibility that play may promote further development by serving as a scaffold within the children's Zone Proximal Development, helping them attain higher levels of functioning.

During the 1980s, the cognitive link between play and literacy in Vygotsky's theory received considerable attention, along with the similar connection in Piagetian theory. During the 1990s, attention shifted to social aspects of Vygotsky's theory, due in large part to the growing influence of the emergent literacy perspective which also places heavy emphasis on the social nature of early literacy learning (Sulzby & Teale, 1991). Play in literacy-enriched settings was viewed as an ideal context for children to learn about literacy from adults and peers.

Using the Zone of Proximal Development as a rationale, several quantitative research studies investigated the effects of teacher scaffolding of literacy during play (Christie & Enz, 1992a; Morrow, 1990; Schrader, 1990; Vukelich, 1991). Results indicated this scaffolding did increase the amount of literacy activity during play. In addition, Vukelich (1991) found that the combination of a literacy-enriched environment and adult mediation led to significant increases in children's knowledge about the functions of writing. Several other studies reported that, when teachers draw children's attention to environmental print in play settings, children's print recognition ability is enhanced (Neuman & Roskos, 1993; Vukelich, 1994).

A qualitative study by Roskos and Neuman (1993) shed light on exactly how teacher's scaffold literacy during play. Findings revealed that teachers assume a variety of roles when interacting with children literacy-enriched play settings, ranging from being an appreciative audience to being an active play leader. Experienced teachers shift between these roles, matching their interaction to children's personalities and ongoing play activities.

Several other studies focused on peer interaction in literacy-enriched play settings. Neuman and Roskos' (1991) ethnographic study revealed that children use strategies such as negotiating and coaching to help each other learn about literacy during play. Christie and Stone (1999) discovered the collaborative literacy interactions that occurred during the play were more complex than the Zone of Proximal Development would lead one to believe. While a majority of the joint literacy learning conformed to the Zone construct, flowing from older or more competent "experts" to "novices," a sizable number of collaborative interactions in both groups were multi-directional in nature, with the "expert" and "novice" roles not firmly set.

These Vygotskian-inspired studies have shed considerable light on what children can take from their encounters with others during play in which literacy is be embedded. At the same time, opportunities exist to take fuller advantage of Vygotsky's theory. Researchers have concentrated their attention on the Zone of Proximal Development as it pertains to social interaction between children and adults or more competent peers. In so doing, researchers have minimized the role of activities and things, in contradiction with the third connection between play and literacy in Vygotsky's theory – the part of his theory that views play *itself* as a Zone of Proximal Development. Play provides its own "Zone" with rules, roles and routines as forms of assistance that can pull literacy development forward. Viewed from this perspective, activity itself can have a key role in what children learn in literacy-enriched settings.

Bronfenbrenner – Linking Child and Activity in the Play World

Both the Piagetian approach and the Vygotskian approach (as applied in play-literacy research) examine the play-literacy interface at the level of interactions among individuals. In these approaches, the central focus is that of child-object or child-other interactions that become internalized and increasingly coordinated as symbolic representations of experience or structures of mind. The goal is to connect what develops as intellectual competencies in the play experience, e.g. role-taking, negotiating, using elaborated language, heeding print cues, etc., with what develops as foundational skills of emergent literacy, e.g. pretend reading, sense of story, letter-name knowledge. Each approach assumes connections between these two forms of intellectual work, although each focuses singularly on different routes taken in building the play-literacy relationship (child-object or child-other).

What this basic scheme overlooks, however, are broader patterns of activation that dynamically arise from the integration of child, immediate setting, and the society at large. What the play experience affords literacy development, some would argue, is both individually determined and contextually driven (Bronfenbrenner, 1995; Cole, 1996; Resnick, 1994; Thelen & Smith, 1995). It is the developing, busy child with her cache of bio-psychological resources and developmentally instigative dispositions (i.e. preferences) who is drawn into play as an activity with its own socio-historical and cultural content implicit in playthings, play spaces, play scripts and play themes. The interface between play and literacy involves the complex interaction of:

- *who* children are,
- *where* they are located,
- *with whom* they are interacting, and
- *when* they are playing – both in real time and historically.

These factors work synergistically to create the conditions for play that are of significance to the emergence of literacy. This symbiosis is a grand puzzle indeed. The union of child and play activity does not yield easily to investigation because both are dynamic systems continually on the move (although they sometimes appear to be stable entities).

Recently a few literacy development researchers have ventured beyond the specifics of child-object and/or child-other interactions that describe either content (e.g. print concepts) or process links (e.g. narration) toward a more integrative view of play-literacy relationships – one that encourages the joint consideration of macro- and micro-level processes at work in play activity. Their work at different levels seeks to describe the *patterns of activity* that arise in the play environment and to infer what these may afford young children's literacy development. Dunn, Beach and Kontos (1994), for example, in a series of studies involving 30 day care programs identified a knot of macro-level factors that appear to constrain the amount of children's literacy interactions in their immediate play environment, including a limited supply of print resources (e.g. books), fewer certified teachers, less conversation and verbal exchange between adults and children, and teachers' general lack of emergent literacy knowledge. In combination these factors created a troubling pattern of very little literacy activity in children's everyday play experiences at day care. The consequences of this low level of literacy exposure in the early years of play is not known. A few recent studies suggest, however, that certain macro-level features in the home, such as parents' entertainment orientation toward early literacy or their problem solving processes in game-like situations, may contribute to children's overall literacy development (de Jong & Leseman, in press; Sonneschein et al., 2000).

Closer to the bone of actual play activity, Neuman and Roskos (1997) investigated children's metacognitive activity or "knowing" in three literacy-enriched play settings (doctor's office, restaurant and post office) in a preschool program over a seven-month period. Observations indicated that children used different forms of knowledge to monitor their understandings of events in these sites as demonstrated in their role-taking, rules for play, and play routines. Moreover, the specific strategies used by children to maneuver the play situation varied according to the activity setting. For example, children chose organizing, arranging, and allocating of materials to enable literacy activity in the post office setting, whereas making transactions between one another and using print as a memory device were selected in the restaurant setting. These differences in strategy use indicated that children's responses were related to the purposes of activity, the physical environment, materials, and the occasion which emerged from the interplay between peers.

This evidence suggests that literacy knowledge in play is dynamic and synergistic, assembled in real time, in context from informational bits at hand that in and of themselves do not hold the resultant knowledge children garner. Literacy knowledge, in fact, does not appear structure-like at all, but rather soft-assembled as a pattern of activity in relation to the play setting. In this sense, the emerging literacy knowledge that play affords is essentially *process* – a network of interactions that varies as a joint function of player, setting and the larger circumstances that give rise to the play environment in time and place (e.g. the early childhood program). It is in the dynamics of this process, therefore, that the developmental power of the play-literacy interface may be found.

Still, joint consideration of the bio-psychological characteristics of the players, of the play setting and of the larger play environment as an ecological context has only recently emerged as a promising direction for literacy-play research. Much needs to be considered and weighed to explain the activity systems that unite literacy development and play. There are many good questions to ask. What, for example, does children's literacy knowledge look like across play settings and over time? What strategies contribute to the development of emergent reading and writing knowledge that predicts literacy acquisition? What is the evidence of developmental change in literacy knowledge and skills in play contexts? How do individual children's dispositional proclivities toward the play environment influence their literacy in play patterns? Such queries invite a dynamic account of play as an environmental opportunity for literacy development and learning – an account that integrates knowing and action.

THEORETICAL IDEAS IN REAL SPACE AND TIME

Working through the thicket of theoretical ideas that seek to describe and explain the play-literacy relationship, we now carry these ideas along into visions of everyday practice in early childhood education. Our plan is to bring each set of ideas to life and to describe it as it might live and look in real classrooms. We highlight the strong features of each and its emphases. We then briefly discuss what each offers the eager, young learner as a swiftly developing writer and reader.

Literacy in Play – the Piagetian Perspective

The Piagetian perspective focuses on interactions between the child and objects in the physical environment. If one's goal is to help children learn about literacy through play activities, it follows that the areas in which they play should contain a rich assortment of literacy materials. By interacting with these material resources during play, children will be able to consolidate knowledge about print, reading, and writing. The practical application of this perspective is the literacy-enriched play center – a play setting that contains a variety of literacy artifacts.

Literacy-enriched play centers typically contain three types literacy artifacts. First, these centers always contain general types of writing and reading materials:

- various types of paper, notepads, post-it notes, and other kinds of things for children to write on;
- a variety of instruments to write with such as pencils, markers, pens, and crayons;
- small "mixed" collections of children's books.

These types of "generic" materials have the advantage of being open-end, flexible, and adaptable to any type of literacy activity during play. For example, if children are acting out a restaurant play episode, the child who is waiting tables may decide that he or she wants to jot down customers' orders. Having notepads and writing instruments available will enable the child to incorporate this familiar literacy routine into the play.

Second, literacy-enriched play centers contain theme-related literacy artifacts. These are literacy items that are found in the real-life counterparts to a play setting. For example, if a play center was set up to represent a pizza restaurant, theme-related artifacts for the setting would include: signs (*Pizza Hut©, Pay at the Register*), menus, coupons (for discounts on pizzas), cookbooks, pizza boxes with company logo, etc. These types of artifacts encourage children to act out literacy routines that they have observed in real-life: finding a restaurant by looking for its sign, ordering from a menu, etc.

No matter how well equipped a literacy-enriched play center is, children will inevitably come up with other literacy items that are needed for their play. This gives rise to the third type of artifact – child-produced. In our pizza shop example, children might decide that they need to make some money to pay for the pizzas. Or they might decide that more signs are needed for their restaurant (*No Smoking!*). In order to encourage this type of print production, teachers often include a small set of art construction materials (scissors, tape, paste) in the play center.

Perhaps an apt metaphor for these Piaget-inspired play setting would be the "cafeteria." These settings offer children a broad selection of literacy artifacts with which they can interact. If a child discovers that something she wants is not on the menu, she can pop back in the kitchen and cook it up herself in the form of a child-produced literacy object. This cafeteria offers children literacy tools, functions, routines, and ideas that enable children to build their concepts of written language. However, the possibility also exists that some children may not know how to incorporate the menu items into their play and, as result, may not be able to profit from what the cafeteria has to offer. So the literacy-enriched play setting, by itself, provides children with wonderful learning opportunities, but not all children will likely take advantage of these – at least not on their own.

A Vygotskian View of Literacy in Play

In the Vygotskian perspective, the focus shifts from child-object to child-other interactions. Parents, teachers and more capable peers are viewed as social resources that can help children learn about literacy through play by extending the Zone of Proximal Development, enabling children to engage in advanced literacy activities that they could not do on their own. This perspective also relies on literacy-enriched play settings. However, now these enriched settings are viewed not as sites for individual discoveries about literacy, but rather as settings in which more knowledgeable "experts" can assist "novices" construct knowledge about literacy through play.

On the surface, Vygotskian- and Piagetian-oriented classrooms look very similar. The play centers contain similar types of generic, theme-related, and child-produced literacy artifacts. Where differences emerge is in the teacher's role during play and in how children are grouped for play.

The teacher's role in a Piagetian classroom is that of an 'appreciative audience'. The teacher sets up literacy-enriched play centers and watches from the sidelines as children engage in play. These play-based observations allow the teacher to assess children's growing concepts about the concepts and structure of print and the "fine tune" the environment to slightly lead children's current level of literacy development.

The teacher is much more actively involved during play in the Vygotskian classroom, using a continuum of play interaction roles to match children's current play interests and activities (Enz & Christie, 1997; Roskos & Neuman, 1993). The 'appreciate audience' is one of these roles, but it is joined by more active methods of play involvement, including:

- stage manager – the teacher watches from the sidelines and, when opportunities arise, provides children with temporary assistance with their play and literacy activities (e.g. helping them make a sign for their supermarket).
- co-player – the teacher takes on a minor role and joins in children's play. While enacting this role, the teacher follows the flow of the dramatic action, letting the children take the lead. When opportunities arise, the teacher models play-related literacy activities (e.g. writing a shopping list, ordering food from a menu).
- play leader – as in the co-player role, the teacher joins in and actively participates in the children's play. However, play leaders exert more influence and take deliberate steps to enrich and extend play episodes. They do this by making suggestions interjecting new elements into the ongoing play. For example, if the teacher is taking part in a domestic play episode in which family members are preparing a meal, she might exclaim, "Oh my goodness, we don't have any meat or vegetables for our soup! What should we do?" This creates a problem for the children to solve and may result in the writing of a shopping list and a trip to a nearby store center.

Another feature of the Vygotskian classroom is that teachers take measures to enhance the collaborative interactions between children. Because a mixture of "novices" and "experts" is conducive to stretching the novices' Zone of Proximal Development, one is likely to encounter some form of multi-age grouping (Christie & Stone, 1999). This can range from full-time multiage classrooms (e.g. a K-2 multiage class) to part-time "play buddy" arrangements in which older students visit periodically to play with younger students. As an alternative, teachers in same-age classrooms can take steps to encourage children of differing abilities to play together. When mixed age/ability play is going well, the teacher's role is primarily that the appreciate audience –

observing from the sidelines while children work together to help each other learn about literacy.

A good metaphor for the play settings in a Vygotskian classroom is that of a "community." The literacy-rich play environment becomes a social resource that invites, shows, and guides children's discoveries about literacy. The teacher devotes a considerable amount of time and effort to creating this community. If push comes to shove, social relationships will receive higher priority than the physical artifacts that inhabit the areas where children play.

Literacy in Play as a Dynamic Activity System

The previous two theories put to work in the play environment show how literacy artifacts and persons as literacy resources can augment everyday play – to enrich it, as it were, with literacy ideas, actions, tools, and functions. Essentially this involves 'adding to' what's already there whether in the form of actual things and/or behaviors (talk and actions) that make literacy more visible and more accessible in play episodes.

From an ecological perspective, however, there is more involved than just what surrounds and infiltrates play as literacy resources. Active and enduring processes that hold literacy potential – those invisible patterns, pervasive structures, and ground rules that emerge in the present yet remain shaped by the past – are also at work. Finding the practical application in this complex theoretical idea is difficult. It helps, though, to envision the literacy-rich environment ecologically as a thick jungle with many sensory, intellectual, and cultural layers that stimulate the young child as a developing writer and reader. What is this 'jungle' classroom like? From what we know and can infer at this point, it is distinguishable by at least four features.

One feature is that such places provide a widespread *presence* of print and print-related activities in the environment in ways that are accessible to children. Placing print at children's eye level, using print in functional ways (e.g. to cook), directing children's attention to print, and extending their knowledge and uses of it in meaningful ways enhances children's use and desire to become literate. Literacy is embedded in multiple settings across the environment and in layered ways so as to 'build up' the potential for literacy interactions. The 'old favorite' house corner, for example, includes (at least) labeled bins for sorting kitchen ware or food coupons by type, grocery packages for reading environmental print, recipe cards and books for pretend reading, posted charts from class experiences for shared reading and word pointing, and supplies for pretend writing of lists, labels, coupons and personal recipes. The resources, in short, are layered to scaffold explorations to higher levels as children combine and re-combine what they know and see others do. As a result children help to create an amazingly complex literacy environment that reflects and stretches their knowing and thinking.

A second is that print is in close *proximity* to children's daily activities and life experiences. Books are within reach of their small hands and nearby in the play setting. Writing supplies are at hand and easy to grip, turn, and hold. Children's culture, experience, language, and interests are also carefully considered so that print and print-related activities in play are within the reach of the mind as well as the hand. Items of cultural meaning are placed in specific play areas to support continuity between home and classroom and stimulate communication. Play roles and routines that incorporate reading and writing are linked to what children are likely to know about, but also pulled into new directions through peer- and teacher-modeling. The riches of social resources in school and community are fully tapped to provide information, show how, extend ideas, and link literacy experiences to future play opportunities.

A third involves support for the *portability* of resources among and between play spaces and activities. What this affords, beyond children's desire to carry things back and forth, is the opportunity to carry literacy activities and ideas from one place to another, which opens up new possibilities for literacy uses and also for play. The 'jungle-like' play environment has 'depots' of literacy supplies here and there for writing, supports the creation of toy-book sets for play-enriched pretend reading and storytelling, encourages movement and flexibility in activity, and includes materials for multi-modal expressions of experience. Familiar things and the chance to move them about by hand and in mind create a flexible environment thick with opportunities to explore print and other media for purposes of meaning-making and communication.

A fourth feature has to do with the environment's *productivity*, which resides in its overall capacity to provide opportunities for selection and adaptation on the part of intelligent and developing children. Literacy in play is not only on offer in the environment, but it also contains surprises (novelty) and challenges that lead children to new ideas about the print-sound code, about getting meaning from texts, and about the habits of readers and writers. This is brought about by the press of larger social forces – believing in the literacy capabilities of young children, for example – in concert with adults' everyday acts of professionalism as children's literacy mentors and teachers. For preschoolers, busily exploring and testing language, scripts and roles, their playing in real time and space may, in fact, be part of the fuel that drives emergent literacy forward. In its richly layered ecology, the uniting of child, the world of objects and symbol (society's meaning systems, e.g. curriculum) creates a proximal environment that allows children to "actively integrate conceptual pieces of knowledge" into more elaborated (and more useful) forms of literacy activity (Bus, deVries, de Jong, Sulzby, de Jong & de Jong, 2001).

These four features – presence, proximity, portability, productivity – characterize literacy in play as an activity system that creates certain kinds of conditions for children's literacy interactions and experiences. In its multi-layered, jungle-like environment the emphasis is on drawing out children's existing abilities so they can make good use of what's there in the environment. The personal construction of categories is encouraged because it is through self-organizing processes that literacy development occurs (Lewis, 2000). There is a push for individual, personal solutions of challenges that stimulate the emergence of new conceptualizations about literacy, which in turn contribute to the overall capacity of the system through joint activity. Linking the child and literacy activity in the play world, in sum, creates a 'web of relationships' that becomes a pattern of activation – a proximal process – that may more or less bless the developing writer and reader. What children gain in this environment, if well maintained, is access to literacy at multiple entry points at multiple times. They also are exposed to adventure, which is to say that they encounter novel situations that they can't explain and intellectual challenges on which they stub their thinking – and these 'bouts' urge them forward in their knowing and action.

TEACHERS – KNOWLEDGE, SKILL AND HEART

The bridge from good theory to practical application is not easily crossed and no where is this more sharply felt than in teaching. If teachers are to understand and support literacy-play relationships to the benefit of all children in their classrooms, what is required of them? We have three suggestions about the pedagogical knowledge base for literacy in play practice that we hope our fellow teacher educators at 2-year and 4-year levels of higher education will contemplate and consider in their early literacy and play courses, early literacy pedagogy curriculums, and professional education programs.

Teachers need theoretical knowledge; it's at the core of good practice. And, in early literacy education, they need to know about children's cognition, their play and their emerging literacy to be able to interconnect basic concepts that integrate all three in worthwhile and significant ways. Acquiring knowledge about children's cognition, learning and play have long been mainstays of the professional education in early childhood. Knowledge of early literacy development, however, is a newcomer and also a demanding one. Teachers need to grasp basic concepts about the emergence of literacy as an individual and social matter, which involves knowing not only about child-world interactions, but also about the biology of the brain and development of mind (Eliot, 1999; National Research Council, 2000). How to incorporate the considerable knowledge base on early literacy into already packed 2-year or 4-year teacher education curriculums poses new challenges for educators. Rather than "add on" literacy coursework, we urge an integrative approach that layers basic concepts about literacy development and "best practice" pedagogy into existing language arts courses and builds connections between literacy and the theoretical foundations of other developmental and curricular domains, such as art, movement, numeracy, and science inquiry. In this same vein, we hope for consensus around a basic set of early literacy education concepts and principles that students may experience at increasingly deeper and more complex levels along a continuum of professional education from the Child Development Award (CDA) through the baccalaureate degree. We advocate, in sum, a spiral curriculum that develops and deepens teachers' conceptual understandings across a range of professional education contexts.

Teachers also need to become skillful at embedding literacy in play so that both literacy and play benefit. Along with good ideas, they need practical knowledge. Instruction, via lecture and textbooks may be adequate for conveying theoretical information and basic instructional strategies, but comprehension alone does not guarantee commitment or competent application (Christie & Enz, 1992b). Teachers need practical knowledge so that they can interact with children spontaneously and effectively. In addition, being an effective play partner for children requires not only knowledge but a playful spirit – a willingness to be a player and a risk taker.

Our own experiences as researchers and teacher educators has led us to believe that a process-simulation approach is the most effective way to help teachers acquire these practical skills and play attitudes. The following set of activities could be employed to help novice teachers learn to interact effectively with children during play:

- Read articles about effective teacher play interaction styles (e.g. Roskos & Neuman, 1993; Enz & Christie, 1997);
- View and critique videotapes of teachers interacting with children during play;
- Conduct a play workshop in which novice teachers have opportunities to engage in different types of play (constructive play, dramatic play, games, etc.);
- Have the novice teachers construct a literacy-enriched play center;
- Using the above play center, conduct simulations in which several novice teachers take on the roles of children engaging in dramatic play. Others, in

the role of teacher, are asked to act out how they would interact with the children.
- Under the supervision of a mentor, novice teachers interact with children at play in a school classroom. Ideally, these sessions could be video-taped and self-critiqued by the novice.

Through this sequence of activities, teachers have an opportunity to mentally rehearse, evaluate, and apply new ideas, and thus should be better able to translate theory into practice.

Teachers need knowledge; they need skill. But they also need a willing heart – one that believes in very young children as emergent writers and readers and believes that children are capable of stretching and growing during their preschool years. To value the literacy-play relationship in their hearts, teachers need many opportunities to examine their own thinking in ways that respect their current knowledge and experience and, at the same time, challenge the status quo. This obligates us as teacher educators to create and support learning contexts wherein teachers can wonder, query, deliberate and share their personal histories openly and wholeheartedly – not only in coursework, but also through study groups, discussion threads, conversation circles, mentoring relationships and ongoing professional development. As a result of these types of activities, teachers are helped to develop a stance of inquiry and critique toward literacy-play applications in their everyday work – a stance that opens minds and wins hearts.

REFERENCES

Bodrova, E., & Leong, D. (1996). *Tools of the mind: The Vygotskian approach to early childhood education.* Englewood Cliffs, NJ: Prentice-Hall, Inc.

Bronfenbrenner, U. (1995). Developmental ecology through space and time: A future perspective. In: P. Moen, G. Elder & K. Luscher (Eds), *Examining Lives in Context: Perspectives on the Ecology of Human Development.* Washington, D.C.: APA.

Bus, A. G., deVries, A. B., de Jong, M., Sulzby, E., de Jong, W., & de Jong, E. (2001). Conceptualizatons underlying emergent readers' story writing. Unpublished manuscript.

Christie, J., & Enz, B. (1992a). The effects of literacy play interventions on preschoolers' play patterns and literacy development. *Early Education and Development, 3*, 205–220.

Christie, J., & Enz, B. (1992b). *Teacher education: A key element in improving school play.* Meeting of the International Council for Children's Play, Paris, France.

Christie, J., & Stone, S. (1999). Collaborative literacy activity in print-enriched play centers: Exploring the "zone" in same-age and multi-age groupings. *Journal of Literacy Research, 31*, 109–131.

Cole, M. (1996). *Cultural psychology: A once and future discipline.* Cambridge, MA: The Belknap Press of Harvard University Press.

de Jong, P. F., & Leseman, P. M. (in press). Lasting effects of home literacy on reading achievement in school. *Journal of School Psychology*.
Dunn, L., Beach, S., & Kontos, S. (1994). Quality of the literacy environment in day care and children's development. *Journal of Research in Childhood Education, 9*, 24–34.
Eliot, L. (1999). *What's going on in there? How brain and mind develop in the first five years of life* (chapter 14). NY: Bantam Books.
Enz, B., & Christie, J. (1997). Teacher play interaction styles: Effects on play behavior and relationships with teacher training and experience. *International Journal of Early Childhood Education, 2*, 55–75.
Lewis, M. (2000). The promise of dynamic systems approaches for an integrated account of human development. *Child Development, 71*(1), 36–43.
Morrow, L. (1990). Preparing the classroom environment to promote literacy during play. *Early Childhood Research Quarterly, 5*, 537–544.
National Research Council (2000). *Eager to learn*. Washington, D.C.: National Academy Press.
Neuman, S., & Roskos, K. (1990). The influence of literacy-enriched play settings on preschoolers' engagement with written language. In: S. McCormick & J. Zutell (Eds), *Literacy Theory and Research: Analyses from Multiple Perspectives* (pp. 179–187). Chicago: National Reading Conference.
Neuman, S., & Roskos, K. (1991). Peers as literacy informants: A description of young children's literacy conversations in play. *Early Childhood Research Quarterly, 6*, 233–248.
Neuman, S., & Roskos, K. (1992). Literacy objects as cultural tools: Effects on children's literacy behaviors during play. *Reading Research Quarterly, 27*, 203–223.
Neuman, S., & Roskos, K. (1993). Access to print for children of poverty: Differential effects of adult mediation and literacy-enriched play settings on environmental and functional print tasks. *American Educational Research Journal, 30*, 95–122.
Neuman, S., & Roskos, K. (1997). Literacy knowledge in practice: Contexts of participation for young writers and readers. *Reading Research Quarterly, 32*, 10–32.
Pellegrini, A., Galda, L., Dresden, J., & Cox, S. (1991). A longitudinal study of the predictive relations among symbolic play, linguistic verbs, and early literacy. *Research in the Teaching of English, 25*, 215–235.
Piaget, J. (1962). *Play, dreams and imitation in childhood*. New York: Norton.
Piaget, J. (1963). *The origins of intelligence in children*. New York: Norton.
Resnick, L. (1994). Situated rationalism: Biological and social preparation for learning. In: L. Hirschfield & S. Gelman (Eds), *Mapping the Mind: Domain Specificity in Cognition and Culture* (pp. 474–493). NY: Cambridge University Press.
Roskos, K., & Neuman, S. (1993). Descriptive observations of adults' facilitation of literacy in play. *Early Childhood Research Quarterly, 8*, 77–97.
Schrader, C. (1990). Symbolic play as a curricular tool for early literacy development. *Early Childhood Research Quarterly, 5*, 79–103.
Silvern, S., Taylor, J., Williamson, P., Surbeck, E., & Kelley, M. (1986). Young children's story recall as a product of play, story familiarity, and adult intervention. *Merrill-Palmer Quarterly, 32*, 73–86.
Sonnenschein, S., Baker, L., Serpell, R., & Schmidt, D. (2000). Reading is a source of entertainment: The importance of the home perspective for children's literacy development. In: K. Roskos & J. Christie (Eds), *Play and Literacy in Early Childhood: Research from Multiple Perspectives* (pp. 107–124). Mahwah, NJ: Lawrence Erlbaum.
Sulzby, E., & Teale, W. (1991). Emergent literacy. In: R. Barr, M. Kamil, P. Mosenthal & P. D. Pearson (Eds), *Handbook of Reading Research* (vol. 2). New York: Longman.

Thelen, E., & Smith, L. B. (1995). *A dynamic systems approach to the development of cognition and action*. Cambridge, MA: MIT Press.

Vukelich, C. (1991). Learning about the functions of writing: The effects of three play interventions on children's development and knowledge about writing. Paper presented at the meeting of the National Reading Conference, Palm Springs.

Vukelich, C. (1994). Effects of play interventions on young children's reading of environmental print. *Early Childhood Research Quarterly, 9*, 153–170.

Vygotsky, L. (1976). Play and its role in the mental development of the child. In: J. Bruner, A. Jolly & K. Sylva (Eds), *Play: Its Role in Development and Evolution* (pp. 537–554). New York: Basic Books.

Vygotsky, L. (1978). *Mind in society: The development of psychological processes*. Cambridge, MA: Harvard University Press.

THE PLAY FRAME AND THE "FICTIONAL DREAM": THE BIDIRECTIONAL RELATIONSHIP BETWEEN METAPLAY AND STORY WRITING

Jeffrey Trawick-Smith

The relationship between play and literacy has been well documented (see Christie, 1998, for a review); a variety of explanations have been offered for why the two are connected. Some researchers have proposed that play involves cognitive processes that contribute to literacy growth. For example, it has been argued that the representational transformations required in play (e.g. using a toy telephone to represent a real one) promote the same symbolic abilities needed to read and write (e.g. representing an object, person, or action in a story, using scribbles, pictures, words, or letters) (Trawick-Smith, 1990). Others have suggested that the rich oral language used in play supports literacy (Levy, Wolfgang & Koorland, 1992). Davidson (1998) has shown that children sometimes use unique, literature-related phrases, intonations, and words as they pretend-the language of books-which are especially useful in learning to read and write.

Another body of work has focused on actual literacy behaviors during play (Christie & Enz, 1992). As children use reading and writing props in pretend enactments, they have been found to practice *literacy routines* and to construct an understanding of the *functional uses of print* (Roskos & Neuman, 1993).

Children using crayons, markers, order pads, and signs while playing in a make believe restaurant, for example, not only acquire understandings of specific features of print, but learn, more generally, how print can be used as a practical tool in life-in this case, to take a customer's lunch order or to post a special of the day.

Much of this previous work has focused on how behavioral or language elements in play directly enhance an understanding of print and how it is used. In this paper a more basic connection between play and literacy is considered- the relationship between certain aspects of make believe and an overall conception of *story*. I will propose here that, when children read and write, they often construct in their minds a fictional world – a vivid, detailed mental construction that includes an imagined physical setting, characters, action, and a full story to go with these. When they hear a story of a train moving through the arctic, for example, they may hear, see, and feel this event. When they write about a firefighter, they may visualize her appearance, the actions she takes in fighting a blaze, and even the anxiety she feels in doing so. I will argue that, when children play, they often construct a similar imagined world, which includes these same elements. When a child enacts the role of a patient receiving an injection in a make believe doctor's office, he can visualize the physical setting, hear the comforting words of the doctor, and feel the discomfort of a medical procedure. The focus of this paper is on play worlds and the worlds of literature-how they are constructed, and the ways they support each other in early childhood.

STORY, PLAY, AND METAPLAY

Lehr (1991) argues that, in order for children to derive significant personal meaning from literature or a piece of their own writing, they must acquire an overall sense of story. As they write, they gradually move beyond simple narratives or the retelling of personal histories (e.g. "This says, 'My grandma picked me up after day care.'"), to consider the feelings, viewpoints, and interrelationships among multiple characters, to understand the problems they face and settings they find themselves in, and even to grasp basic themes – the larger messages to be conveyed to their readers. When reading or being read to, children must contemplate these elements in another author's work; they must come to understand a whole story and even *enter* the fictional world that has been created (Lehr, 1991).

Young children's abilities to conceptualize a full story – to create a mental fictional world – vary considerably, as illustrated by the two examples below:

The Play Frame and the "Fictional Dream"   341

Literacy Example 1: "My Mom"

Two four year olds sit at the writing table in their classroom, working on separate stories, both using unique forms of scribble writing to represent their ideas.

Child A: (Points to Child B's story.) What's that say?
Child B: It's my mom.
Child A: No, but what does it say?
Child B: The story?
Child A: Yeah.
Child B: (Runs a finger along the "text," and uses an adult-sounding voice). 'This is my mom.'
Child A: (Points to the drawing.) Her?
Child B: Yeah.

The two children resume writing independently.

Literacy Example 2: "The Monster Who Eats Up Little Boys"

A four year old child sits alone at the writing table. He has drawn on several pages of a blank book, and is now adding "text" – a series of jagged lines – underneath his illustrations. A teacher moves over and asks him about his story.

Teacher: Tell me about what you're writing.
Child: It's going to be a monster.
Teacher: Really? Is it a nice monster?
Child: No, silly, 'cause it's a monster who's bad, 'cause he *loves* little boys to eat up. And it's almost his supper time.
Teacher: Oh.
Child: See? (Points to his drawing). He's got sharp teeth, like those points? See? And these are the claws. And he's going eat the boys. (Laughs.)
Teacher: Oh, my.
Child: And, know what?
Teacher: What?
Child: (Turns the page.) It's the little boy's birthday and the monster's going to eat up the birthday presents.
Teacher: Oh no.
Child: Yeah and he's five years old, too.
Teacher: Wow. So, how's the little boy going to feel?
Child: Well . . . (pauses) . . . I think he's going to cry.
Teacher: Right.
Child: See? (Points to another drawing.) The little boy's going so say, "My birthday presents! Stop it!" And then the monster eats them and he doesn't have any more presents. And then he's going to be sad and that's the end.

Children in both examples display an impressive understanding of print. However, the child of example two has constructed a fuller story with multiple characters, dialogue, and a plot. Stated another way: The second child has created a more vivid fictional world. What specific elements has he included in his story to accomplish this? What early experiences have contributed to his ability to create such a world? A question to be raised in this paper is whether play experiences have influenced this child's abilities to conceptualize such a full, detailed story.

Norout (1998) has proposed that in play children engage in a process very similar to writing, often creating compelling imaginary worlds and constructing elaborate stories within these. As they engage in make believe, she notes, children invent characters, plot, and setting, not unlike authors do when writing fiction. Play may be useful, then, in helping children to understand and construct key elements of a story. (I will argue here that writing imaginative stories will, in turn, contribute to more complex play. A purpose of this paper is to elevate the status of play as an important *outcome*, a goal of meaningful writing experiences, just as writing is an outcome of play.)

Norout (1998) suggests that a distinct process in play – *metaplay* – is most similar to story writing. In metaplay, children suspend their make believe enactments momentarily and step out of their imaginary settings to think about or negotiate play themes from outside of actual role playing. When children do this, they become scriptwriters, directly inventing new characters, planning actions, or negotiating the next steps in a plot line. Children plan for the pirate ship to sink, the baby to stop crying and finally go to sleep, or the customer at the restaurant to complain about the high price of a meal. Norout notes that these outside-of-play interactions are somewhat different from actual role playing, because, as children engage in these, they must, "draw on skills at coordinating multiple perspectives. A child must represent not only her own perspective as a player ... but the complementary perspectives of other characters" (p. 380). So, in metaplay, a child creates a story from an omniscient viewpoint, as good fiction writers do (Gardner, 1991).

In metaplay, the child is not just enacting a script from the perspective of a single character, then, but planning out actions and events that may be unknown to the characters, themselves, or situations that have not yet occurred. A group of four year olds, for example, decide, outside of play, that a hurricane is coming and agree that the "children," but not the "grown ups," will be afraid and cry. They plan for the house to get knocked down, but, "let's say no one gets blown to bits." Such interactions are more planful than role playing, itself, and more external to the thoughts and feelings of individual make believe characters.

The Play Frame and the "Fictional Dream" 343

It is argued here that, in metaplay, children most directly plan a story. Through metaplay interactions, some children create rich, authentic play worlds; however, others do not, as the following examples illustrate:

Play Example 1: "Driving"

Child A: (Sitting on a chair and manipulating a plastic steering wheel.) We're driving the car, alright?
Child B: Yeah. (Makes engine noises.) B-r-r-r . . .
Child A: (Makes engine noises.) B-r-r-r . . .
Child B: But I drive now, alright?
Child A: Alright. (Hands the steering wheel to Child B, continues to make noises) B-r-r-r . . .)

This goes on for several minutes until both children leave the play area.

Example 2: "Drinking and Driving"

Child C: (Sits on a "car" built from blocks.) Let's say I turn on the radio. (Makes a gesture as if turning on a car radio.)
Child D: Turn it 98.6, all country all the time, okay?
Child C: Alright. And you be the baby.
Child D: No, I'm not a baby. We're the grown up mothers that live together, alright? And let's say these are our babies. (Retrieves two dolls from the nearby dramatic play area and places them in block "seats" behind them.) And let's say we need to go to New York City and take them to the show, alright Sonia?
Child C: No, let's say it's too crowded there today. We have to go to McDonald's instead.
Child D: Okay, maybe McDonald's. But then we go to New York to the show, right?
Child C: Alright. (Pretends to drive the car) Let's say we have beer now, okay? (Makes a gesture and a noise, *psst*, to indicate she is opening a beer can.)
Child D: No, Cheryl, not beer. Mommies don't drink beer.
Child C: They do too. My mommy drinks beer. She drinks it all the time.
Child D: No, beer makes you drunk. Let's say we're mommies that don't drink beer.
Child D: No, I'll be the Mom that drinks the beer.
Child C: But not really beer, okay? Like not the beer that makes you drunk.
Child D: No, it needs to be, because let's say the mommies had a really hard day, alright? Like the babies are being too loud?
Child C: (In a puzzled tone.) Okay. But . . . I don't . . .
Child D: You open a can like this. (Makes another beer opening gesture and noise) And then you drink it. But let's say the mothers don't get sleepy.
Child C: (Still with a puzzled expression.) Okay.
Child D: The mothers aren't mad at the babies, because let's say it's the kind of beer that doesn't make you mad.
Child D: Alright. But I'm not the mother who drinks it, okay?
Child C: You be the mother who says, "Honey, you just drink too much cans of beer."

The two then resume their roles, one of them drinking and driving, the other complaining about this. The play theme goes on for many minutes and contains much language and role enactment.

In the first play example, there are few metaplay negotiations and the play world that is created is somewhat sparse. In the second, metaplay discussions lead to the invention of elaborate characters and relationships among them, a detailed physical setting (e.g. a car with a radio playing country music), and several plot lines (e.g. driving babies to New York City or McDonald's and beer drinking). Because of metaplay interchanges, a far more authentic play world has been constructed in example two – a world in which there likely will be greater make believe, language, and social interaction, a world in which children will nearly live the experiences of their characters. What prompts children to add this authenticity to their play worlds? How do young children acquire metaplay abilities? A questions to be raised in this paper is whether literacy experiences-particularly story writing-can enhance these competencies.

THE FICTIONAL DREAM AND THE PLAY FRAME

In further examining the relationship between metaplay and story writing, two phenomena are considered here – *the fictional dream* and *the play frame*. One is a concept drawn from the field of adult fiction writing, the other is a classic notion about play which has guided play research for decades.

The Fictional Dream

Gardner (1991) has described a mental state which writers and readers of fiction often enter – "a kind of dream, a rich and vivid play in the mind, which engages us heart and soul" (p. 30). Full entry into this fictional dream is required, he argues, if one is to experience what characters do, if a piece of writing is to hold true meaning. In Gardner's view:

> We not only respond to imaginary things – sights, sounds, smells – as though they were real, we respond to fictional problems as though they were real: We sympathize, think, judge. We act out, vicariously, the trials of the characters and learn from the failures and successes of particular modes of action, particular attitudes, opinions, assertions, and beliefs exactly as we learn from life (p. 31).

Even among very young authors, there may be a point in the creation of a story – after it has been conceived and plotted, possibly through negotiation with peers – when it becomes an imaginary world in its own right – a fictional dream. As the child plans the story, "The Monster," in Literacy Example 2, above, he appears almost to step into the world he has created, to feel the sadness of a child whose birthday presents have been gobbled up and, perhaps, the joyful power of the monster who is doing the gobbling.

How is this fictional dream established and maintained? What elements lead the story of the monster, above, to be more vivid in the minds of children than the simpler first story, "My Mom?" From Gardner's view, it is *detail* that makes the difference. Detail is, "the life blood of fiction, which draws readers and writers in, causes them to worry about the characters, listen in a panic for some sound behind the fictional door, and exult in the characters' successes and bemoan their failures" (p. 31). When the identities, feelings, and behaviors of a story's characters are specified, and settings and situations in which they find themselves elaborated, a fictional world comes to life, and the story being written or read becomes nearly true.

It is proposed here that children acquire an ability to add such detail to their stories in a variety of ways: Through shared book reading, story telling, and writing at home and school they learn how authors use detail to add authenticity and emotional power to their fiction (Lehr, 1991; McGill-Franzen & Lanford, 1994). It may be that play experiences – particularly metaplay negotiations-also help children learn to add detail to their stories, in order to make them more real in one's own mind and more clearly understood by others.

The Play Frame

Bateson (1972) proposed that children, in their spontaneous activities, often create and enter *play frames* – intellectual, emotional, and behavioral contexts in which they act and think in an *as if* manner. Within a play frame, children come to an agreement with others – or on their own, in the case of solitary play – that interactions are not to be taken literally. A significant feature of play, according to Bateson, is the process by which children initiate, maintain, and elaborate on the frame. Sometimes they communicate the *as if* features of play with smiles or silly voices (e.g. a child assumes the deep, adult voice of a firefighter). Often, they step out of the play frame in order to negotiate symbolic meanings-that is, they engage in metaplay. In order to maintain an authentic play-fighting episode on the playground, for example, a child might briefly suspend the pretend battle to say, "You hit me now, okay? But not too hard, but lct's say I still cry, because I'm a little kid, alright?" A number of studies have shown that metaplay negotiations contribute to the organization and duration of play (Corsaro, 1992; Farver, 1992; Goncu, 1993; Reifel & Yeatman, 1993; Trawick-Smith, 1994).

It is proposed here that the play frame is quite similar to Gardner's fictional dream. Both are states of mind that allow entry into an imaginary world and cause the fiction or play to hold great personal meaning. Reifel and Yeatman (1993), in fact, might argue that the two are really identical-both instances of a play frame, since they each have an *as if* quality. I propose that, not only are the fictional dream and the play frame similar states of mind, but they are established in a similar way-through detail. I have argued in previous work that metaplay negotiations are what lead play frames to become most detailed, socially organized, verbal, and sustained (Trawick-Smith, 1994). In metaplay, children can create sufficient detail in play to provoke strong feelings and to cause play worlds to become nearly (though not completely) real. In a pirate play episode, a child may establish enough detail to truly feel – to a safe extent – the loss of her pretend ship when it sinks. When a child, through metaplay, creates a detailed and authentic make believe family, he might experience real, if minor, exasperation that his imaginary children are throwing plastic fruit all over the floor. (This simultaneous coordination of the feelings of make believe characters with a knowledge that "this is all just pretend," make play, at once, emotionally-charged and safe. When the child in the second play episode, "Drinking and Driving," above, elaborates on the effects of beer on parenting, she is confronting true feelings, while at the same time controlling and manipulating make believe to maintain emotional safety. The simple driving play of the first example, which contains little detail, does not appear to have an equivalent emotional impact.)

Metaplay Behaviors and Detail in the Play Frame

In a recent study of preschool children's symbolic play, I described and categorized 40 specific metaplay behaviors which were used by children to establish, maintain, and elaborate symbolic play frames (Trawick-Smith, 1998). One key finding of this study was that these metaplay behaviors were regularly used to achieve greater detail in play. A discussion of each behavior identified in the study is beyond the scope of this paper; a summary of eight broad categories of metaplay interactions are presented in Table 1.

As shown in the table, most of these metaplay behaviors had the purpose of adding detail to play. A number of these were found to establish or clarify the identities and traits of pretend characters (e.g. "Am I the older sister? Let's say I'm a teenager, alright?"), and another set to specify their make believe actions (e.g. "You're just sweeping and sweeping the kitchen to clean up the baby's mess, right Sara?"). Some metaplay initiatives were aimed at describing characters' internal mental states (e.g. "Let's say the mother's really angry,

The Play Frame and the "Fictional Dream" 347

Table 1. Categories of Metaplay Behavior Performed by Four and Five Year Old Children.

Metaplay Category:	Examples:
1. Establishing/clarifying identities and traits of pretend characters.	"Am I the older sister? Let's say I'm a teenager, alright?"
	"Will you be a hunter?"
2. Narrating/clarifying make believe actions of characters.	"You're just sweeping and sweeping the kitchen to clean up the baby's mess, right?"
	"Let's say I'm baking these pies."
3. Announcing/clarifying characters' internal mental states.	"Let's say the mother's really angry, okay?"
	"The hunter doesn't know I'm hiding here, okay?"
4. Transforming objects.	"This'll be the spider web, alright?"
	"Is this going to be the telephone?"
5. Announcing/clarifying make believe situations.	"Is the carnival starting today, James? Today at twelve o'clock, when it's dark?"
	". . . so it's windy and stormy out."
6. Relating the play theme to real life.	"My mother has these tools, but they're too sharp."
	"This is how my teacher writes it."
7. Constructing the play setting.	"I'm putting the house over here. A big house, not an apartment."
	"I just have to set up the spook house."
8. Agreeing/disagreeing, answering, clarifying in response to a peer's initiative.	"There's no tornado. It's sunshine."
	"No, let's say it's not time yet for the party."

okay?"); several others served to transform real objects or props into imaginary ones (e.g. A child holds up a rubber net and states, "This'll be the spider web, alright?"). As shown in the table, other metaplay behaviors were focused on creating or clarifying specific, make believe situations ("Is the carnival starting today, James? Today at twelve o'clock, when it's dark?"). A few metaplay negotiations were found to connect elements of make believe with real life events ("My mother has these tools, but they're too sharp.") One category, shown in the table, involved physically arranging or constructing space and materials to create a make believe setting (e.g. A child placing boxes together says, "I'm setting up the house over here. A big house, not an apartment.") Finally, a variety of responses to the above metaplay initiatives were identified. Children agreed or disagreed, answered, or offered clarification in response to the metaplay initiatives of peers (e.g. "There's no tornado. It's sunshine.") All of these appeared to lead to the construction of a more detailed play frame. With each additional metaplay pronouncement, an imaginary world came into sharper focus. These metaplay behaviors are not unlike the mental and social actions of young authors, described previously, who are creating characters, action, and plot. It is argued here that metaplay and story writing are, in many cases, nearly the same thing.

A BI-DIRECTIONAL RELATIONSHIP BETWEEN METAPLAY AND STORY WRITING

In this paper, I propose a bidirectional relationship between metaplay and the writing of stories in early childhood. The relationships between several selected categories of metaplay – as delineated in my previous research (Trawick-Smith, 1998) – and specific examples of story writing are illustrated in Figs 1 and 2.

Figure 1 shows the association between several types of metaplay interactions that are related to *character* and a young author's character development in story writing. As shown in the figure, some interchanges from the "Drinking and Driving" play example, presented above, are aimed at establishing detailed descriptions of pretend characters. These include initiatives to clarify the identities of characters (e.g. "You're the baby," and "We're the grown up mothers that live together . . ."), and to announce a character's internal mental state (e.g. "The mothers aren't mad at the babies . . ."). There are also character-related responses to a peer's initiatives that include disagreeing with a counter proposal (e.g. "No, beer makes you drunk. Let's say we're mommies that don't drink beer.") These behaviors appear to serve two very important functions in play: (1) In these pronouncements, children guide their own internal constructions of play elements-that is, they create detailed, personal, mental representations of pretend characters. (2) These behaviors also serve to convince, clarify, and/or establish mutual agreement with peers about the traits of characters. As shown in the figure, the overall effect of these interchanges is to establish a detailed play frame.

What parallel processes can be found in the literacy example – "The Monster" – presented above? Figure 1 shows that as the child discusses his writing plans with his teacher, he also establishes the identities and traits of his story's characters (e.g. "He's got sharp teeth, like those points?" and ". . . he's

The Play Frame and the "Fictional Dream" 349

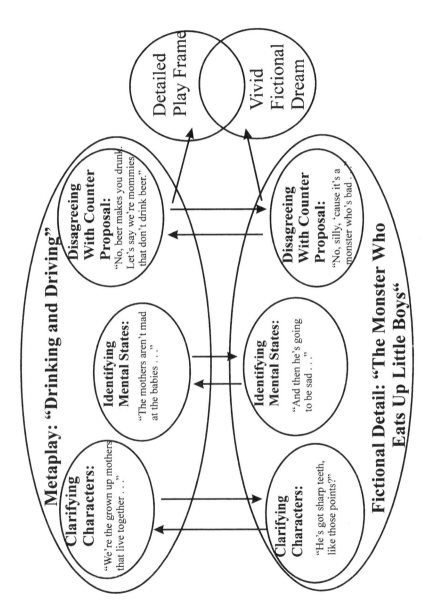

Fig. 1. Relationships Between Metaplay and Fictional Detail in Regard to *Character*.

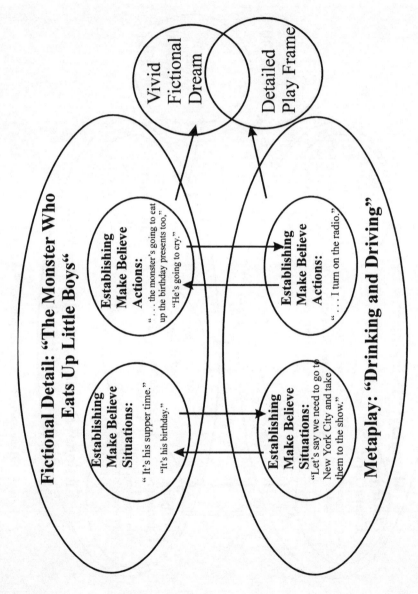

Fig. 2. Relationships Between Fictional Detail and Metaplay in Regard to *Make Believe Actions and Situations*.

The Play Frame and the "Fictional Dream" 351

five, too."). Prompted by a teacher's question, he contemplates internal mental states of at least one character (e.g. "And then he's going to be sad ..."), and perhaps another's (e.g. ... 'cause he *loves* little boys to eat up."). The child disagrees with a contributing author – in this case, a teacher – and offers a counter argument (e.g. "No, silly, 'cause it's a monster who's bad ...") Are these interactions and mental processes substantially different from the metaplay negotiations shown in the figure? They appear to serve the same functions: They lead to the creation of a detailed mental construction of characters, and, at the same time, establish a mutual understanding of story elements with another author. As shown in the figure, these interchanges lead to the creation of a more detailed fictional dream.

Arrows in figure 1 show the bidirectional relationship between each metaplay behavior and its analog in writing. Adding detail to character in play may enhance character planning in story writing and vice versa. Both metaplay and writing experiences lead to rich, imaginary worlds – a detailed play frame or vivid fictional dream. These are proposed to be nearly the same phenomenon, as indicated by the intersection between their two ellipses.

Figure 2 shows a similar bidirectional relationship between metaplay behaviors aimed at establishing *make believe actions and situations* and writing activities that are intended to clarify these same elements in a story. As shown in the figure, the author of "The Monster" establishes fictional situations (e.g. "... it's his supper time," and "It's his birthday.") and also plans specific actions for his characters (e.g. "... the monster's going to eat up the birthday presents too," and "... he's going to cry."). In the "Drinking and Driving" play theme, analogous interactions take place. The children create make believe situations (e.g, "... let's say we need to go to New York City and take them to the show," and "Let's say we have beer now, okay?") and negotiate specific character actions ("... I turn on the radio," and "You be the mother who says, 'Honey, you just drink too much cans of beer.'") Arrows in Figure 2 show the bi-directional relationship between these metaplay and writing behaviors. It is proposed here that each supports the other. Also shown in the figure is the contribution that each makes to the creation of an imaginary world – a fictional dream or play frame, which intersect, since they are hypothesized to be similar mental constructs.

CONCLUSIONS AND IMPLICATIONS

In summary, it is argued that, through the process of establishing and negotiating detail, children create authentic imaginary worlds in both literacy and play. This authenticity is what leads play themes or stories to seem real, allows children to experience the emotions of their characters, and prompts them to communicate and negotiate more fully with peers. I propose that practice at creating detailed play frames, through metaplay, contributes to greater competence in establishing detail within stories. Likewise, when children negotiate detail in stories, they may be better able to do so in play.

The model presented here requires additional empirical support. Specific metaplay behaviors that establish play detail have been identified in research (Trawick-Smith, 1998) (see Table 1). Can these same eight categories of interaction be observed in young children's collaborative writing with peers? Do children, in their story writing, create and negotiate character identities, traits, and mental states, as well as make believe situations, actions, objects, or physical settings, as they do in play? If so, research is needed to verify and further describe the relationships among these writing and metaplay interactions. Do children who engage in more frequent and elaborate metaplay include more fictional detail in their writing or more often negotiate with peers about specific elements of a story? Do children who write in more detail also engage in more metaplay and construct more detailed play frames?

Several questions are more difficult to address empirically, yet are at the heart of the theoretical model proposed here: How vivid and authentic *are* play frames and fictional dreams in the minds of children? In what ways are these influenced by metaplay and story writing behaviors? To answer these questions, researchers might use interview strategies, in which preschool or primary children are asked directly to reflect on their play and literacy experiences. During or after a play episode, for example, children might be asked open-ended questions about their subjective experiences: "Tell me all about what you were playing." "What did it feel like to be the mother?" "Did you feel like you were a pirate or still a child when you were playing on the boat?" Such questions relate to how real the play frame had become for players. Although these would seem to be extraordinarily complex questions, Fein & Wiltz (1998) found that children as young as four were quite competent in describing feelings and internal mental processes in play.

Similar interview methods have been used to study young children's responses to literature and writing and their overall *sense of story* (Applebee, 1978; Clay, 1979; Lehr, 1991). In these investigations children as young as four were found to construct schemas for stories, including a knowledge of characters-their feelings, goals, and actions-the sequence of events, the physical setting, and interrelationships among these. In other words, young children displayed a growing ability to conceptualize a full and vivid story. In future research, interview methods might focus on children's emotional reactions to reading and writing events, on the extent to which children can identify with characters, and on the degree to which stories seem real.

Researchers could examine how children's responses to these play and literacy interviews vary as a function of the frequency and complexity of their metaplay and story planning behaviors.

This proposed relationship between play and story has a number of classroom implications. It has been recommended that teachers spontaneously scaffold children's reading and writing throughout the classroom (Neuman & Roskos, 1991). A focus of some interventions might be on the elaboration of fictional detail. Questions related to character traits and actions, the situations in which they find themselves, or descriptions of physical setting, might be asked (e.g. "Is it a nice monster?" or "How's the little boy going to feel about that?"). It is proposed here that such strategies will enhance, not just writing, but play.

Similar teacher interventions might occur in the play area. Although most recommendations for play intervention, found in the literature, focus on enhancing make believe, itself (Trawick-Smith, 1995), some strategies might focus on facilitating metaplay. During a pause in play activity, a teacher might ask questions which prompt metaplay discussions about the details of a play theme (e.g. "Where are you pretending to drive in your car today?" "How's the weather for your trip?" or "How does it feel driving in all this traffic?"). (Elsewhere, I have urged that great caution be exercised in such interventions, so that play will not be disrupted – Trawick-Smith, 1995). One strategy which holds promise in enhancing metaplay-and thus, perhaps, story writing-is the provision of non-realistic play objects (e.g. nondescript materials, such as cardboard pieces or boxes, which suggest no specific use). It has been found that four and five year olds engage in the most detailed metaplay discussions when using these non-realistic items (Trawick-Smith, 1990).

Finally, based on the proposed connections between play and literacy, teachers might consider strategies which more directly link the two. Books might be made available in the dramatic play center to inspire more detailed play, as proposed by Neuman and Roskos (1991). Thematic fantasy play, studied in the classic work of Salz and Johnson (1974) might be implemented, in which children are told or read stories, then provided with props and guidance to playfully re-enact these. Paley's (1981) strategy of encouraging young children to dictate or write their own stories and then enact these in the dramatic play center might be considered. Based on the present work, teachers might even implement strategies for guiding children in writing what they play. After a play episode, for example, children might be encouraged to reconstruct a story they have just enacted by dictating or writing about it in the writing center. These strategies may cause fictional worlds and play worlds to be fully integrated.

REFERENCES

Applebee, A. (1978). *The child's concept of story: Ages two to seventeen*. Chicago: University of Chicago Press.

Bateson, G. A. (1972). *Steps to an ecology of mind*. New York: Ballantine.

Christie, J. F. (1998). Play as a medium for literacy development. In: D. P. Fromberg & D. Bergen (Eds), *Play from Birth to Twelve and Beyond: Contexts, Perspectives, and Meanings* (pp. 37–55). New York: Garland.

Christie, J. F., & Enz, B. (1992). The effects of literacy play interventions on preschoolers' play patterns and literacy development. *Early Education and Development*, *3*, 205–220.

Clay, M. (1979). *What did I write?* Portsmouth, NH: Heinemann.

Corsaro, W. (1992). *Interpretive approaches to children's socialization*. San Francisco: Jossey-Bass.

Davidson, J. I. F. (1998). Language and play: Natural partners. In: D. P. Fromberg & D. Bergen (Eds), *play from Birth to Twelve and Beyond: Contexts, Perspectives, and Meanings* (pp. 175–183). New York: Garland.

Farver, J. (1992). Communicating shared meaning through in social pretend play. *Early Childhood Research Quarterly*, *7*, 501–516.

Fein, G. G., & Wiltz, N. W. (1998). Play as children see it. In: D. P. Fromberg & D. Bergen (Eds), *play from Birth to Twelve and Beyond: Contexts, Perspectives, and Meanings* (pp. 37–49). New York: Garland.

Gardner, J. (1991). *The art of fiction*. New York: Random House.

Goncu, A. (1993). Development of intersubjectivity in dyadic play of preschoolers. *Early Childhood Research Quarterly*, *8*, 99–116.

Lehr, S. S. (1991). *The child's developing sense of theme: Responses to literature*. New York: Teachers College Press.

Levy, A., Wolfgang, C., & Koorland, M. (1992). Sociodramatic play as a method for enhancing language performance among kindergarten age children. *Early Childhood Research Quarterly*, *7*, 245–262.

McGill-Franzen, A., & Lanford, C. (1994). Exposing the edge of the preschool curriculum: Teachers talk about text and children's literacy understandings. *Language Arts*, *71*, 264–273.

Neuman, S. B., & Roskos, K. (1991). Play, print, and purpose: Enriching play environments for literacy development. *The Reading Teacher*, *44*, 214–221.

Norout, P. M. (1998). Sociodramatic play: Pretending together. In: D. P. Fromberg & D. Bergen (Eds), *Play from Birth to Twelve and Beyond: Contexts, Perspectives, and Meanings* (pp. 378–391). New York: Garland.

Paley, V. (1981). *Wally's stories*. Cambridge, MA: Harvard University Press.

Reifel, S., & Yeatman, J. (1993). From category to context: Reconsidering classroom play. *Early Childhood Research Quarterly*, *8*, 347–367.

Roskos, K., & Neuman, S. B. (1993). Descriptive observations of adults' facilitation of literacy in play. *Early Childhood Research Quarterly*, *8*, 77–97.

Salz, E., & Johnson, J. (1974) Training for thematic fantasy play in culturally disadvantaged children: Preliminary results. *Journal of Educational Psychology*, *66*, 623–630.

Trawick-Smith, J. (1990). Effects of realistic versus non-realistic objects on young children's symbolic transformation of objects. *Journal of Research in Childhood Education, 5,* 27–36.

Trawick-Smith, J. (1994, April). A qualitative study of metaplay. Paper presented at the annual meeting of the American Educational Research Association, New Orleans.

Trawick-Smith, J. (1995). *Interactions in the classroom: Facilitating play in the early years.* Columbus, Ohio: Merrill.

Trawick-Smith, J. (1998). An analysis of metaplay in the preschool years. *Early Childhood Research Quarterly, 13,* 433–452.

AUTHOR INDEX

Abbott, L., 170, 172, 198, 200
Abbott-Shim, M., 67, 74, 76, 82, 83, 89, 102, 106, 107, 110, 111, 113
Adamson, P., 4, 5, 51
Adler, L., 299, 307, 314, 317
Ahmeduzzaman, M., 303, 304, 317
Alexander, L., 186, 200
Alexander, P. A., 205, 223
Almqvist, B., 280, 292
Althoff, W., 229, 254
Anderson, B-E., 87, 88, 90, 93, 101, 107
Anderson, C. W., 88, 107
Anderson, R. C., 223, 236, 239, 242, 243, 253
Antler, J., 206, 223
Apple, M. W., 6, 51
Applebee, A., 352, 354
Arbuthnot, M. H., 210, 223
Arnett, J., 68, 69, 74, 79, 100, 107
Aroni, R., 186, 200
Ayers, W., 264, 273
Azer, S., 122, 168

Bailey, C. D., xi, xiv
Bairrao, J., 25, 55
Baker, L., 315, 336
Baker, T., 307, 315
Bakhtin, M. M., 216, 221, 223
Baptiste, N., 117, 167
Barbour, N. E., 117, 167
Bates, J. E., 88, 107
Bateson, G. A., 260, 262, 273, 345, 354
Bauersfeld, H., 230, 232, 253–255
Bauman, Z., 265, 273
Baydar, N., 91, 107
Beach, B. B., 171, 198, 224
Beach, S., 326, 336
Belenky, 196
Belsky, J., 60, 74, 78, 88, 107, 170, 198
Bennett, D. S., 88, 107

Bennett, N., 263, 273
Bereiter, C., 250, 254
Berk, L. E., 216, 223, 232, 233, 245, 247, 248, 253, 254
Berkowitz, M. W., 229, 254
Bernas, R. S., 229, 255
Bernhard, J., 12, 13, 55
Bertram, T., 37–39, 42, 51
Biklen, S., 171, 198
Bjorkman, S., 84, 89, 107
Blau, D. M., 86, 100, 104, 107
Bloch, M., 299, 307, 314
Blow, S. E., 206, 285, 288–290, 292
Bodrova, E., 233, 254, 324, 335
Bohlin, G., 79, 84, 91, 96, 99, 109
Bookstein, F. L., 78, 111
Bornstein, M. H., 199, 306, 318
Boutte, G., 312–314
Bowman, B., 310, 318
Boykin, A. W., 298, 314
Bradley, R. H., 78, 108
Bransford, J. D., 213, 218, 223
Brazelton, T. B., 53
Bredekamp, S., xi, xiv, 5, 7, 51, 54, 64, 66, 75–77, 82, 107, 122, 167, 213, 214, 218, 223, 262, 273
Brennan, D., 14, 48, 51
Bretherton, I., 306, 317
Britain, L., 37, 214–216, 221, 223
Broberg, A. G., 5, 16, 53, 78, 111
Broman, I. T., 49, 52
Bronfenbrenner, U., 60, 107, 125, 167, 170, 198, 325, 326, 335
Brooks-Gunn, J., 91, 107
Brophy, H., 170, 173, 200, 305, 317
Brosterman, N., 278, 279, 281, 282, 291, 292
Brown, J., 259, 293, 304, 317
Bruner, J., 204, 213, 219, 221–223, 261, 273, 274, 301, 306, 314, 337

Bryan, Y., 75, 109
Bryant, D. M., 66, 90, 93, 107, 108
Buell, M. J., 108
Burbules, N. C., 6, 7, 51
Burchinal, M. R., 25, 52, 63, 65, 66, 68, 83, 84, 90, 92–94, 96, 97, 107, 108, 111, 112
Burk, C. F., 275, 283–287, 292, 293
Burk, F., 275, 284, 292
Burts, D. C., 78, 82, 91, 108, 318
Bus, A. G., 333, 335
Buzzelli, C. A., 245–249, 254
Byler, P., 112

Caffarella, R., 121, 167
Cai, W., 12, 52
Caldwell, B. M., 60, 78, 108
Calicchia, D. J., 229, 255
Campbell, J. J., 87, 108
Campbell, S., 102, 109
Cannella, G. S., xi, xiv, 261, 273
Caplan, M., 60, 111
Carnoy, M., 6–9, 51
Carr, M., 21, 51
Casper, V., 184, 198
Cassells, C., 250, 254
Cassidy, D. J, 69, 108
Caughy, M. O., 63, 94, 96, 108
Chaille, C., 214–216, 221, 223
Chang, J., 236, 253
Chard, S. C., 212, 224
Charlesworth, R., 78, 83, 108, 283, 292, 318
Chase-Lansdale, P. L., 63, 108
Chevalier, M., 271, 273
Children's Defense Fund, 123, 124, 167
Chin, J., 306, 315
Chinn, C., 236, 253
Chodorow, N., 172, 173, 176–179, 182, 187–189, 191–199
Choi, S.-H., 18–21, 51
Christie, J. F., 269, 274, 303, 316, 317, 321, 323–325, 330, 334–336, 339, 354
Chung, A., 124, 167, 168
Clandinin, D. J., 171
Clarke, L. J., 260, 261, 273
Clarke-Stewart, K. A., 72, 86, 88, 97, 108
Clay, M., 352, 354

Cleghorn, A., 5, 17–19, 23, 51
Clements, M., 110
Clifford, R. M., 72, 76, 80, 82, 90, 101, 107, 109, 112, 127, 129, 130, 138
Clinchy, B. M., 196
Clinton, W. J., 268, 273
Clyde, M., 13, 51
Cmic, K., 315
Cobb, P., 230, 232, 234, 245, 250, 254–256
Cochran, M., 5, 15, 51, 52
Cocking, R., 299, 315
Cohen, D. H., 112, 212, 223
Cole, M., 55, 199, 254, 293, 326, 335
Coll, C. G., 298, 300, 315
Collins, M., 170, 199
Cone, J., 113
Connelly, F. M., 171
Copley, J. V., 7, 51
Copple, C., xi, xiv, 7, 54, 64, 66, 107, 213, 214, 218, 223, 262, 273
Corasaniti, M. A., 88, 91, 92, 96–99, 102, 104, 113
Corbett, S. M., 266, 267, 273
Corsaro, W. A., 262, 271, 273, 345, 354
Costley, J., 117, 122, 168
Coughlin, P., 31, 51
Courtney, G., 55
Coutsoukis, P., 48, 51
Cox, S., 322, 336
Crane, J., 22, 52
Cryer, D., 25, 52, 55, 60, 64, 65, 76, 82, 108, 109, 112
Cuffaro, H. K., 184, 198, 220, 221, 223
Culkin, M., 112
Curry, N., 311, 315
Cyphers, L., 306, 318

Dahlberg, G., 5, 44, 45, 52, 265, 273
Damon, W., 229, 254
Daniels, D., 68, 92, 113
Darling, M. A., 11, 14, 55
Davidson, D. H., 307, 318
Davidson, J. I. F., 339, 354
Davies, R., 5, 10, 17, 18, 20, 23, 52
Davis, M. D., 276, 293
Daycare Trust, 39, 40, 52
de Jong, E., 333, 335

Author Index

de Jong, M., 333, 335
de Jong, P. F., 326, 336
de Jong, W., 333, 335
Deater-Deckard, K., 65, 66, 68, 75, 83, 92, 108, 113
Degotardi, S., 244, 254
Deich, S. G., 63, 109
DeMarrias, K., 307, 315
Derman-Sparks, L., 311, 315
Desai, S., 63, 96, 99, 108
Dettling, A. C., 92, 108
deVries, A. B., 333, 335
DeVries, R., 83, 106, 108, 214–216, 221, 223, 224
Dewey, E., 205, 207, 208, 223
Dewey, J., 205, 207, 208, 210, 223, 289, 292
DiPeitro, J. A., 63, 108
Dixon, D., 301, 318
Dixon, S., 5, 53
Dockett, S., 227, 231, 238, 244, 252, 254, 255
Dodge, K. A., 88, 107
Doherty-Derkowski, G., 65, 66, 108
Donovan, B., 303, 317
Doucet, A., 182, 200
Draper, P., 298, 300, 315
Dresden, J., 322, 336
Dunn, J., 229, 234, 244, 251, 254, 255
Dunn, L., 60, 68, 71–73, 75, 79, 80, 83, 87, 97, 108, 229, 234, 244, 251, 254, 255, 306, 319, 326
Dunn, T., 306, 319
Durkin, D., 213, 224

Edwards, C., 217, 218, 221, 224
Edwards, R., 182, 200, 201
Eisenberg, A. R., 229, 254
Eisenberg, M., 60, 65, 66, 111, 113
Eliot, L., 334, 336
Elkind, D., 270, 273
Elliot, K., 168, 199
Emblem, V., 21, 52
Engelsmann, F., 109
Enz, B., 323, 324, 330, 334–336, 339, 354
Erikson, E. H., 261, 271, 273
Essa, E. L., 59, 69, 91, 109

Evans, R. I., 292

Farquhar, E., 66, 111
Farver, J., 306–308, 315, 345, 354
Farver, T. A. M., 306, 308, 315
Fasoli, L., 291, 293
Favre, K., 91, 109
Federal Register, 64, 109
Feeney, S., xi, xiv, 13, 52, 55, 261, 265, 273
Feiler, R., 92, 113
Fein, G. G., 203, 293, 303, 313, 318, 319, 352, 354
Fenstermacher, G. D., 264, 274
Fernald, A., 305, 315
Ferrier, A., 205, 224
Field, T., 66, 68, 70, 71, 91, 93, 109
Fiene, R., 72, 96, 109, 111, 115, 118, 159, 166–168
Fiese, B. H., 306, 315
File, N., 66, 67, 69, 75, 77, 79, 83, 89, 109
Fitzgerald, L. M., 97, 108
Flavell, J. H., 236, 254
Fleege, P. O., 78, 108
Fogel, A., 305, 315
Forman, G., 223, 224
Fortner-Wood, C., 86, 113
Fosnot, C., 245, 250, 254
Foster, J. C., 208, 211, 224
Frank, A. L., 88, 112, 113
Franks, J. J., 213, 223
Freeman, E. B., 213, 224
Freeman, N. K., xi, 259, 260, 264, 265, 273, 274
Friendly, M., 17, 18, 23, 52
Froebel, F., 205, 224, 262, 276–284, 286–293
Fromberg, D., 314, 315, 354
Frost, J. L., xi, xiv, 260, 261, 263, 271, 274

Galda, L., 322, 336
Galinsky, E., 75, 98, 110, 116, 124, 164, 167
Gallagher, J., 101, 102, 109
Gallagher, M., 217, 224
Galluzzo, D. C., 68, 88, 110

Gammage, P., 170, 199
Gandini, L., 224
Garcia, H. V., 298, 315
Gardner, H., 268, 274
Gardner, J., 342, 344, 354
Garvey, C., 229, 254
Gaskell, J. S., 170, 199
Gelman, R., 213, 224, 336
Genser, A., 117, 168
Gerber, E., 71, 113
Gerber, M., 261, 274
Gestwicki, C., 262, 274
Ghazvini, A. S., 69, 83, 100, 109
Gibson, M., 299, 315, 317
Gill, P., 305, 317
Gilligan, C., 171, 178, 180–182, 187, 194, 195, 198, 199
Glickman, C. D., 291, 293
Glucksmann, M., 184, 199
Goelman, H., 96, 109
Goffin, S. G., 212, 213, 224
Goldberg, M., 102, 112
Goldberger, W. R., 196, 198
Goldstein, L., 170, 171, 199, 292
Goncü, A., 248, 254, 345, 354
Goodlad, J. I., 212, 224, 260, 264, 274
Goodman, I., 117, 118, 122, 168
Grajek, S., 86, 111
Greenfield, P., 299, 315
Greeno, J. G., 219, 224
Grice, H. P., 243, 254
Grieshaber, S., xi, xiv, 169
Grubb, W. N., 45, 52
Gruber, C. P., 97, 108, 255
Grumet, M. R., 171, 172, 199
Guild, D. E., 20, 52
Gulcur, L., 110
Gunnar, M. R., 92, 108
Gunnarsson, L., 32–34, 36, 52
Gutek, G. L., 276, 277, 279, 281, 293

Hagekul, B., 79, 84, 91, 96, 99, 109
Hall, G. S., 224, 225, 283–285, 287, 288, 293
Halliwell, G., 13, 55, 171, 199
Hamilton, C. E., 60, 71, 87, 88, 91, 110
Harkness, S., 298, 299, 315

Harms, T., 72, 74, 76, 80, 82, 109, 127, 129, 130, 136, 138
Harpending, H., 298, 300, 315
Harris, P., 303, 316
Harrison, E., 279, 280, 282, 293
Hart, C. H., 78, 108, 318
Hartup, W. W., 244, 255
Hasselbring, T., 218, 223
Hatch, J. A., 213, 224
Hausfather, A., 83, 91, 96, 109
Hayden, J., 20, 52, 53
Hays, S., 174, 175, 176, 189, 191, 199
Headley, N. E., 208, 211, 224
Heath, S. B., 297, 309, 316, 318
Helburn, S., 63, 66, 71, 72, 86, 90, 92, 96, 100, 108, 109, 111, 112, 116, 124, 167
Henderson, V. K., 66, 113
Hendley, S., 312, 314
Hennessy, E., 60, 111
Henry, M., 170, 199
HERA, 52
Hernandez, S., 108
Hestenes, L. L., 75, 76, 80, 81, 84, 90, 99, 109
Hewes, D. W., 205, 207, 224, 291–293
Hewitt, J., 250, 254
Hewlett, B., 298, 304, 316
Hill, P. S., 112, 206, 209, 222, 224, 275, 293
Hirsh-Pasek, K., 82, 83, 92, 111, 113
Hite, S., 170, 173, 176, 177, 191, 199
Hofferth, S. L., 63, 66, 109, 111
Holloway, S. D., 68, 76, 77, 87, 89, 110
Holm, G., 11, 14, 15, 55
Holt-Reynolds, D., 121, 167
Honig, A. S., 170, 173, 182, 196, 199, 200, 307, 316
Hooper, F., 302, 314–317
Hoot, J. L., 12, 52
Horn, I. S., 230, 255
Hossain, Z., 299, 303–305, 317
Howe, N., 17, 18, 55
Howes, C., 60, 63–71, 75, 76, 80, 81, 83, 87–93, 95–100, 102, 104, 106, 109–113, 124, 164, 167, 168, 303, 306, 316
Hubbs-Tait, L., 89, 113

Author Index

Hughes, F. P., 260, 263, 274, 303, 316
Hutchins, T., 170, 200
Hwang, C.-P., 5, 16, 53, 78, 87, 108, 111
Hymes, J., 261, 274
Hyson, M. C., 82, 83, 92, 111, 113
Hyun, E., 311, 316

Inhelder, B., 229, 255

Jagadish, S., 71, 91, 109
Jain, D., 304, 317
Jalongo, M. R., 12, 17, 21, 22, 52
Jambor, T., 307, 318
James, A., 44, 55, 115, 261, 295, 321, 347
Jarolimek, J., 212, 224
Jenkins, R., 298, 315
Jessup, P., 3, 7, 53
Jipson, J., xi, xiv, 310, 316
Johnson, D., 299, 318
Johnson, J., 115, 119, 166–168, 269, 274, 295, 299, 301–303, 308, 309, 313–318, 353, 354
Johnson, P., , 11, 17–20, 22, 52
Johnson, R., xi, xiv
Jones, E., 72, 112, 122, 167, 251, 255, 289, 293
Jorde-Bloom, P., 121, 127, 150, 165, 168
Joshi, P., 304, 317

Kabiru, M., 17, 55
Kagan, S. L., 112, 164, 168
Kahn, A. J., 12, 23, 52
Kallós, D., 49, 52
Kamerman, S. B., 12, 23, 52
Kamii, C., 214–216, 224, 271, 274
Karns, J., 305, 315
Katz, L. G., 8, 13, 53, 116, 168, 212, 224, 310, 316
Kavanaugh, R., 303, 316
Keefe, N., 67, 111
Keefer, C., 5, 53
Kelley, M., 322, 336
Kelly, T., 88, 107
Killen, M., 229, 254
Kilpatrick, W., 208, 211, 224
Kim, Y. K., 306, 315
Kipnis, K., 261, 273

Kisker, E., 66, 68, 91, 96, 111, 112
Klugman, E., 291, 293, 315, 317
Kohlberg, L., 214–216, 221, 223
Kontos, S. J., 60, 63, 66, 67, 69, 71, 72, 75, 77, 79, 80, 83, 84, 86, 87, 89, 95–98, 101, 109, 111, 113, 159, 167, 168, 326, 336
Koorland, M., 339, 354
Korpi, B. M., 33, 52
Kritchevsky, S., 72, 112
Krogh, S. L., 280, 289, 293
Kruger, A. C., 229, 255
Krummheuer, G., 228, 230–232, 234–236, 238, 255

Lamb, M. E., 5, 16, 53, 60, 78, 87, 89, 108, 111, 316
Lamberty, G., 298, 315
Lande, J., 102, 112
Lane, S., 92, 108
Lanford, C., 345, 354
Langlois, J., 89, 113
LaRoche, C., 83, 109
Lasater, C., 309, 316
Lasker, J., 303, 317
Lawrence, J. A., 53, 108, 199, 246, 254, 255, 293, 336
Layder, D., 185, 200
Lazerson, M., 209, 224, 283, 293
Leal, T., 25, 52
Leavitt, R. L., 170, 200
Lee, G., 5, 21, 22, 53
Lee, J., 19, 27, 53
Lee, M., 94, 107
Lee, Y., 306, 315
Lehr, S. S., 340, 345, 352, 354
Leiderman, P. H., 5, 53
Leong, D. J., 233, 254, 324, 335
Leseman, P. M., 326, 336
Levin, B., 12, 53
Levin, D., 268, 274
LeVine, R., 5, 11, 53
LeVine, S., 5, 53
Levy, A., 339, 354
Lewin, K., 125, 168
Lewis, M., 256, 333, 336
Lilliard, A., 301, 316
Lim, S. E., 307, 316

Logie, C., 5, 19, 20, 22, 23, 53
Lombardi, J., 117, 168
Love, J. M., 106, 111, 117, 124, 168
Lu, M., 304, 317
Lubeck, S., 3, 7, 15, 53, 310
Lyons, L., 20, 52

Macdonald, K., 303, 305, 315, 316
MacNaughton, G., 244, 255
Main, N., 244, 255
Maioni, T., 308, 318
Malaguzzi, L., 170, 200
Marshall, D., 56, 311, 316
Martin, J., 170, 171, 182, 200
Martini, M., 307, 316
Marvinney, D., 88, 107
Matheson, C. C., 71, 110, 303, 316
Mauthner, N., 182, 200
May, H., 21, 51
McAdoo, H. P., 298, 315
McAllister, J., 116, 168
McCarthy, J., 110
McCartney, K., 66, 67, 69–72, 75–77, 83, 86, 87, 89–92, 97, 100–102, 106, 111–113
McGill-Franzen, A., 345, 354
McGimsey, B., 117, 118, 122, 168
McGurk, H., 60, 111
McLean, V., 13, 55
McLoyd, V. C., 303, 316
Meacham, J., 310, 316
Meade, A., 18, 19, 22, 53
Meckstroth, A., 117, 124, 168
Meckstroth, A. L., 106, 111
Melhuish, E. C., 11, 13, 20, 53, 54, 60, 111
Meljeteig, P., 10, 53
Mellor, E. J., 5, 17–19, 21, 23, 53
Melnick, S., 159, 167
Melton, G. B., 10, 53
Mertens, D. M., 297, 309, 317
Michael, R. T., 63, 108
Milburn, S., 68, 92, 113
Miller, C. A., 229–232, 242, 251, 255
Miller, E., 170, 199
Miller, P. H., 229, 236, 254, 255
Miller, S. A., 236, 254
Miller, T., 182, 200

Milner, V., 5, 17, 19, 20, 22, 23, 54
Minichiello, V., 186, 200
Ministry of Education and Science, 36, 53
Mischler, E. G., 186, 200
Mocan, H. N., 67, 69, 70, 111
Modigliani, K., 168
Moe, T., 4, 25, 26, 43, 53
Moely, B., 88, 113
Monighan-Nourot, P., 311, 317, 319
Morgan, G., 117, 119, 122, 168, 254
Morris, J. R., 111
Morrison, G. S., 276, 278, 282, 293
Morrison, J. W., 5, 17, 19, 20, 22, 23, 54
Morrow, L., 323, 324, 336
Mosley, J., 78, 108
Moss, P., 5, 13, 41, 44, 52–54, 60, 111, 265, 273
Moylett, H., 170, 172, 198, 200
Munirathnam, G., 17, 19, 23, 54
Munro, J. H., 276, 293
Murphy, P. K., 205, 223
Myers, L., 88, 110

Nabors, L. A., 63, 66, 107
Nagle, R. J., 88, 107
Nakagawa, K., 299, 318
Nash, R. J., 260, 264, 274
National Association for the Education of Young Children, 5, 54, 83, 109, 112, 122, 129, 223, 254
National Research Council, 7, 54, 334, 336
Neebe, E., 93, 108
Nelson, P., 307, 315
Neuman, S. B., 7, 54, 323–325, 327, 330, 334, 336, 339, 353, 354
Newport, S. F., 17, 19, 20, 54
NICHD Early Child Care Research Network, 75, 92, 111, 112
Nicolopoulou, A., 252, 255
Nimmo, J., 251, 255
Nitsch, K. E., 213, 223
Nixon, M., 72, 109
Njenga, A., 17, 55
Noddings, N., 180–182, 187, 193–195, 198, 200, 260, 274
Noirin, H., 18, 20, 54

Nordenstam, U., 33, 52
Nourot, P. M., 342, 356
Nwokak, E., 305, 315

O'Connell, B., 306, 317
O'Neill, D., 305, 315
Oberhuemer, P., 36, 37, 54
OECD, 4, 8, 16, 23, 25, 26, 28–34, 37, 39, 43, 48, 50–54, 199
Ogbu, J., 296, 298, 299, 315, 317
Ogino, M., 306, 318
Olenick, M., 66, 68, 87, 90, 99, 110
Olmstead, P., 13, 23, 53, 54
Oloman, M., 17, 18, 23, 52
Osborn, D., 205, 207, 224
Oser, F., 229, 254

Paciorek, K. M., 276, 293
Palacios, J., 25, 52, 55
Paley, V., 314, 319, 353, 354
Pan, W. H. L., 209, 306, 307, 317
Park, S., 12, 52, 107, 254, 293
Parke, R., 303, 316
Parker, S. C., 207, 210, 224
Parker, S. W., 92, 108
Parmenter, G., 13, 51
Pascal, C., 37–39, 42, 51
Pattnaik, J., 12, 52
Pearson, P. D., 217, 224, 336
Peisner-Feinberg, E. S., 84, 87, 90, 92, 93, 97, 107, 112
Pellegrini, A. D., 293, 301, 317, 322, 336
Pence, A. R., 5, 44, 52, 96, 109, 265, 273
Penn, H., 5, 24, 51, 54, 55, 119, 293
Perry, B., 227, 231, 238, 255
Peters, D. L., 60, 111, 117, 167
Petrogiannis, K., 20, 54
Petrovski, P., 244, 254
Pettit, G. S., 88, 107
Philipsen, L. C., 110
Phillips, D., 63–66, 68, 71, 77, 80, 83, 86, 96, 98, 101, 102, 108–113, 124, 125, 138, 159, 168
Phillipsen, L. C., 64–68, 70, 71, 112
Phoenix, A., 172–174, 176, 191, 200, 201
Piaget, J., 204, 213–216, 219, 221, 229, 232, 255, 256, 271, 274, 292, 293, 314, 317, 321–323, 336

Pinkerton, R., 66, 108
Piscitelli, B., 13, 55
Polakow, V., 18
Poteat, G. M., 84, 107
Power, M. B., 170, 200
Powers, C. P., 66, 68, 70–72, 87, 89, 113
Pratt, C., 205, 206, 210, 216, 218, 222, 224
Prawat, R. S., 217, 224
Prescott, E., 72, 112
Prior, G., 12, 14, 40, 185
Prochner, L., 5, 12, 13, 17–19, 23, 51, 55
Prout, A., 44, 55
Puch-Hoese, S., 108

Qualifications and Curriculum Authority, 7, 42, 43, 50, 55

Raggozzine, D., 168
RameyRamey, C., 94, 107
Razey, M., 170, 199
Readdick, C. A., 69, 83, 100, 109
Reguiero-de-Atiles, J. T., 102, 113
Reichhart-Erickson, M., 68, 74, 76, 77, 87, 89, 110
Reifel, S., ix, xi, xiv, 103, 112, 167, 168, 225, 273, 274, 303, 306, 310, 315, 317, 345, 346, 354
Rescorla, L., 79, 83, 92, 111, 113
Resnick, L., 326, 336
Reynolds, G., 289, 293
Ribbens, J., 182, 200, 201
Richman, A., 5, 53
Riegraf, N., 303, 304, 317
Rieser-Danner, L., 89, 113
Riggins, R., 93, 108
Roberts, J. E., 66, 93, 107, 108
Roberts, W. A., 88, 107
Robinson, H. B., 11, 14, 15, 55
Robinson, J. F., 212, 224
Robinson, N. M., 11, 14, 15, 55
Rocheleau, A., 111
Rodd, J., 13, 51
Rodning, C., 88, 110
Rogers, S., 263, 273
Roggman, L., 89, 113
Rogoff, B., 216, 224, 246, 255
Rolfe, S., 13, 51

Rooney, R., 102, 109
Roopnarine, J. L., 295, 299, 300, 302–307, 314–317
Rosaldo, R., 296, 318
Rosegrant, T., 214, 218, 223
Rosenthal, R., 106, 111
Rosenthal, S., 67, 111
Roskos, K., 317, 321, 323, 324, 327, 330, 334, 336, 339, 353, 354
Ross, E. D., 115, 276, 293
Rovers, F., 244, 254
Rovine, M., 88, 107
Rubenstein, J. L., 100, 110, 113
Rubin, K., 303, 307, 308, 318
Ruddick, S., 175, 189, 201
Rudolph, M., 212, 223
Ruh, J., 67, 111
Rumelhart, D. E., 217, 225
Russell, S., 108
Rustici, J., 112
Ryan, S., 213, 214, 225

Sacks, M., 303, 317
Sakai, L. M., 71, 98, 110, 113
Saltz, E., 301, 313, 318, 353, 354
Saracho, O. N., 261, 262, 274, 276, 289, 293, 317
Scardamalia, M., 250, 254
Scarr, S., 60, 65–68, 70, 71, 81, 83, 86, 95, 102, 104, 105, 108, 111–113, 138
Schindler, P. J., 88, 89, 113
Schmidt, D., 326, 336
Schochet, P., 117, 168
Schochet, P. Z., 106, 111
Schoonmaker, F., 213, 214, 225
Schrader, C., 324, 336
Schultz, S., 184, 198
Schwartz, J. C., 86, 111
Schwartzman, H., 309, 318
Schweder, R., 310, 318
Scott, F., 166, 310, 316, 318
Sebanc, A., 92, 108
Seefeldt, C., 113, 291, 293
Segoe, M. V., 303, 318
Serpell, R., 326, 336
Shantz, C., 229, 255
Shapiro, M., 275, 276, 293
Sharma, A., 17, 18, 19, 23, 55

Sheerer, M., 168
Sherwood, R., 223
Shin, Y. L., 299, 300, 303, 308, 315, 317
Shinn, M., 75, 98, 110, 111, 167
Sibley, A., 76, 83, 106, 107, 110
Siegal, M., 243, 255
Silin, J., 184, 198
Silvern, S. B., 262, 273, 322, 336
Singer, E., 17, 20, 22, 55
Sirotnik, K. A., 260, 274
Slaughter-Defoe, D., 299, 318
Smith, E., 67, 69, 75, 80, 81, 83, 110, 124, 164, 167
Smith, J. W., 88, 107
Smith, L. B., 322, 326, 337
Snell-White, P., 304, 317
Snider, D. J., 276, 281, 293
Snow, C. W., 84, 102, 107, 113
Soder, R., 260, 274
Soltis, J. F., 260, 274
Sonnenschein, S., 326, 336
Sperry, L., 229, 255
Spodek, B., 110, 171, 201, 212, 213, 224, 225, 261, 262, 274, 276, 289, 293, 317
Srivastav, P., 304, 317
Staley, L., 217, 221, 225
Starnes, L., 5, 21, 23, 55
Stein, N. L., 229–232, 242, 244, 245, 249, 251, 255
Sternberg, K. J., 5, 16, 53, 60, 89, 111
Stewart, P., 76, 95, 96, 110
Stipek, D., 68, 69, 75, 78, 80, 81, 92, 113
Stone, S., 325, 330, 335
Stoney, L., 119, 124, 167, 168
Stores, M., 303, 317
Streitz, R., 210, 225
Stremmel, A., 314, 318
Strike, K. A., 260, 274
Strobino, D. M., 63, 108
Sulzby, E., 323, 324, 333, 335, 336
Sun, L., 304, 318
Super, C., 298, 299, 315, 318
Suppal, P., 303, 317
Surbeck, E., 322, 336
Sutton-Smith, B., xi, xiv, 260, 261, 269, 274, 290, 291, 293, 305, 309, 318
Swadener, E. B., 17, 21, 23, 55
Szarkowicz, D., 244, 254

Taharally, L. C., 307, 318
Takanishi, R., 299, 318
Talukder, E., 304, 317
Tamis-Lemonda, C. S., 306, 318
Tarule, J. M., 196
Taylor, J., 199, 322, 336
Teale, W., 323, 324, 336
Teleki, J. K., 102, 113
Temple, A., 210, 224
Tesla, C., 229, 255
Thelen, E., 322, 326, 337
Thomas, R. M., 283, 293
Thomasson, R. H., 78, 108
Thweat, G., 91, 109
Tietze, W., 25, 52, 55
Timewell, E., 186, 200
Tinworth, S., 51, 252, 256
Tobin, J. J., 200, 261, 267, 269, 274, 307, 308, 318
Toda, S., 306, 318
Toharia, A., 83, 109
Tomasello, M., 229, 255
Torres, C. A., 6, 7, 51
Toulmin, S., 232, 236, 238, 242, 256
Trabasso, T., 229, 255
Trawick-Smith, J., 207, 214, 215, 225, 339, 345, 346, 348, 352, 353, 355
Trevarthen, C., 248, 256
Tudge, J. R. H., 248, 256
Tulananda, O., 304, 306, 307, 319
Tulviste, P., 245, 247, 256

U.S. Department of Education, 116, 168
Ulich, M., 37, 54
UNESCO, 4, 8, 11, 18, 50, 51, 55
Ungerer, O. A., 316
UNICEF, 4, 8, 10, 11, 18, 32, 37, 55

Valsiner, J., 246, 255
Van Hoorn, J., 305, 319
Van Manen, M., 182, 184, 201
Van Scoy, I., 312, 314
Vandell, D. L., 60, 66, 68, 70–72, 87–89, 91, 92, 96–99, 102, 104, 105, 113
Vandenberg, B., 262, 274, 303, 318
VanderVen, K., 122, 168
Vega-Lahr, N., 71, 91, 109
Verba, M., 301, 319

Voran, M., 96, 112
Vukelich, C., 324, 337
Vygotsky, L. S., 170, 201, 204, 213, 216, 219, 222, 223, 225, 246, 254–256, 261, 271, 274, 298, 306, 314, 319, 323, 324, 337

Waggoner, M., 236, 253
Walker, A., 74, 78, 107
Walsh, D. J., 213, 225
Waniganayake, M., 51
Washington, V., 101, 113
Wasik, B. H., 298, 315
Watson, K., 307, 318
Waugh, S., 91, 109
Wearing, B., 172–174, 176, 191, 201
Webb, W., 304, 317
Weber, E., 205–207, 209, 211, 225, 276, 277, 279, 280, 282, 285, 287, 288, 293
Webster, F., 6, 55
Wechsler, S., 164, 168
Weikart, D., 13, 23, 53, 54
Wertsch, J. V., 216, 223, 225, 245, 247, 256
Wetzel, G., 25, 55
Whiley, J., 20, 52
White, C. S., 213, 225
White, M., 5, 11, 53
Whitebook, M., 63, 70, 71, 80, 91, 96, 98, 100, 102, 110, 112, 113, 124, 168
Wickens, E., 184, 198
Wilcox-Herzog, A., 86, 94, 105, 106, 113
Willer, B., 122, 123, 167, 168
Williamson, P., 322, 336
Wilson, K. S., 66, 113
Wiltz, N. W., 313, 319, 352, 354
Winsler, A., 216, 223, 247, 253, 254
Winterhoff, P. A., 248, 256
Wolfe, B., 60, 105, 113
Wolfgang, C., 339, 354
Wood, L., 263, 273
Wood, T., 232, 254
Woodhead, M., 5, 44, 55
Woodill, G., 4, 12, 55
Woods, B. S., 223
Woodson, B., 279, 280, 282, 293
Woollett, A., 172–174, 191, 200, 201
World Bank, 8, 9, 11, 18, 26, 55, 56

Wortham, S., xi, xiv, 209, 225
Wright, A., 229, 255
Wu, D. Y., 318

Yackel, E., 230, 232, 234, 250, 254, 256
Yandell, K., 297, 319
Yawkey, T. D., 269, 274, 293, 303, 316
Yazejian, N., 112
Yeatman, J., 345, 346, 354

Yi, H., 236, 253
Yin, R., 185, 186, 201
Youngblade, L. M., 306, 319

Zan, B., 83, 108, 216, 223
Zaslow, M. J., 60, 113
Zeisel, S. A., 93, 108
Zelazo, J., 112

SUBJECT INDEX

administration, 30, 100, 135
argumentation, 229, 242

caring, 170, 172, 180, 186, 198
child study, xii, 283
cognition, 92, 205, 213, 214, 321
community, 101
comparitive analyses, ix, 11, 23
context, 296, 326
culture, 9, 44
curriculum, x, xi, 20, 31, 36, 42, 80, 155, 212, 221, 252, 266, 277, 311
curriculum theory, 207, 213, 264

developmental theory, xii, 204, 209, 217, 298
Developmentally Appropriate Practice (DAP), 5
diversity, xii, 261, 301, 303
discourse analysis, x, 245

ecology, 60, 125, 301
economics, 7, 45, 96
emotions, 90, 193
environments, 70, 131, 150, 155
ethnicity, 97

feminism, 182
funding, 18

gender, 99
globalization, 6

history, 205, 276

language, 92, 233

materials, 278
measures, 65, 74, 76, 82, 127, 130, 138

mothering, x, 97, 170, 173, 186

narrative, 219, 340

parents, 22, 29, 31, 34–35, 37–39, 42, 95, 303
play, x, 89, 220, 244, 260, 289, 296
    culture, xii, 297
    ethics, xi, 265, 269
    freedom/democracy, xii, 271, 284, 312
    history, 205
    literacy, xii, xiii, 322, 328, 339, 348
    teachers, 263, 267, 289, 307, 313, 333, 353
    theory, 261, 263, 328, 344
policies, 30, 35, 40, 101, 162
post-modernism, xi, 296
process, 73
program structure, 66

quality, ix, 61, 72, 81, 85, 120, 124, 159

research, 31, 37, 42, 119, 142, 283

social relations, 86, 176, 192, 230, 244, 323
staff, 21, 30, 36, 41, 121

teacher characteristics, 77, 131, 249, 276, 310
teacher education/training, 68, 117, 136, 141, 145, 197, 250
thematic units, x, 203, 222

values, 153